AF412671

Optoelectronics for Low-Intensity Conflicts and Homeland Security

Anil K. Maini

ARTECH HOUSE

BOSTON | LONDON
artechhouse.com

Library of Congress Cataloging-in-Publication Data
A catalog record for this book is available from the U.S. Library of Congress.

British Library Cataloguing in Publication Data
A catalogue record for this book is available from the British Library.

Cover design by John Gomes

ISBN 13: 978-1-63081-570-7

10 9 8 7 6 5 4 3 2 1

To my parents,
the late Smt. Vimla Maini and
the late Sh. Sukhdev Raj Maini

Contents

Preface

Low-intensity conflicts in the present day environment are an area of grave concern to the entire international community. These conflicts pose an alarming threat to national security. The scope of countering these conflicts extends from emergency preparedness and response to domestic intelligence activities to riot and mob control; from combating illegal drug trafficking to protection of critical infrastructure; from handling counterinsurgency and antiterrorist operations to detection of nuclear threats; and from detection and identification of chemical and biological warfare agents and explosive materials to detection of concealed weapons. Technology plays a major role in countering threats originating from these conflicts. The low-intensity conflict and homeland security operations under discussion in this book mainly include detection, observation, and surveillance operations, riot/mob control operations, counterinsurgency and antiterrorist operations, detection and neutralization of explosives, detection of chemical, biological, radiological, and nuclear (CBRN) agents, and operations relating to protection of critical infrastructure. There are various technology alternatives including electrical, chemical, kinetic, acoustic, laser-based, and optoelectronic available for most operations.

Laser and optoelectronics technologies play an important role in handling low-intensity conflicts and homeland security related situations. The key advantages of use of laser technology in such applications are near-zero collateral damage, speed-of-light delivery, and potential for building nonlethal—or to be more precise—less-lethal weapons. Different technologies and devices covered in the book include less-lethal laser devices, sighting, observation and surveillance devices, night vision technologies and devices, directed energy lasers, and technologies for detection and identification of hazardous chemical, biological, explosive, and radiological materials.

Chapter 1 provides an overview of nonoptical and optoelectronic technologies with applications in homeland security. This is followed by Chapter 2, which presents an overview of laser fundamentals and devices and photosensors that would be particularly beneficial to a large cross section of working professionals in the defense industry and the armed forces. Chapter 3 comprehensively covers less-lethal laser weapons intended for antipersonnel operations. The chapter begins with an overview of less-lethal nonoptical technologies for similar application scenarios and then moves on to discuss at length operational aspects of laser dazzlers. Salient features of some representative laser dazzler systems designed for various deployment modes are also briefly discussed. Chapter 4 focuses on use of directed energy lasers for neutralizing optoelectronic sensors of adversary, ordnance disposal, and satellite debris removal. Chapter 5 discusses sighting, observation, and surveillance devices including different types of weapon sights, surveillance cameras, laser microphones used for covert listening, optical sniper locators, laser range finders, and detectors of concealed weapons. Chapter 6 comprehensively covers night vision technologies and devices including image intensification and thermal imaging technologies. Chapter 7 covers different nonoptical and optical technologies for detection and identification of explosive materials with a focus on the latter. Chapter 8 focuses on CBRN threats including the types of CBRN threats, challenges involved in their detection, and technologies for their detection and identification. All chapters are suitably illustrated with a large number of tables, photographs, and practical information and end with a comprehensive list of references that will be of particular interest to researchers and working professionals.

The book is intended to be a ready reference for graduate students specializing in defense electronics, lasers, and optoelectronics, R&D scientists and engineers engaged in R&D and manufacturing of defense electronics and optoelectronics equipment in the defense industry, professionals in the armed forces engaged in operation and maintenance activities, military planners, amateurs, and hobbyists. The book provides information in a single volume on operational aspects, deployment scenarios, actual usage, and the state of the art in the spectrum of laser- and optoelectronics-based technologies for low-intensity conflicts, enabling readers to fully understand the utility of each technology/device for a given operational requirement. It will allow personnel in paramilitary forces and the armed forces to gain insight into technological aspects rather than only an overview of a wide spectrum of laser and optoelectronic sensor systems designed for homeland security applications, which will allow them to evaluate the pros and cons of various technologies. I hope that the book will be well received by the readers. Constructive suggestions toward making the book more useful in future editions are highly appreciated.

1

Optoelectronics for Homeland Security: An Overview

This chapter introduces different laser and optoelectronics technologies with the potential to be employed in low-intensity conflict and homeland security applications. Beginning with an overview of low-intensity conflicts and its associated security implications, the chapter briefly describes both laser and non-laser technologies having applications in low-intensity conflicts. Major topics introduced in the chapter include less-lethal weapons including different variants of laser dazzlers for riot/mob control, countering insurgents, terrorists, and asymmetric threats, surveillance technologies, night vision devices, devices for detection of chemical, biological, and explosive agents, directed energy lasers for neutralization of threats, and technologies for detection of concealed weapons. The topics discussed in this chapter are described at length in subsequent chapters.

1.1 Low-Intensity Conflicts

A low-intensity conflict is the most common form of warfare today and is likely to be so in foreseeable future. Data suggests that more than 75% of the armed conflicts since World War II have been of the low-intensity variety. Low-intensity conflict operation is a military term used for deployment and use of troops and/or assets in situations other than conventional war. Compared to a conventional war, in the case of low-intensity conflict operations, armed forces engaged in the conflict operate at a greatly reduced tempo, perhaps with fewer soldiers, reduced range of tactical equipment, and limited scope to operate in a military manner. The use of artillery is also avoided in the case of conflicts in urban territories and use of air power is often restricted to surveillance and transportation of personnel and equipment. Low-intensity conflicts pose an alarming threat to national security. It is an area of concern for the international community today. Its scope extends from emergency preparedness and response to domestic intelligence activities to riot and mob control, from combating illegal drug trafficking to protection of critical infrastructure, from handling counterinsurgency and antiterrorist operations to detection of nuclear threats, and from detection and identification of chemical and biological warfare and explosive agents to detection of concealed weapons.

1.2 Technology for Homeland Security

The low-intensity conflict and homeland security operations under discussion in this book mainly include detection, observation, and surveillance operations, riot/mob control operations, counterinsurgency and antiterrorist operations, detection, and neutralization of explosives, detection of chemical, biological, radiological, and nuclear (CBRN) agents, and operations relating to protection of critical infrastructure.

There are various technology alternatives available for most operations. For example, there are a number of different less-lethal technologies and devices other than optical technologies such as use of laser dazzlers available for riot/mob control operations. These include use of rubber bullets, blunt impact projectiles, foam batons, pepper spray, and conducted energy devices (CEDs), use of chemicals such as oleoresin capsicum (OC) or chlorobenzalmalononi-trile (CS) gas. As another example, there are laser as well as RF technologies available for the purpose of detection of explosives. Ground penetration radar (GPR) technology, X-ray backscatter technology (XBT), vibrometry and laser bathymetry are potent technologies for detection of buried mines. Ion mobility spectrometry (IMS), electronic noses, millimeter-wave imaging, terahertz spectroscopy, and laser spectroscopy based devices are also available for the purpose of detection of overground explosives. As yet another example, there are host of technologies including microwave, infrared, and laser technologies available for protection of critical infrastructure such as military establishments, nuclear installations, commercial hubs, and industrial plants.

Laser and optoelectronics technologies play an important role in handling low-intensity conflict situations. The key advantages of use of laser technology in such applications are near-zero collateral damage, speed-of-light delivery, and potential for building nonlethal weapons. Some of the well-established laser devices in low-intensity conflict (LIC) applications include laser dazzlers for close combat operations, mob/riot control, countering asymmetric threats, and use of laser-based intrusion detection systems for protection of critical infrastructures from aerial threats; lidar sensors for detection of chemical, biological, and explosive agents; terahertz imaging detection of concealed weapons; laser devices such as optical target locator for sniper and gunfire location identification; and so on. Use of laser vibrometry and electron speckle interferometry techniques for detection of buried mines and high-power lasers for disposal of unexploded ordnances are emerging applications of laser technology for homeland security.

1.3　Less-Lethal Laser Weapons

Nonlethal, or more appropriately, less-lethal laser weapons act as a force multiplier enabling friendly forces to discourage, delay, or prevent hostile action. They are particularly effective in situations where use of lethal force is not preferred, such as when limiting escalation and temporarily disabling facilities and equipment. Laser-based nonlethal weapons such as laser dazzlers can be used for counterinsurgency, antiterrorism, unruly riot/mob control, and infrastructure protection applications. Laser dazzlers are emerging internationally as a new nonlethal alternative to lethal force for law enforcement, homeland security, border patrol, coastal protection, infrastructure protection, and a host of other low-intensity conflict scenarios due to their proven efficacy and significantly minimized collateral damage.

1.3.1　Laser Dazzler

A laser dazzler emits a high-intensity laser beam in the visible band, usually in the blue-green region, to temporarily impair the vision of the adversary without causing any permanent or lasting injury or adverse effect to the subject's eyes. The blue-green region is the chosen wavelength band as a human eye's response is the highest in this band. Laser dazzlers typically employ diode-pumped solid-state lasers. Nonlinear optical devices are employed for shifting the laser wavelength to the visible part of the spectrum. Semiconductor diode lasers emitting in the visible spectrum, generally in the form of an array of laser diodes, have also been used in some cases. For achieving a temporary blindness effect, any visible wavelength can be used. However, the green wavelength, falling at the peak response of an eye, is found to be more effective. Here a widely diverging laser beam is pointed toward the object of interest. The diverging output also results in a larger spot at the intended target, making aiming the device over long distances or at multiple subjects much easier.

A laser dazzler system typically uses a laser power of a few hundreds of milliwatts for short-range handheld devices to a few tens of watts for long-range (up to 10 km) platform-mounted devices. The most commonly employed wavelength for the purpose is 532 nm, usually generated as mentioned earlier by using either laser diodes or frequency-doubled neodymium: yttrium-aluminum-garnet (Nd-YAG) laser modules. A laser dazzler is a nonlethal weapon specifically designed for applications where subject vision impairment is to be achieved at a distance ranging from a few tens of meters to several kilometers in bright ambient conditions.

1.3.2   Applications of Laser Dazzlers

Lasers dazzlers have shown lot of promise in close quarter battle (CQB) operations including counterinsurgency and antiterrorist operations, riot control, preventing potentially hazardous vehicles from getting too close to security forces, warning potentially malicious intruders in no-fly zones, protection of critical assets from aerial threats, countering asymmetric threats on sea by using laser dazzlers on friendly naval vessels, and countering man-portable air-defense systems (MANPADS) using dazzlers on friendly aerial platforms.

1.3.3   Choice of Parameters

The choice of performance parameters such as laser power, spot size at the target, or laser power density are driven by the nature of deployment. The beam-shaping and directing optics are designed to achieve the desired value of nominal ocular hazard distance (NOHD) and a laser power density that does not exceed the maximum permissible exposure (MPE) figure dictated by American National Standards Institute (ANSI) standards for eye safety. The MPE is dependent on wavelength and exposure time. At 532 nm wavelength, the MPE value is 2.5mW/cm^2 for an exposure time of 0.25 second and 1.0 mW/cm^2 for an exposure time of 10 seconds. These devices usually produce a randomly pulsed output in the range of 10 to 20 Hz riding a DC level for better overall effect. DC level is usually kept at 30% to 50% of the peak intensity level. Nighttime maximum operational range is typically three to four times the maximum daytime operational range.

1.3.4   Representative Laser Dazzler Devices

A large number of companies are offering short- to medium-range laser dazzlers with operational ranges from tens of meters to a few kilometers for various scenarios. These devices are particularly suited to counterinsurgency and homeland security applications. These are available in a variety of package and mounting configurations. Most devices are configured around frequency doubled Nd-YAG modules operating at 532 nm and are designed to produce power density in the range of $0.2 - 0.6 \text{ mW/cm}^2$ at the target plane. Advanced versions of these devices have built-in safety measures that virtually eliminate the possibility of causing any lasting injury to the target's eyes in case it gets too close. Two types of safety measures are available in these devices. In one type, laser spot diameter is controlled in accordance with the target distance to keep the laser power density at the target plane within prescribed safe limits. In the other type of safety, the laser is switched off in case the target comes within the NOHD.

The first reported use of laser dazzlers was from Royal Navy warships by the British during the Falkland Islands War of 1982. Russians have also been reported to use laser dazzlers against American aerial platforms in the straits of Juan de Fuca. One popular family of short range laser dazzlers including GLARE MOUT (Figure 1.1), GLARE MOUT PLUS, GLARE GBD-IIIC, GLARE LA-9/P, and GLARE RECOIL have been developed by B.E. Meyers to provide a nonlethal deterrent weapon which temporarily interferes with a suspect's vision without causing any ocular damage. The GLARE MOUT is a nonlethal visual disruption laser with an effective range of 150m to 2 km. The device is ideally suited for small arms integration as well as mobile crew-served applications. GLARE GBD-IIIC emits twice the power of GLARE MOUT and also produces a more concentrated beam, thereby offering a maximum range of 4,000m. Laser dazzlers type GLARE LA-9/P and GLARE RECOIL from B.E. Meyers have a built-in safety that switches off the laser beam in case of a target coming within the specified NOHD. Laser output is resumed once the target gets out of NOHD. Yet another short-range laser dazzler is the Compact High Power (CHP) laser dazzler from LE Systems, developed with the sponsorship of DARPA. The dazzler emits a 500-mW flashing green dazzling laser beam and is capable of engaging targets up to 200m. Some of the other popular portable laser dazzler devices with comparable specifications include Dazer Laser from Laser Energetics, which is available in two variants, namely Defender and Guardian, Saber-203, Chinese JD-3 laser dazzler, Threat Assessment Laser Illuminator (TALI) from Wicked Lasers, United States, Medusa

Figure 1.1   B. E. Meyers' GLARE MOUT laser dazzler. (Courtesy: Wikimedia Commons.)

and Hydra laser dazzlers from Passive Force, United Arab Emirates, and Green Light Optical Warner (GLOW) from Thales Group, United Kingdom.

Laser dazzler systems for riot, mob, and crowd control applications are also under development. Several crowd control laser dazzler systems with day and night time operational ranges of the order of 250 to 500m have been developed. Figure 1.2 shows one such system. These are generally vehicle-mounted systems with the laser integrated into a gimbal platform and equipped with a joystick control, usually provided in front of the codriver's seat for viewing, scanning, and controlling laser beam position on the target scene. An electronically controlled laser beam projector assembly usually controls the laser beam spot size at the target.

Long-range dazzler systems with maximum operational range approaching 10 km are also commercially available. The ZM-87 Portable Laser Disturber is a Chinese electro-optic countermeasure laser device. It can blind enemy troops at up to a 2- to 3-km range and temporarily blind them at up to a 10-km range. The ZM-87 has been widely deployed, including on naval vessels. Long-range laser dazzlers intended for protection of critical infrastructure and strategic assets are usually integrated with a 360° gimbal mount and an electro-optic tracker. Such a system would be capable of providing effective means to thwart suspect aircraft and other aerial vehicles from violating no-fly zones and protected airspace. The dazzler can also provide effective means for intention assessment of the suspect aircraft, thereby facilitating quick decision for restoration of lethal forces.

While short- and medium-range handheld and weapon integrated versions of laser dazzler are in widespread use for antiterrorist and counterinsurgency applications; vehicle-mounted systems designed for operations against unruly and violent crowds are beginning to appear on the scene. Maritime port security, exclusion zone enforcement, nuclear plant perimeter security, aviation

Figure 1.2   Vehicle-mounted laser dazzler.

airport perimeter security, oil and gas critical infrastructure protection, and industrial petrochemical plant perimeter security are other applications. Another interesting development in the recent past has been a keen interest in long-range laser dazzler systems mounted on gimbal platforms and equipped with an electro-optic tracker and also integrated with a network of radars to provide 24/7 protection to strategic assets from airspace violation by aerial platforms. In a not-too-distant future, one would see deployment of laser dazzlers with global coverage. These systems are proposed to use a remotely controlled membrane reflector to receive the dazzling laser beam from the source station and guide it to the intended target location. Figure 1.3 illustrates the concept.

1.4　Directed Energy Lasers

Directed energy laser systems primarily use directed energy of the coherent laser beam in the targeted direction to cause intended damage to the object, which could be an enemy's electro-optic systems, facilities, and even personnel. The targeted electro-optic systems could include sighting and observation devices, such as night vision devices, thermal imagers, charge-coupled device (CCD) or complementary metal-oxide semiconductor (CMOS) sensor-based surveillance cameras, and laser range finders. Laser-based directed energy weapons (DEW) could cause structural damage to an enemy's weapon systems and platforms.

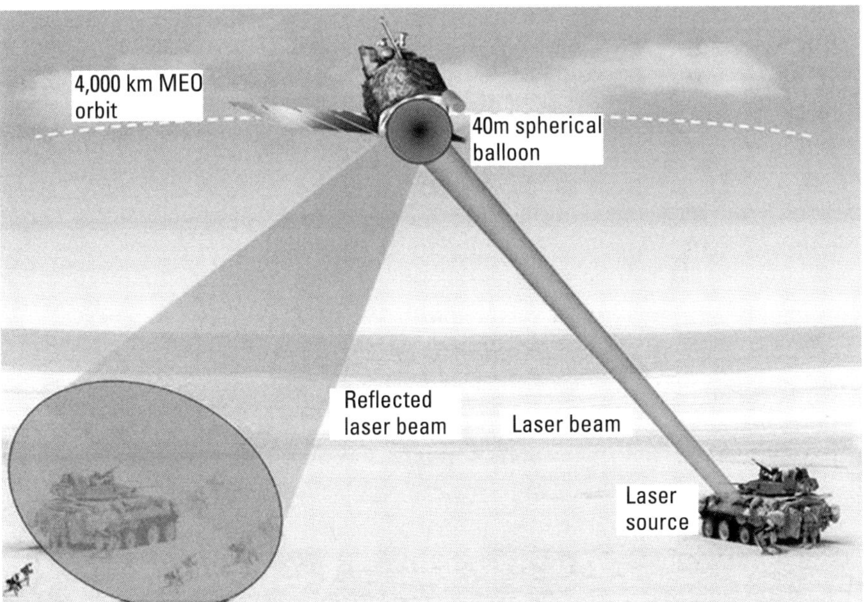

Figure 1.3　Laser dazzler for global coverage.

Though high-power directed energy laser weapons may not be of great significance for low intensity conflict operations, there could be a role for relatively lower-power laser DEW, as briefly outlined in subsequent paragraphs, in homeland security applications. A laser dazzler, briefly discussed in the previous section, also uses directed energy property of the laser beam for the intended effect. From the viewpoint of countermeasures terminology, it may be viewed as a type of antisensor laser only where the target sensor is the human eye. They are discussed further in detail in Chapter 4.

1.4.1 Laser Countermeasures

In the category of antisensor systems, we have systems capable of causing only a temporary disability of electro-optic devices and optoelectronic sensors deployed by the adversary and systems that are capable of inflicting permanent damage. In both cases, the target is the front-end optics and optoelectronic sensors. These are sometimes referred to as *soft-kill systems*. These systems have, by no means, capability of inflicting a physical or structural damage to the platform carrying weapons. There are *hard-kill systems* that are capable of inflicting physical damage to the front-end optics of any electro-optic system. These systems are usually vehicle mounted, are large in size, and weigh more than their soft-kill counterparts. The pulse energy level in such lasers is of the order of several kilojoules, as compared to a few hundreds of millijoules to few joules in the case of soft-kill systems.

No military platform, be it land-based, aerial, or shipborne, today is free from the risk of being exposed to laser radiation. The activities of these platforms are under constant surveillance by various kinds of electro-optic devices and optoelectronic sensors. Rendering these devices and sensors ineffective in the battlefield therefore makes a huge difference to the battlefield competence of a nation. Deployment of EOCM devices and systems designed to incapacitate or neutralize the more conventional laser devices and systems would act as a force multiplier, as a platform incapacitated in the enemy camp is a platform added to your own. This goes a long way in enhancing the survivability quotient of armed forces equipped with such a capability. In the context of low-intensity conflicts or homeland security applications, neutralization of electro-optic devices and optoelectronic sensors, such as sighting and observation devices including night vision devices, thermal imagers, CCD and CMOS sensor-based surveillance devices, laser range finders used by insurgents, terrorists, and other rogue elements, can be a huge tactical advantage for the security agencies and paramilitary forces. It may be mentioned here that it could be possible to inflict an irreversible damage to photodiode, CCD, or CMOS-sensor based observation and surveillance devices by using small arms mounted laser modules. As an example, a Q-switched Nd-YAG laser module generating 1mJ, 10-ns pulses, or a 1-W CW laser at 1,064-nm wavelength when focused to a 100-micron spot on the photodiode sensor would produce energy density or power density greater than the damage threshold. CCD/CMOS sensor arrays and image intensifier sensors used in night vision devices have much lower damage thresholds.

1.4.2 Lasers for Ordnance Disposal

In the category of the DEW class of laser systems, disposal of unexploded ordnances including surface-laid mines, improvised explosive devices (IED), grenade shells, artillery or mortar rounds, cluster bombs, and so forth, from safe stand-off ranges using a high-power laser beam is an emerging application of directed energy laser systems in homeland security related applications. Earlier, disposal of unexploded ordnance necessitated that highly trained bomb disposal specialists wore body armor and protective suits to ensure operational safety. Also, a large part of the adjoining area needed to be cleared for safety reasons before the operation could be executed. Subsequently, robotic platforms were developed for explosive ordnance disposal. These robots offer high reliability and excellent maneuverability and can be used to identify and disarm booby traps, fireworks, improvised explosive devices, and other dangerous objects in closed areas, buildings, and vehicles. They also perform reconnaissance, monitoring, and investigation of objects in exceptionally dangerous conditions. Figure 1.4 shows a photograph of one such explosive ordnance disposal robotic platform named EOD-5.

The laser ordnance neutralization system is being considered for a range of application scenarios. One possible application relevant to homeland security is disposal of improvised explosive devices, also called roadside bombs. Laser-based disposal system offers a quick, safe, and reliable method of neutralizing these roadside bombs from a safe stand-off distances in excess of 150m. Another

Figure 1.4　EOD-5 robot. (Courtesy: Wikimedia commons.)

possible application is in disposal of friendly explosives that have outlived their shelf life. Since full-scale conventional wars are not there to be fought every now and then, there is a large inventory of ammunition stored in the ammunition depots that have exceeded their shelf lives and are waiting to be neutralized. Neutralization of such large quantities of ammunition is not only a cumbersome exercise, it also poses a safety hazard.

The laser in this case is a kilowatt-class laser with CW output power in the range of 1 to 5 kW as compared to 100 kW to a megawatt required in the case of any realistic directed energy laser weapon system. Ordnance is disposed of by focusing a high-power laser beam on the ordnance casing, thereby heating it until the temperature of the backplane of the casing exceeds the ignition temperature of explosive filler. The explosive filler ignites and begins to burn. The process is independent of the type of fusing used by the target explosive. This leads to a low-level detonation or deflagration rather than full-power detonation. The advantages of laser energy usage for ordnance disposal include a large magazine, high precision, controllable effects with reduced collateral damage, and assured and fast disposal from safe stand-off ranges.

The concept of neutralization of live ordnances using laser energy was first demonstrated in the field in 1994 with the development and field testing of the mobile ordnance disrupter system (MODS) that employed a 1.1-kW arc lamp driven solid-state laser mounted on an M113 A2 armored personnel carrier. Some better-known laser ordnance neutralization systems in use today include ZEUS-HLONS (HUMMWV Laser Ordnance Neutralization System) configured around a 2-kW fiber laser-mounted on top of an adapted Humvee platform, the Laser Avenger system from Boeing Combat Systems using a 1-kW solid-state laser mounted on a converted antiaircraft vehicle, and the THOR system from Rafael. The United States Army and Air Force have also tested a similar high-power laser system named Recovery of Airbase Denied by Ordnance (RADBO) for rapid clearance of airbases of unexploded ordnances (Figure 1.5).

1.5   Protection of Critical Infrastructure

Every sovereign and independent country or state has assets that need to be protected against natural disasters and intrusion by rogue elements and terrorist attacks for strategic, economic, and social reasons. These assets are referred to as the critical infrastructure. Laser and optoelectronics technologies play an important role in providing critical infrastructure protection, particularly to physical infrastructure. There are a host of technologies and devices available for the purpose. Some of the important ones include infrared imaging cameras, such as short-wave infrared (SWIR), midwave infrared (MWIR), and long-wave in-

Figure 1.5 RADBO UXO clearing system.

frared (LWIR) cameras for surveillance applications, laser fencing for perimeter protection, laser scanners or lidar sensors creating customized infrared barriers for intrusion detection, laser-based directed energy weapons for protection against rocket, artillery, mortar (RAM) threats, precision strike weapons and small unmanned aerial vehicles (UAVs) and *laser debris removal* for protection of space assets.

SWIR cameras are used to detect covert battlefield lasers (laser rangefinders, laser designators, and laser pointers); MWIR cameras can penetrate fog and smoke, see through camouflage colors and patterns, detect gun flashes and long-range surveillance applications; and LWIR cameras are used for thermal imaging applications. While SWIR cameras look through fog, haze, and smoke and can identify humans or objects during night and day, thermal imaging cameras can detect at very long distances and in pitch darkness. These cameras can be installed on mobile platforms, such as vehicles, vessels, and airborne platforms, and also as fixed monitoring positions on roofs and posts.

Laser fencing, also called laser wall, uses multiple pairs of laser sources and sensors to create an invisible laser fence along the perimeter of the infrastructure to be protected. Intrusion of this line-of-sight between source and sensor is used to activate an alarm or relay the information immediately to a nearby post for action. The concept can be used by security forces to thwart infiltration along the border.

Laser scanners, also called lidar sensors, can be configured to create customized infrared (IR) barriers or follow the contours of some landscape or surroundings. Lidar sensor technology is highly suited to the task of securing critical infrastructure such as power plants, nuclear reactors, nuclear material and waste disposal sites, bridges and dams, drinking water and water treatment facilities, chemical plants, and arms and ammunition depots. These scanners, employing the time-of-flight principle, create an invisible IR barrier and trigger an alarm or a specified event once the IR plane is broken. Multiecho technology and high-tech filters provide enhanced security system reliability even in harsh and inclement weather conditions. In harsh weather conditions such as rain, fog, or snow, part of the energy from a pulse of the laser scanner may be reflected by nearby objects, like rain, while the remainder of the beam continues to propagate and is reflected by the actual obstacle. A laser scanner with multiecho technology evaluates multiple echoes and ignores the closer, weaker reflections caused by the environmental factors. Laser scanners, when used together with other security apparatus, provides timely and automated response to security threat. For example, when integrated with autotracking cameras and GPS, the system allows tracking of an intruder's location and movement. Laser-Guardian Intrusion Detection from M/s Sick, Inc. in the United States is one such system that employs laser scanning technology combined with security cameras and software tools to provide reliable intrusion detection and security for any facility.

Directed energy weapons capable of neutralizing RAM threats, drones, MANPADS, guided weapons, and ballistic missiles in their flight paths have been traditionally considered as weapons of deployment in conventional wars and rarely the weapons for homeland security or low-intensity conflict operations. However, emergence of global terror networks and sophisticated smuggling techniques has allowed terrorists and other rogue elements an access to missile technologies and precision strike weapons, thereby making critical infrastructure highly vulnerable to precision missile attacks. Directed energy weapons therefore can be very effective in countering short-range missile threats to critical infrastructure. Although there are other options for protection of critical infrastructure from missile threats, in the long run, none will be as cost-effective, precise, safe, and swift as a directed-energy weapon-based defense system. The technology of directed energy weapons is yet to come to a level to be deployable for protection of critical infrastructure. A well-known laser-based DEW system that has been successfully field tested is Northrop Grumman's Tactical High Energy Laser (THEL). The laser is built in two configurations; baseline static HEL (Figure 1.6) and Mobile-THEL. THEL systems are point defense weapon systems designed to engage and destroy artillery rockets, artillery shells, mortar rounds, and low-flying aircraft. The system uses a deuterium fluoride (DF) laser operating at 3.8 μm. Although DEW system prototypes configured around chemical lasers such as DF and chemical oxygen iodine laser (COIL) lasers and high-power solid-state and fiber lasers have been built and field-tested, the deployable DEW systems are going to be the ones configured around high-power solid-state lasers and high-power fiber lasers.

Space assets mainly including different categories of satellites constitute one of the most vital and critical infrastructures whose uninterrupted

Figure 1.6   Static Tactical High Energy Laser (THEL). (Courtesy: Wikimedia Commons.)

operability is very crucial for national security. These space assets provide support for navigation, communications, reconnaissance and surveillance, weather forecasting, and many other functions key to operations carried out by armed forces. Therefore, they must be protected against various possible threats. There are intentional threats such as those from use of antisatellite weapons, natural threats including magnetic storms, high-energy particle radiation and collision with meteors, and artificial threats posed by space debris mainly caused by decommissioned satellites and rocket stages and their parts. It is the protection of critical infrastructure from space debris where the use of laser technology holds lot of promise. More than 200 million objects with diameters greater than 1 mm accounting for 6,500 tons constitute space debris. Close to 80% of this is in low-earth orbit with high concentrations at 900 and 1,400 km. The debris is growing exponentially. While small-sized debris with diameters up to 1 cm can be countered by making satellites more robust and using other risk mitigation measures, large-sized debris with diameters greater than 10 cm can be tracked by ground stations that enable use of collision avoidance maneuvers. It is the medium-sized debris that is hard to counter. It is here that use of laser technology becomes relevant. Laser technology can be used to clean the low-earth orbit of the hazardous debris.

1.6   Sighting, Observation, and Surveillance Devices

Sighting, observation, and *surveillance devices* such as different types of sights, night vision devices, laser range finders, optical target locators, and laser-based covert listening devices and surveillance cameras constitute an essential component of the arsenal of the armed and paramilitary forces fighting riotous crowds, insurgents, terrorists, and other rogue elements. Sighting devices including different types of sights are used for alignment and aiming of weapons from small arms such as assault rifles, snipers, and carbines to machine guns, antiaircraft guns, mortars, armored carriers, and main battle tanks. Different types of sights include the simplest-of-all mechanical sights such as iron sights and optical sights such as telescopic and reflector or reflex sights, holographic sights, and laser sights.

Iron sights comprise a system of shaped alignment markers having a front sight, usually a block or post, and a rear sight with a notch. While aiming, the front sight and the notch in the rear sight need to be aligned and also be on the target. *Telescopic sights* (Figure 1.7) generally comprise a combination of lenses and reflective surfaces within a metal sleeve. The ocular lens that the shooter looks through is smaller and has lower magnification. The lens at the telescope's far end called objective increases magnification. There are telescopic sights with fixed as well as variable magnification. Although magnification and diameter of objective lens are the important specifications of telescopic sights, they are also characterized by the coatings used for the glass both to protect the glass itself and to alter the way the light enters the optics, levels of adjustment for windage, and elevation and types of reticles. Some telescopic sights also include electronic elements to assist in ballistics calculations.

In a *reflector* or *reflex sight,* the shooter looks through a partially reflecting glass element to see at infinity a projection of a reflective or luminous aiming point or some other image placed at the focus of a lens or image forming curved mirror with the result that anything at the focus will look as if it is sitting in front of the viewer at infinity. When red-light-emitting diodes are used to create illuminated reticle, it is called red dot sight.

Figure 1.7   Swift 687M telescopic sight. (Courtesy: Wikimedia Commons.)

In the case of a *holographic weapon sight* or simply a *holographic sight*, which is a nonmagnifying gun sight; the shooter looks through a glass window having a laser transmission hologram of a reticle image built into it. The shooter sees a reticle image superimposed on the field of view at a distance when the recorded hologram is illuminated by a collimated light beam from a laser diode. It is a nonmagnifying gun sight. Unlike conventional telescopic sights where the shooter must have his or her eye aligned with the telescope in order to place the reticle over the target, with a holographic sight, the reticle is always on the spot on the target where the weapon is pointed. In telescopic sights, if the shooter's head isn't perfectly aligned; the scope's reticle is not pointing to the same spot on the target as the gun. In contrast, with holographic sights, movement of the weapon toward left or right moves the reticle as well, with the result that the reticle continues to remain on aim point. One often compares red dot sights and holographic sights vis-à-vis their preferred applications. Range distance does not limit dot visibility; they are popular for use on pistols, in shooting sports, and even in military applications. In comparison, holographic sights can be used with extreme accuracy from distances of up to 300m and they perform best in closed-quarter battle (CQB) operations where speed is critical and where peripheral vision for dealing with multiple-threat situations is important.

Laser sight also known as *laser aiming aid* is an aiming device that uses small, low-cost, low-power semiconductor diode laser modules for target aiming and pointing particularly during nighttime operations. This increases weapon effectiveness by improving the single-shot hit probability and reducing collateral damage. Both visible and infrared emitting laser diodes have been used for this purpose. Visible laser sights emitting in red and green wavelength are available with the former being more common. Red laser sights require relatively much lower electrical energy for operation that makes them more compact. As well, these are less expensive. The shortcoming is that red laser spot cannot be seen in bright sunlight beyond eight to 10m. Green laser sights offer a brighter, easier-to-see aiming point. Due to peak response of human eye to green wavelength, some green lasers can be seen in bright sunlight out to 100m. However, green lasers consume more power and therefore drain battery power faster. As a result, they tend to be bulkier and more expensive. In the case of infrared laser sights, the laser spots are invisible and can be seen through a night vision device. The module can be integrated with the weapon through Picatinny or universal rails. Laser sights generally have provision of precise X-Y movement to enable alignment of laser beam with the barrel of the weapon. Laser sights are often available as an integrated package of one or more sighting and observation devices such as optical sight, flash light, and laser sight. Figure 1.8 shows a photograph of one such combo of laser sight, flash light, and tactical sight.

The most common usage of laser technology in tactical military operations has been as *laser rangefinders* for surveillance and fire control applications

Figure 1.8   BSA laser sight-tactical sight-flashlight combo.

and *laser target designators* for munitions guidance. One cannot imagine a modern armored fighting platform or a battlefield tank whose fire control system does not utilize the services of a laser rangefinder. Range finders employ various techniques for the purpose of determining range. These include the time-of-flight technique, phase shift technique, triangulation technique and frequency modulated–continuous wave (FM-CW) technique. Short-range semiconductor diode laser-based rangefinders are finding use on squad weapons such as assault rifles and light machine guns. Laser range finders find extensive use as stand-alone devices for the purpose of sighting, observation and surveillance, reconnaissance, target acquisition, and situational awareness of an adversary's movement of personnel and military assets. Figure 1.9 shows of a handheld lightweight all-weather, day and night system with built-in GPS, thermal imaging, and laser range finder of up to 5 km. This laser range finder, which is called Surveillance System and Range Finder (SSARF) from Thales, UK, allows a soldier to quickly establish the exact location and distance of enemy forces and determine the most appropriate and accurate mortar or artillery firepower

Figure 1.9   SSARF. (Courtesy: Wikimedia Commons.)

to use. Compact laser range finder modules are available as OEM products for integration with other systems. While the basic range finder can be used to find target range, when combined with a digital magnetic compass and inclinometer, it can also be used to determine target coordinates. The other important application of a laser range finder is in integrated fire control system of armored fighting platforms. Laser range finder when interfaced with fire control computer significantly enhances target accuracy. When interfaced with night vision, thermal, and daytime optical aids, laser range finders lead to many useful and effective battlefield applications for observation, surveillance, and situational awareness. High-repetition rate laser range finders in a target designator are used for munitions guidance and are also at the heart of gap measuring devices, proximity sensors, lidar sensors, and laser trackers.

With laser-guided munitions becoming as narrow as large-caliber ammunition and a laser target designator becoming as small as a pistol, the munitions can be delivered to its intended target with the pinpoint accuracy once only possible with mammoth precision-guided missiles and hulking bombs, thereby bringing guided munitions technology to low-intensity conflict operations. Pike munition is an example. Pike is a 40-mm guided munition designed by Raytheon that uses a digital semiactive laser seeker. It can be fired from the barrel of a Heckler & Koch M320 grenade launcher module and enhanced grenade launcher module (EGLM) like a standard 40-mm grenade. While one soldier designates the target; the other fires the munition. The munition is powered by a rocket motor that propels it to an extended range of 2,000m.

Another device without which security apparatus would be incomplete is the *surveillance* or *security camera*. Security cameras provide round-the-clock surveillance for our homes, offices, and critical civil, industrial, and military infrastructures such as highways, ports, airports, power plants, oil refineries, dams, bridges, nuclear centers, and ammunition depots, and public places like parks, shopping malls, and restaurants. A strategically placed camera can provide vital evidence against thefts, crimes, and other unlawful activities, and untoward incidents and insurgency. A complete security system would often be comprised of one or more security cameras, a digital video recorder (DVR), and a monitor. There is a wide variety of security cameras depending on size, shape, and mounting arrangement and major performance specifications including resolution, field-of-view, optical zoom, digital zoom, wired or wireless, capability to operate in low light or nighttime conditions, capability to operate in harsh environmental conditions, and remote operation. There are bullet cameras (Figure 1.10), spy cameras, and dome cameras (Figure 1.11). Bullet cameras are particularly suited to homes and small offices. Spy cameras are usually fitted in an object of regular use such as shirt buttons, calculators, pens, car keys, or even watches, which allows them to be carried along without being noticed. Dome cameras, named after their shape, come with cameras fitted in

Figure 1.10   Bullet camera. (Courtesy: Wikimedia Commons: Amirecoy-Own Work.)

Figure 1.11   Dome camera. (Courtesy: Wikimedia Commons: StephaneLL-Own Work.)

the dome and are either mobile or mounted on walls or ceilings. There are also high-definition (HD) cameras, pan-tilt-zoom (PTZ) cameras, wireless cameras, Intenet Protocol (IP) cameras, and thermal cameras.

High-definition security cameras provide the video quality required to identify vital details should an incident occur. In addition to providing better clarity of the scene, the higher amount of pixels also offers greater digital zoom abilities allowing you to see further into the distance and zoom in on distant objects without drastically reducing image quality. High-definition cameras with 4K and 8K horizontal resolution specification are available; 4K and 8K resolution imply number of pixels in a horizontal line of the order of 4,000 pixels and 8,000 pixels, respectively. A typical 4K high-definition camera would offer a resolution of 3,840 pixels × 2,140 lines or 8.3 megapixels. An 8K camera would offer 7,680 pixels × 4,120 lines or 33.2 megapixels.

PTZ cameras (Figure 1.12) with their pan/tilt and zoom features allow them to monitor large areas with a single security camera. The pan and tilt feature, along with continuous 360° rotation, enables PTZ cameras to quickly move to an intended area and object in the scene. The optical zoom feature provides the ability to focus on fine details like faces or license plates. These cameras are generally available as bullet- and dome-shaped cameras. These cameras are available with high-definition resolution and various optical zoom specifications. *Wireless security cameras* offer flexibility of installation. Most wireless cameras are available with all the advanced features of wired cameras including HD resolution, infrared night vision, and motion detection. *IP cameras* allow for viewing the intended object or area remotely from any part of the world by using easy-to-use IP camera software. Thermal cameras see heat as opposed to visible light. Most objects, human beings in particular, tend to give off enough heat to be detected by thermal cameras during daytime or nighttime and in different weather conditions. Thermal cameras make hiding by camouflage extremely difficult and also have a much reduced false alarm rate of objects in motion, such as those from trees blowing in the wind, as they only see moving sources of heat and not objects themselves.

Yet another emerging application of laser technology is in the detection and identification of battlefield optoelectronic sighting systems, which includes weapon sights, binoculars, night vision devices, thermal imagers and datum optical markers, leveling rods equipped with optical reflectors, laser range finders, and target designators. These devices are commonly called sniper detection systems and operate on the principle of the cat-eye effect. Target optical devices are illuminated by a laser beam, in which the optical system returns a fraction

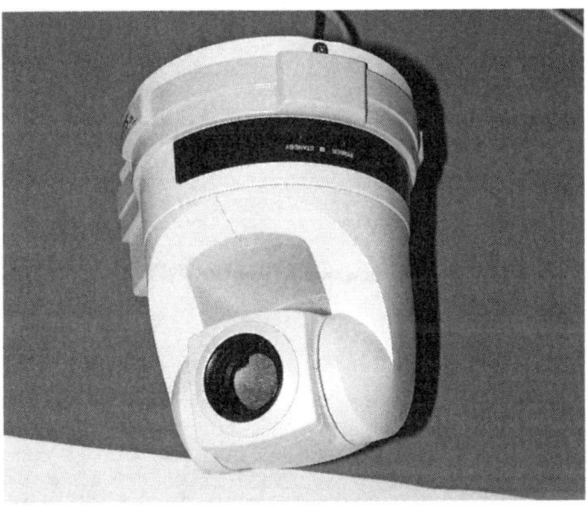

Figure 1.12 PTZ camera. (Courtesy: Wikimedia Commons.)

of it as backscattered energy. Backscattered energy is received by a sensitive receiver. With particular reference to homeland security, such a device could be very useful for detection of optical scopes employed by snipers. Another homeland security related application could be surveillance of sensitive areas particularly in urban territories. Figure 1.13 shows typical deployment of such a device in an urban environment when used for area sanitization. The user illuminates the suspect area, a multistory building in this case, with a widely diverging laser beam emitted by the transmitter portion of sniper detector. If there were an optical device looking toward the laser beam within its field of view; a retro-reflected laser beam would travel back toward the laser transmitter and reveal the location of the optical device.

A large number of portable sniper detector systems with varying specifications of operating or detection range are commercially available. Some of the better-known systems include *laser sniper detector* type SLD-500 from CILAS, France, with a maximum operational range of greater than 2,000m, antisniper scope type Mirage-1200 from Torrey Pines Inc. with a maximum operational range up to 1,200m, LAS-1000 from Newcon Optik with 1,000m maximum operational range, *optics detector* types GCU-OCD10 (Figure 1.14)

Figure 1.13   Sniper detector deployment.

Figure 1.14   Optics detector type GCU-OCD10.

and GCU-OCD20 laser sniper detectors from JSC Sekotech with a maximum operational range of 1,000m, and OCELOT-3 from FidusCrypt GmbH capable of detecting 100-mm optics up to 2,000m. Most devices in this category are capable of detecting optics in their line of sight even if they were covered behind bushes, windowpanes, and windshields. These systems find widespread applications in sniper detection, perimeter intrusion detection and border patrol, protection of critical infrastructure, countersurveillance and counterintelligence operations, and Special Forces combat operations.

A laser microphone is another potent surveillance device. It can be used to eavesdrop on suspect and rogue elements from a distance of few hundred meters. It is a covert listening device that operates by transmitting an invisible infrared beam to the targeted room through one of its windows (Figure 1.15). The device operates as follows. The pressure waves produced due to conversation happening inside the room causes minute vibrations in the windowpane. A part of the laser beam energy bouncing off the windowpane picks up these vibrations. The reflected laser energy is collected by an optical receiver. The receiver converts minute differences in time of travel caused by vibrations in windowpane into intensity variations using principles of interferometry. The device allows simultaneous real-time audio monitoring and recording. Practical devices capable of eavesdropping from as far as 1,000m line of sight under ideal conditions and 500m under practical conditions are available. Laser EMAX-3500, Laser EMAX-3100 and Laser EMAX-2510 from Electromax International, Inc. are some such devices.

1.7 Night Vision Technologies

Night vision devices are extensively used by the military for location of enemy targets, surveillance, and navigation and thereby play a crucial role in enhanc-

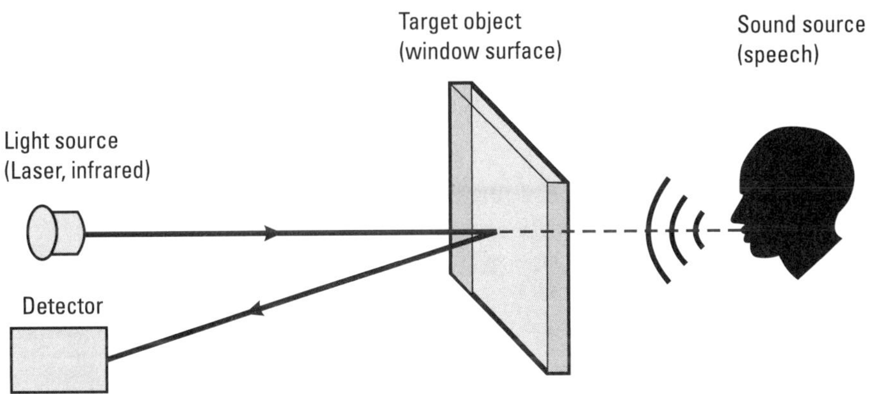

Figure 1.15 Laser microphone—principle of operation.

ing night fighting capability of armed forces. They are also used by law enforcement and security agencies for surveillance. Night vision enabled cameras are even being used by private businesses and civilian homes in addition to military establishments to monitor surroundings of their critical assets. Night vision technologies and the associated night vision devices enable the users see in low light conditions. Contemporary devices allow viewing even in near total darkness. The ability to see in low light conditions is governed by two basic requirements: sufficient spectral range and sufficient intensity range. Low values of spectral range and intensity range in the case of human eyes therefore become the limiting factors for their ability to see with an acceptable level of contrast in low light conditions. Use of technology to enhance both these parameters makes night vision possible. The two basic and widely different approaches to night vision include image intensification (or enhancement) and thermal imaging.

1.7.1 Basic Approaches to Night Vision

Image intensification or *enhancement* works on the principle of collecting small quanta of light reflected off the target scene to be viewed in visible and near-infrared bands of electromagnetic spectrum in low light conditions. The collected photons are amplified through the processes of photon-electron conversion, electron multiplication, and electron-photon conversion. These processes take shape in what is called an image intensifier tube. The other important constituent parts of an image intensifier tube based night vision device include the objective lens used for collection of photons, an eyepiece for viewing an intensified image, and a power supply that generates required DC voltages for electron acceleration. Active illumination is often used in conjunction with image intensifier tube in what is known as active night vision technology to enhance image resolution in very low-level light conditions. Active night vision technology has the disadvantage that it can be prone to giving away the location of the user, which is particularly undesirable in tactical military operations. A variation of the conventional night vision device is the digital night vision device. While in a conventional night vision device, available light is collected through the objective lens and focused on an intensifier; most digital night vision devices process and convert the optical image into an electric signal through a highly sensitive CCD image sensor. Digital night vision devices are free from image distortions on a photo-cathode and blemishes on a phosphorescent screen, are immune to damage by bright light exposure, and offer image recording facility.

Thermal imaging night vision technology works on the principle of detecting temperature difference between the objects in the foreground and those in the background. A thermal imaging device is essentially a heat sensor capable of detecting tiny differences in temperature of different points on the surface of the object to be viewed. The information on the temperature difference available in the form of infrared energy is collected by the thermal imaging device and converted into an electronic image. The ability to detect tiny temperature differences that always exist not only between the desired object and the surroundings but also between different points of the object itself coupled with emission in the infrared region allows a thermal imaging device see in near total darkness. Thermal imaging night vision devices are extensively used by military and law enforcement agencies for the purpose of target detection and acquisition, surveillance and monitoring, search and rescue operations, firefighting, and so forth.

1.7.2　Different Generations of Night Vision Technologies

Night vision technology has undergone substantial changes during its existence of more than 40 years. These changes have led to major improvements in performance standards of night vision devices. Each substantial change in technology is associated with a generation. Both image intensifier and thermal imaging technologies have seen several generations of evolution. Beginning with generation 0, we are currently in the fourth generation in both image intensifier and thermal imaging technologies.

Generation 0 devices were based on image conversion rather than image intensification. *Generation 1 night vision devices* were an adaptation of generation 0 technology in the sense that both used a photo cathode and an anode, the former for photon-to-electron conversion and the latter to accelerate photo electrons toward it. A major deviation in generation 1 night vision devices from generation 0 devices is in the absence of an infrared source used in the case of the latter devices to provide scene illumination. *Generation 2 night vision devices* introduced in 1970s were the first to use microchannel plate (MCP) for electron multiplication leading to significant increase in device sensitivity. As compared to generation 2 devices, *generation-3 night vision devices* had two distinctive changes. These included use of a gallium arsenide photo cathode and an ion barrier coating on the MCP. A gallium arsenide photo cathode enables target detection at longer ranges and in darker conditions. *Generation 3+ devices* offer improved performance specifications over generation 3 devices. Two important features associated with generation 3+ night vision devices are an automatic gated power supply system and a thinned ion barrier layer. The absence of an ion barrier layer or its thinning improves luminous sensitivity even though it comes at the cost of slight reduction in the life of the tube.

There have been different generations of thermal imaging sensors. Each successive generation has incorporated not only a major change in the type of detector but also a major change in optical systems used to image the target onto the detector. Four distinct generations of thermal imagers have been designed based on IR detector technologies developed during the last 35 years, and classified according to the number of elements contained in each group. *First generation* thermal imagers contain single-element detectors or detectors with only a few elements (1 × 3). A two-dimensional mechanical scanner was usually used in order to generate a two-dimensional image. Sensitivity of thermal imaging sensors of the first generation was limited by background radiation. This problem was overcome in *second generation* thermal imaging sensors by using modified front-end optics that reduced unwanted flux. However, this resulted in a fixed f-number for all fields of view. Second generation thermal imagers are vector detectors usually containing 64 or more elements. The two-dimensional scanner was somehow simplified in the vertical direction to include only the interlace motion. *Third generation* thermal imagers contain dual-band two-dimensional arrays with several columns of elements and dual/variable f-number optical system. These thermal imagers still scan in one direction and perform time delay integration of the signal in the scanning direction in order to improve the signal-to-noise ratio. *Fourth generation* thermal imagers contain two-dimensional array detectors (160 × 120, 320 × 240, 680 × 480) called focal plane arrays that do not require any scanning mechanism for acquiring the two-dimensional picture.

1.8 Detection and Identification of Explosives

Terrorism today undoubtedly poses the biggest challenge to survival of humanity on Earth. Though the threat spectrum includes use of conventional explosives, chemical, biological, radiological, and nuclear agents, the most common form of terrorism uses only conventional explosives and more so IEDs. There has been a huge loss of life and extensive damage to public property caused by explosive blasts in the terrorist acts witnessed around the world in the last 10 to 15 years. According to a U.S. Department of State report, while only a few tens of people have been killed due to chemical, biological, and radiological attacks, on the other hand, tens of thousands of people have been killed due to explosive attacks. Some representative examples of explosions include those onboard TWA flight 841 on September 8, 1974, killing all 80 passengers and crew members and Pan Am flight 103 over Lockerbie on December 21, 1988, which left 270 people dead, the World Trade Center bombing of February 26, 1993, killing six and injuring over a thousand, the Oklahoma City bombing on April 19, 1995, leaving 170 dead and 680 injured, the Chechnya border attacks on March 24, 2001, leaving 20 dead and 100 injured, blasts on three commuter trains in Madrid on March 11, 2004 that left 190 dead and 1500 injured, a series of coordinated suicide bomb attacks on July 7, 2005, in the London underground trains, killing 52 people of 18 different nationalities and injuring hundreds, the Mumbai terror attacks of November 26, 2008, killing 164 and wounding 308, and three coordinated suicide bombings in Brussels on March 22, 2016, killing 32 civilians and injuring 300. Equipment needed for detection, identification, and neutralization of explosive threats, therefore, becomes an important component of providing protection to our public transport systems, airports, ports, critical infrastructure and other strategic assets, and protection against roadside bombs and monitoring of container transportation. While timely detection and subsequent neutralization of an explosive saves many lives and prevents damage to public property, knowing the type of explosive even postdetonation helps in identifying the source of origin of the explosive.

There are two broad categories of equipment used for detection of explosives. These include those employing bulk detection methods and those employing trace detection methods. While bulk detection methodology looks for the presence of large amounts of explosives in the explosive device, trace detection methods on the other hand require only trace amounts of explosives either in gas phase or particle form for detection. X-ray scatter, neutron- and γ-based techniques, magnetic techniques, millimeter-wave imaging, and terahertz spectroscopy are important bulk detection methods. Important trace detection methods include IMS, electronic noses, cavity ring down spectroscopy, surface plasmon resonance (SPR) and surface-enhanced Raman spectroscopy (SERS).

Detection of explosives from safe stand-off distances is particularly important when it comes to real operational scenarios. While bulk detection methods have been tried for stand-off detection of explosives, it is trace detection that has been mainly exploited for the purpose. In particular, stand-off detection of explosives using lasers using trace detection method is one of the most widely researched technologies internationally. It is a great technological challenge for the technology to mature to an extent where it can be transformed into a product usable in the kind of environment and field conditions usually encountered in homeland security related applications. One of the major problems comes from a decrease in the intensity of backscattered light signal due to wavelength dependent absorption and scattering losses and the intensity decreasing inversely with distance squared. The problem is compounded by the fact that the trace levels associated with common explosive agents are extremely low, in the range of fraction of parts per billion (ppb) to few parts per million (ppm) in the case of common explosives. The second major problem pertains to unique identification of targeted explosive agents in a background of interferents. Many chemical agents have atomic compositions including sulfur, phosphorus, fluorine, and chlorine in addition to nitrogen, oxygen, hydrogen, and carbon present in organic molecules. Thus the detection methodology needs to be highly sensitive and selective. Laser-based spectrometric methods have the potential of being fast, sensitive, selective, able to detect and identify wide range of explosive agents, and upgradeable to handle new threats.

Atmospheric transmission at the wavelengths concerned is an important factor while assessing the suitability of a given stand-off detection methodology. Commonly used technologies are laser-induced breakdown spectroscopy (LIBS), Raman spectroscopy and its variants; laser-induced fluorescence (LIF) spectroscopy, and IR spectroscopy. These are all trace detection methods, although there are bulk detection methods such as *millimeter-wave imaging* and *terahertz spectroscopy.*

Laser-induced breakdown spectroscopy focuses a high-energy laser beam on the trace sample to break down a small part of the sample into plasma of excited ions and atoms. The plasma emits light that is characteristic of emissions from ionic, atomic, and small molecular species. These light emissions are detected by a spectrometer to identify the elemental composition (Figure 1.18). One of the challenges in the use of LIBS is to assess its efficacy to detect and identify explosive species in a real environment that is replete with many interfering substances. One way to get the desired selectivity is to use double-pulse LIBS. In double-pulse LIBS, the first pulse is used to create a laser-generated vacuum and the second pulse transmitted a few microseconds later generates the return signal. Double-pulse LIBS is also observed to improve sensitivity in addition to enhancing selectivity. Selectivity can be further improved by adding temporal resolution to the LIBS emission analysis. The wavelength of 1,064 nm has been widely employed for LIBS systems. Due to serious eye hazards posed by 1,064 nm, scientists have also tried 266 nm, which also allows the designer to build Raman capability into the system. Note that 266 nm has a 600 times higher MPE limit as compared to 1,064 nm.

Raman spectroscopy offers another method for stand-off detection of explosive agents. It has been extensively used for many years as a standard analytical tool for identification of chemical agents in the laboratory environment. The basis of detection in this case is the shift in the wavelength caused by inelastic Raman scattering by the target molecule. The inelastic scattering of impinging photons where some energy is lost to (or gained from) the target molecule returns scattered light with a higher (or lower) wavelength depending on whether energy was lost to (or gained from) the target molecule. The difference is dictated by the energy of vibrational modes of the target molecule and therefore constitutes the fingerprint or the basis of identification. Complex mixtures are identified using algorithms for pattern recognition. Figure 1.16 shows a Raman explosive sensor type G-SCAN Pro from Laser Detect Systems, Israel. LDS G-SCAN Pro is a sensitive, low-false positive and low-false negative handheld Raman spectroscopy detector. It is designed to be used by law enforcement, the military, and security agencies for detection of a wide variety of explosive substances and suspicious drugs in forms of liquid, bulk, or powder.

Figure 1.16　Raman explosive sensor, Type G-SCAN Pro.

A major drawback of the Raman technique is its extremely poor sensitivity caused by the fact that Raman scattering occurs for one in about 10^7 photons impinging on the sample. Weak return signal intensity of Raman spectroscopy limits its use for trace detection as it also makes it sensitive to ambient light and fluorescence from the sample itself or other chemicals in the vicinity. The fluorescence masks the Raman signal. These problems are overcome by use of resonant Raman spectroscopy. With a tunable laser, the wavelength can be chosen to match or nearly match a resonant absorption in the target molecule leading to intensity enhancement of the order of 10^6. The problem of a fluorescence masking Raman signal can be overcome by use of either infrared or ultraviolet radiation. Infrared radiation does not have sufficient energy to cause fluorescence, and ultraviolet radiation will cause fluorescence in visible wavelength bands, which is well separated from Raman signal.

LIF is another important tool for similar applications. Although a very valuable tool in combustion diagnostics and for studying decomposition of explosives, it has not been found very useful for detection of explosive agents. When combined with pulsed laser photo (PLP) dissociation also called photo fragmentation (FP), it can be used to detect nitro compound based explosives through detection of nitric oxide (NO) radicals produced as a result of photo fragmentation. LIF spectroscopy is used to monitor NO by probing its vibrational levels in ground electronic state. A wavelength of 248 nm is used for both fragmentation and LIF spectroscopy.

Stand-off explosive detection techniques based on IR-laser spectroscopy represent another promising technique. Almost all explosive materials typically exhibit strong, characteristic absorbance patterns in the mid-IR spectral range. Also, the atmosphere is fairly transparent to the mid-infrared spectral region, which is an important factor for any stand-off detection technique if it were to work over distances at least up to few meters or preferably few tens of meters.

Infrared laser spectroscopy makes use of a tunable mid infrared laser source. The laser energy backscattered from the sample is detected as a function of wavelength to provide fingerprinting of the trace explosive. In a variant of infrared laser spectroscopy called infrared photothermal spectroscopy, resonant absorption by the explosive traces causes a thermal contrast that is captured by an infrared camera.

Another stand-off detection technique is that of nonlinear wave mixing. In this, two laser beams are made to overlap in the region of the presence of an explosive agent. Molecules present in the overlapping region interact with laser beams and the chemical information is transmitted to a detector as a laserlike beam. It has the potential of being a highly sensitive technique capable of detection of parts per quadrillion (ppq) levels. However, it is still in the R&D stage.

Almost all the laser-based stand-off detection methods have detection limits too high to make them suitable for detection of explosives in field conditions, which requires the capability to detect traces in the vapor phase or in the form of particles. Another problem is insufficient selectivity to identify target explosive in the presence of interferents. In fact, none of the laser-based spectroscopic methodologies available today are ready for full functionality prototype manufacturing. There is lot of scope of further research to improve the performance of the more mature technologies in terms of both detection sensitivity and selectivity. The final solution would probably lie not in one type of sensor but integration of several technologies to derive the benefits of the best of each. One such example is a stand-off explosive sensor concept that combines Raman hyperspectral imaging and LIBS. This approach trades off the high sensitivity of LIBS elemental analysis with the high specificity of Raman molecular analysis to provide a potential technique for high sensitivity, low false alarm rate detection, and identification of explosives on surfaces.

1.9 Detection and Identification of CBRN Agents

CBRNs are weaponized or nonweaponized chemical, biological, radiological, and nuclear materials, which if accessible to terrorists and other rogue elements could pose the gravest danger to national security. Weaponized materials can be delivered using conventional bombs, improved explosive materials, and enhanced blast weapons. Nonweaponized materials are traditionally referred to as hazardous materials. Though response to intentional CBRN incidents caused by weaponized forms of CBRN materials and accidental CBRN incidents occurring due to human error or natural or technological reasons is the same; the former can have unique implications relating to national security and international relations. Ease with which CBRN materials can be produced and the fact that CBRN incidents have the potential for causing mass casualties in the short term and creating an extremely hazardous environment with long-term effects, there is need for specialized detection equipment and decontamination and countermeasure systems need to be in place in short time frames.

The spectrum of CBRN materials includes chemical and biological warfare agents and radiological and nuclear materials. Under chemical materials, we have sodium or potassium cyanides, blister agents such as mustard gas, nerve agents such as tabun, sarin, and VX, toxic industrial chemicals including chlorine and phosgene, and organophosphate pesticides such as parathion. Common biological agents include *Bacillus anthracis* that causes anthrax, *Botulinum toxin*, and Ricin. Radiological dispersal devices (RDDs) are conventional bombs designed to disperse radioactive material to cause contamination, destruction, and injury from the radiation produced by the material. There are different variants of RDDs including passive RDD in which an unshielded radioactive material is dispersed or kept manually at the target, explosive RDD that uses the explosive force of detonation to disperse radioactive material around the targeted area, and atmospheric RDD in which radioactive material is in such a form as to be easily transported by air currents. Common radioactive materials used in RDDs include Cesium-137, Strontium-90, and Cobalt-60. As far as use of nuclear devices is concerned, use of an improvised nuclear device (IND) by terrorists cannot be entirely ruled out. The threat does exist. Improvised nuclear devices, however, require fissile material like highly enriched uranium or plutonium to produce nuclear yield unlike radiological dispersal devices that can be produced with any radioactive material.

1.9.1 Detection and Identification of Chemical Agents

There are host of techniques that can be used for detection and identification of chemical warfare agents and toxic industrial chemicals, including point detection techniques and stand-off detection techniques, and both laser-based techniques as well as nonoptical methods. Common techniques extensively explored for detection of chemical warfare agents and toxic industrial chemicals include Infrared spectroscopy, Raman spectroscopy, IMS, flame photometry, surface acoustic wave (SAW) technology, colorimetric technology, photo ionization detection (PID) technology, and flame ionization detection (FID) technology. An overview of these technologies is presented next. Optoelectronic technologies are discussed in greater detail in Chapter 8.

In the case of *infrared spectroscopy*, infrared radiation, generally in the mid-IR region of electromagnetic spectrum of 2.5–15 μm, is passed through the sample. The radiation is partly absorbed and partly transmitted. The wavelength spectrum of detected radiation represents the molecular absorption or transmission and hence the fingerprint of the sample. Infrared spectroscopy is employed in several point and stand-off detectors of chemical agents. Different approaches used in infrared spectroscopy include photo acoustic spectroscopy, passive infrared detection, infrared hyperspectral imaging, and differential absorption lidar (DIAL).

Raman spectroscopy detects the presence of chemical agents by way of spectral fingerprinting. In Raman spectroscopy, a sample is illuminated with a monochromatic laser light and the scattered light is then detected as a function of wavelength. A Raman spectrum is a plot of the intensity of Raman scattered radiation as a function of its frequency or wavelength difference from the incident radiation. This difference is called the Raman shift, which is independent of the frequency of the incident radiation.

IMS is one of the most commonly used techniques for building point detectors of narcotics, drugs, and chemical warfare and explosive agents for law enforcement and security agencies. It is an analytical technique that allows ionized analyte molecules to be distinguished on the basis of their mass, charge, and mobility in the gas phase. A small sample of air containing the suspected species is periodically taken into the IMS system where a radioactive source ionizes the molecules in the sample. As a result of the charge, the ionized analyte molecules drift into an electric field inside what is termed a *drift cell*. Each type of molecule has a typical drift velocity in the air that can be used to identify the molecule.

Flame photometry is an atomic spectroscopy technique that uses the characteristic emission spectrum of the atoms for fingerprinting as they return to lower energy states. Flame photometry is an important chemical warfare agent detection technique that has been successfully used by armed forces and civil agencies worldwide. A flame photometric detector (FPD) is more commonly found integrated with a gas chromatograph (GC) in the laboratory. A GC is an analytical instrument that measures the concentration of various components in a sample. GC-FPD has been one of the most useful methods in identifying chemical warfare agents and determining their concentrations in samples sent to a laboratory for confirmatory analysis.

SAW devices constitute an important category of chemical warfare agent sensors. A SAW device employs an acoustic wave guided along the surface of a piezoelectric crystal, in which the stresses and strains of the wave are coupled to electric fields. An AC signal applied to the input interdigital transducer (IDT) initiates the surface acoustic wave and the output IDT converts it back to electrical signal. In a typical SAW device used for detection of chemical warfare agents, the piezoelectric crystal plate is coated with a chemically selective polymer. In general, a sample is introduced into a SAW sensor via a preconcentrator, which adsorbs the test vapors for a given period of time and is then heated to release the vapors over a much shorter time span, thereby increasing the effective concentration of the vapor. The chemicals present in the vapor undergo sorption by the polymer on the surface of the piezoelectric substrate, which alters the surface wave propagation on the substrate in terms of its amplitude and frequency. The signals are processed to identify chemicals sorbed from the sample and also determine the concentration of the chemical by assuming that sorption equilibrium is reached. The detector is made to go through a purge cycle to ensure that the sorbed chemicals are effectively released.

A *colorimetric chemical agent sensor* uses a sorbent substrate such as paper or a paper ticket to which a reagent has been applied. When the targeted chemical agent comes in contact with the substrate, it reacts with the reagent to produce a distinctive color change. This allows identification of chemical agent. It also allows determining concentration of the targeted chemical from the intensity of the developed color over a given exposure time.

PID technology relies on the ionization of molecules for detection of chemical agents. A source of ultraviolet radiation is used to ionize molecules in the sample being analyzed. For ionization to take place, the photon energy in the ultraviolet radiation must be greater than the energy needed to remove electrons from the species. The positive ions in the ionized molecules are attracted to a negatively charged electrode where they release their charge and become neutralized. As a consequence, a measurable electrical current is generated. The magnitude of current produced is proportional to the concentration of the target analyte. PID sensors are typically used to give preliminary information about a variety of chemicals as they can detect vapors given off by certain inorganic compounds that other detectors may not.

FID technology in principle is similar to PID technology in the sense that analyte in both cases is ionized with the difference that FID sensors use a hydrogen flame as the ionization source rather than the UV radiation. In an FID sensor setup, the sample can be introduced either directly or via a GC column. The sample vapor is mixed with hydrogen and air in the combustion chamber and burned. This causes decomposition of organic substances in the vapor into fragments to be subsequently ionized. The ions moving along the electrical field toward the electrodes generate electrical current signal. The signal is processed to produce the desired response. Since FID sensors respond to any molecule containing carbon-hydrogen bonds, they are nonselective and therefore do not have the ability to identify the detected compounds.

1.9.2 Detection and Identification of Biological Agents

In the context of CBRN threats, biological warfare agents pose an alarming concern to the security of a nation, not only because of its destructive potential but also the psychological, economic, and social impact it can have on the entire population. Timely detection and identification of a biohazard, intentional or accidental, is therefore paramount to countering it with correct and effective response measures. There are a number of detection and identification techniques for point and stand-off detection of biohazards. There are specific point detection devices that can both detect and identify the biological warfare agents and nonspecific point detection devices that detect only the presence of biological agents but do not identify them. As well, there are stand-off detection systems capable of detecting the presence of biological warfare agents at a stand-off distance from the point of release.

Some of the better-known nonspecific point detection techniques include particle sizers, fluorescence-based systems, viable particle size samplers, and virtual compactors. Important specific point detection techniques include molecular biology, flow cytometry, and mass spectrometry and immunoassay technologies. Stand-off detection systems are based on lidar technology. Again, as in the case of chemical warfare agents, an overview of various optoelectronic and nonoptoelectronic technologies is presented next, while optoelectronic technologies are discussed at length in Chapter 8.

The operational principle of a *particle sizer* is based on determining the relative number of particles in a predetermined size range. In one of the types of a particle sizer, particles within the aerosol under investigation are exposed to a constant flow of concentrated air. Depending upon their size, the particles accelerate with different rates with smaller particles experiencing higher acceleration. A suitable laser-based device is used to measure number, size, and distribution of particles, thereby providing information on presence or absence of a biological agent. Particle sizers cannot discriminate between biological and nonbiological aerosols. *Fluorescence technologies* involve excitation of molecular components of a biological agent with light, usually in the ultraviolet (UV) region of the spectrum. Fluorescence-based devices exploit the properties of endogenous fluorophores (i.e., fluorophores growing or originating from within the organism) to detect biological agents through bioluminescence. Different fluorescence-based biosensors differ in the nature of source of light, wavelength of operation, number of measuring channels, and so forth. *A viable particle size sampler* or *impactor* operates by accelerating an airflow through a nozzle before deflecting it against an impact surface maintained at a fixed distance. Particles of different sizes get separated as they pass through different stages in the sampler that allows diffusion and collection on a specific surface of particles of different sizes. Small particles exit the sampler. After an incubation period, which is typically 24 to 48 hours, the number of colonies grown on each plate is evaluated to give information on the presence or absence of certain specific biological agents. *Virtual impactors* also belong to the broad category of viable particle size samplers. A virtual impactor separates particles by size into two airstreams. The impaction surface of a conventional impactor is replaced with a virtual space of stagnant or slow-moving air. Large particles are captured in a collection probe rather than impacted onto a surface.

Polymerase chain reaction (PCR) is one of the most commonly used molecular biology techniques in clinical laboratories for identification of microorganisms, thereby allowing detection of biological agents such as bacteria and bacteria spores or viruses. A PCR amplifies a single copy or a few copies of a segment of DNA across several orders of magnitude, generating thousands to millions of copies of a particular DNA sequence enabling investigators to obtain the large quantities of DNA that are required for various investigative procedures in molecular biology, forensic analysis, and medical diagnostics. The main limitation to this technique is the requirement for a prior knowledge of the biological agent being analyzed due to the need of specific primer sequence for the nucleic acid amplification. There is also a specific reaction for each agent, although multiplex PCR allows concurrent analysis of several agents. Portable devices based on the PCR principle and suitable for operation in field conditions are commercially available.

Flow cytometry is a technology that is widely used to analyze the physical and chemical characteristics of particles in a fluid as it passes through at least one laser beam. Cell components are fluorescently labeled and then excited by the laser to emit light at varying wavelengths. A flow cytometer counts and measures the size of particles dispersed after liquid phase concentration using a laser diffraction system. An optoelectronic conversion system is used to record the way in which the particle emits fluorescence and scatters incident light. A flow cytometer instrument is comprised of the fluidics, the optics, and the electronics. The purpose of the fluidics system is to transport the particles in a stream of fluid to the laser beam where they are interrogated. The optics system comprises lasers that illuminate the particles present in the stream as they pass through scattering the laser beam in the process. Optical filters and beam splitters then direct the light signals to the relevant detectors, which produce electronic signals corresponding to optical signals. Characteristics of particles are determined based on their fluorescent and light scattering properties.

Mass spectrometry is an analytical technique that provides information about structure and chemical properties of biological agents requiring minimal sample amounts of the order of nanograms. A mass spectrum is a plot of the ion signal as a function of the mass-to-charge ratio. The complete process involves the conversion of the sample into gaseous ions, with or without fragmentation, which are then characterized by their mass-to-charge ratio and relative abundances. A mass spectrometer consists of three major components: an *ion source* that is used to produce gaseous ions from the sample under analysis, an *analyzer* for resolving the ions into their characteristic mass components according to their mass-to-charge ratio, and a *detector system* used to detect ions and record the relative abundance of each of the resolved ionic species.

An *immunoassay* is a test that relies on biochemistry to measure the presence and/or concentration of an analyte. Immunoassays technology allows detection and identification of biological agents using the principle of specific antigen/antibody interaction. The antibodies used in immunoassays must be carefully selected because their affinity and specificity represent the limiting factors for these technologies. One common device based on this technology is the *immunochromatographic assay* using colorimetry to show their results. These devices have found important application during the anthrax emergency in 2001, demonstrating their efficacy in screening practices. There are other immunoassays technologies that utilize fluorescence properties to detect biological agents.

Lidar technology is the most commonly used technology for detection of biological warfare agents from safe stand-off distances with respect to point of release. Lidar systems used for stand-off detection of chemical warfare agents and toxic industrial chemicals use infrared wavelengths. Use of the infrared spectrum does not allow discrimination of biological agents from nonbiological aerosols. (UV-LIF) is a promising technique that allows fast stand-off detection of biological warfare agents. *UV-LIF lidar* allows gross discrimination between biological agents and background noise, taking advantage of the intrinsic fluorescence of biological molecules. However, the problem of discrimination of hazardous biological agents from naturally occurring biological agents and the issues of distinction between organic and inorganic aerosols does affect the stand-off systems that are currently available or under development.

1.9.3 Radiation Detectors

Radiation detectors are devices or instruments that can detect and identify presence of specific types of radiation including alpha, beta, gamma, and neutron radiations. There are specific devices to detect radiation in the environment, on the surface of people in case of external contamination, inside people in case of internal contamination, and also radiation received by people as exposure. *External contamination* results when radioactive material is deposited on skin, hair, eyes, or other external structures, much like mud or dust. External contamination may cover the full or partial body. It may permeate through wound contamination with radioactive shrapnel. Full and partial body contamination can be eliminated by removing the material by shedding contaminated clothes and/or completely washing off the contamination. *Internal contamination* is caused by taking radioactive material into the body via inhalation or ingestion or open wounds. Internal deposition of radioisotopes in organs results in local exposure at that location. Internal contamination continues until the radioactive material decays or until it is flushed from the body by natural processes or is removed by medical countermeasures. A person gets *radiation exposure* when the whole or part of the body absorbs penetrating ionizing radiation from an external radiation source and it stops when the person leaves the area of the source or the source is shielded completely or the process causing exposure ceases.

Radiation exposure also occurs after internal contamination. Exposure due to internal contamination stops only if the radionuclide is totally eliminated from the body, with or without treatment.

There are many types of radiation detectors, each better suited to a specific situation. There are devices suitable for detection of specific radiation such as alpha, beta, gamma, or neutron radiation. There are devices designed for specific levels or ranges of radiation energy. Some devices measure exposure to X-rays or gamma radiation as Roentgens per unit time. Other devices measure accumulated dose and current dose rate, respectively, in gray and gray per unit time.

Two common categories of radiation detectors are radiation survey meters and dosimeters. *Radiation survey meters* are portable radiation detection and measurement devices used to detect and measure external or ambient ionizing radiation fields. They find extensive application in monitoring personnel, equipment, and facilities for radiation and radioactive contamination. Most radiation survey meters are designed and built to be handheld, battery-powered devices for easy deployment and use. Common types of radiation survey meters include *scintillation counters* for measurement of alpha, beta, and neutron particles, *Geiger counters* commonly used for detection of beta particles and gamma rays, and *ion chamber* used for beta, gamma, and X-rays. A *scintillation counter* comprises a scintillator that generates photons in response to incident radiation, a photomultiplier tube that converts photons into an electrical signal and processing electronics that extracts the desired result. A *Geiger counter* is comprised of a tube filled with an inert gas. When exposed to an ionizing radiation, the high-energy particles penetrate the tube and collide with the gas thereby releasing more electrons. The electrons get attracted to a high-voltage middle wire. The electron buildup on reaching a certain threshold value generates electrical pulse that is registered on a meter. An *ionization chamber* type of radiation detector is similar to a Geiger counter and is comprised of a gas-filled cylindrical container with two electrodes: an anode and a cathode. In most cases, the wall of the cylindrical container serves as cathode. The anode is a wire along the axis of the cylindrical container. An electric field is maintained between the two electrodes by impressing a voltage that keeps the wall negative with respect to the wire. When a photon or a charged particle enters the chamber; it converts some of the gas molecules to positive ions and electrons. The positive ions and the electrons migrate to the wall and the wire, respectively, producing an observable pulse of current to flow through the circuit joining these elements.

Dosimeters measure the amount of energy deposited by ionizing radiation over a given period of time and therefore are used to estimate the effective dose received by the human body through exposure to external ionizing radiation while working in radiation area. Note that dosimeters cannot assess the radiation dose received due to internal exposure from incorporation of radioactivity.

There are personal dosimeters, also known earlier as film badges, operational dosimeters, and extremity dosimeters. Personal dosimeters are also called legal dosimeters, passive dosimeters, and individual dosimeters. Personal dosimeters are for personal use and are not transferable. There are three types of personal dosimeters: film badge dosimeters, thermoluminescent dosimeters, and ion chamber dosimeters. *Film badge* is comprised of a photographic or dental X-ray film wrapped in light-tight paper and mounted in plastic, which is checked periodically to determine the cumulative amount of radiation to which the wearer has been exposed by the degree of exposure of the film. *Thermoluminescent dosimeters* are nonmetallic crystalline solids that trap electrons when exposed to ionizing radiation, which can be calibrated to estimate the radiation level. The *ion-chamber dosimeter* is reusable, but it is self-reading for immediate determination of exposure. The *operational dosimeter* is sometimes referred to as active dosimeter, electronic dosimeter, or DMC.

The operational dosimeter is generally used while working in *limited stay* and *high radiation controlled radiation areas.* It may be used outside these areas also for specific activities or for optimization purposes if so recommended by the Radiation Protection Group. An operational dosimeter features a direct dose display, audible indication of the radiation level and alarm functions when thresholds for dose or dose rates are exceeded. It can either be assigned permanently to a person or can be used in pool mode. *Extremity dosimeters* are generally used when high local doses to the extremities such as hands, feet, or eyes are a distinct possibility.

1.10 Detection of Concealed Weapons

A number of terror attacks around the globe have brought to public awareness the extent of security deficits at airports and other public spaces and the need for introducing body and baggage scanners as additional security measures. With suicide missions becoming a common mode of terrorist attacks, equipping security agencies with the capability to detect concealed weapons can be an important force multiplier. The need to enhance the security infrastructure in these places and also at the sites of critical infrastructure and strategic assets has resulted in strong interest from the science and technology community in sensor technologies capable of detecting concealed weapons and explosive devices. The objective of the sensor technology is to facilitate law enforcement agencies in noncontact, reliable, and rapid screening of individuals and groups without any significant disruption of their activities.

Guns, knives, and explosives are common potential threats usually carried by terrorists and antisocial elements who conceal them underneath their clothing or hidden in baggage. The solution lies in the use of advanced sensor technologies that would allow individual and baggage screening as well as area screening.

Increased threats from terrorists and antisocial elements in recent years have led to the development of many technologies intended for detection of concealed weapons, contraband, explosives, and other such items. These include X-ray scanners, trace detection systems including both point and stand-off detection systems capable of detecting explosive vapors or particles left behind while handling, neutron activation, and other systems based on energetic radiation. Millimeter-wave (MMW) imaging and terahertz (THz) spectroscopy are two very potent techniques relevant to detection of concealed weapons and explosives. The MMW region of electromagnetic spectrum (1–10 mm, 30–300 GHz) penetrates clothes and many other materials allowing imaging of objects under clothes. In addition, MMW radiation is also safe at moderate intensities to use on people. However, at these wavelengths, the spectra do not have any features that can be used to identify materials.

The THz band lies between the infrared and microwave bands typically extending from 0.1 to 1.0 mm. At shorter wavelengths, (0.1–1.0 mm, 300 GHz–3.0 THz), one can get spectroscopic information. THz radiation being nonionizing in nature is also very safe at moderate intensities for use on people. Frequencies above 3.0 THz yield still better spectroscopic information but these are not suitable for stand-off detection due to heavy absorption in moist air. In fact, moist air is nearly opaque to electromagnetic radiation above 3.0 THz. Another significant challenge to the use of THz imaging technology for detection of concealed weapons is the required level of imaging resolution that would enable discrimination between handguns, knives, or explosive belts from commonly carried objects such as cell phones, wallets, and pens. Note that the technology of THz imaging has advanced to a level that allows imaging of concealed objects with acceptable resolution. Figure 1.17 shows an illustrative security scan with THz imaging.

THz can penetrate many nonmetallic materials such as paper, cloth, and leather, which allows it to be used effectively in building imaging systems suitable for screening individuals and baggage for any weapons, contraband, explosives, and so forth at airports, railway stations, and other similar public places. Figure 1.18 shows one such THz imaging security camera—type T5000 from M/s ThruVision. The camera can detect weapons, drugs, or explosives hidden under people's clothes from up to 25m and is effective even when people are moving. It does not reveal physical body details and the screening is safe for use on human beings. The technology has a range of military and civilian applications.

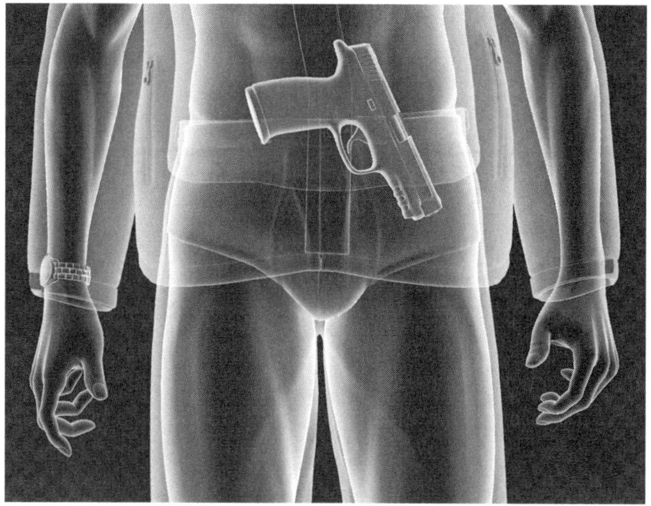

Figure 1.17 Simulated security scan with THz imaging. (Courtesy of NASA.)

Figure 1.18 THz security camera type T5000.

Selected Bibliography

Accetta, J. S., and D. L. Schumaker, in *The Infrared and Electro-optic Systems Handbook,* Volume 7, J. S. Accetta (ed.), Bellingham, WA: SPIE International Society for Optical Engineering Optical Engineering Press, 1993.

Baudelet, M. (ed.), *Laser Spectroscopy for Sensing: Fundamentals, Techniques and Applications,* Waltham, MA: Woodhead Publishing Limited, 2014.

Cremers, D. A., and L. J. Radziemski, *Handbook of Laser Induced Breakdown Spectroscopy,* Hoboken, NJ: Wiley-Blackwell, 2006.

Demtröder, W., *Laser Spectroscopy Volume 1: Basic Principles,* Berlin: Springer, 2008.

Demtröder, W., *Laser Spectroscopy, Volume 2: Experimental Techniques,* Berlin: Springer, 2008.

Hecht, J., *Understanding Lasers: An Entry Level Guide,* Third Edition, Piscataway, NJ: IEEE Press, 2011.

Lee, Y., *Principles of Terahertz Science and Technology,* New York: Springer, 2009.

McAulay, A., *Military Laser Technology for Defence,* Hoboken, NJ: Wiley-Interscience, 2012.

Perram, G., *An Introduction to Laser Weapon Systems,* Albuquerque, NM: Directed Energy Laser Society, 2009.

Sayeedkia, D., *Handbook of Terahertz Technology for Imaging, Sensing and Communications,* Cambridge, UK: Woodland Publishing Limited, 2013.

Waynant, R., and M. Ediger, *Electro-optics Handbook,* Second Edition, New York: McGraw-Hill, Inc., 2000.

Webb, C. E., and J. D. C. Jones, *Handbook of Laser Technology and Applications,* Volume III3, Boca Raton, FL: CRC Press, 2003.

Weitkamp, C., *Lidar: Range Resolved Optical Remote Sensing of the Atmosphere,* New York: Springer, 2005.

Wilson, C., *Improvised Explosive Devices (IEDs) in Iraq and Afghanistan: Effects and Countermeasures,* Congressional Research Service Report for Congress, 2007.

Woolard, D. L., J. O. Jens, R. J. Hwu, and M. S. Shur, *Terahertz Science and Technology for Military and Security Applications,* Hackensack, NJ: World Scientific, 2007.

2

Lasers and Optoelectronics Fundamentals

This chapter presents an overview of laser fundamentals and devices and photo sensors aimed at laying the foundation before moving on to discuss actual laser and optoelectronic devices and systems with applications in low-intensity conflicts and homeland security operations. This would be particularly beneficial to a large cross section of working professionals in the defense industry and armed forces. The chapter briefly covers laser fundamentals, characteristics, and types, particularly those relevant to LIC operations. Operational basics of photo sensors and the different types used in various optoelectronic sensor systems including imaging sensors of relevance to security and surveillance cameras are also discussed.

2.1 Laser Basics

Operational fundamentals of lasers with a necessary dose of quantum mechanics are briefly discussed in this section. The topics introduced in this section include the principle of operation of lasers and related concepts.

2.1.1 Operational Principle

The operational principle of a laser device is evident from the expanded form of the acronym LASER, which says that it produces a light output due to stimulated emission of radiation. In the case of ordinary light such as that from the sun or an electric bulb, different photons are emitted spontaneously due to various atoms or molecules releasing their excess energy on their own. In the case of stimulated emission, an atom or a molecule holding excess energy is stimulated by another photon emitted earlier to release that energy in the form of a photon. *Population inversion* is an essential condition for the stimulated emission process to take place. To understand the process of population inversion subsequently leading to stimulated emission and laser action, a little briefing on quantum mechanics and optically allowed transitions would be well worth its place here. An *optically allowed transition* between two energy levels is the one that involves either absorption or emission of a photon satisfying the resonance condition of $\Delta E = h\nu$, where (ΔE) is the difference in energy of the two involved energy levels, (h) is Planck's constant (= $6.6260755 \times 10^{-34}$ joules/s or $4.1356692 \times 10^{-15}$ eV/s), and (ν) is frequency of the photon emitted or absorbed. Absorption, spontaneous emission, and stimulated emission are common optically allowed transitions.

2.1.1.1   Absorption, Spontaneous Emission, and Stimulated Emission

In an *absorption transition*, an electron, atom, or molecule makes a transition from a lower level to a higher level. The necessary conditions to be satisfied for an absorption transition to occur include (a) the particle that must make the transition should be in the lower energy level and (b) the incident photon should have energy (= $h\nu$) equal to the transition energy, which is the difference in the energies of the two involved energy levels (i.e., $\Delta E = h\nu$). If the above conditions are satisfied, the particle may make an *absorption transition* from the lower level to the higher level (Figure 2.1(a)). The probability of occurrence of such a transition is proportional to both the population of the lower level and also the related Einstein coefficient indicating probability of absorption.

There are two types of emission processes: *spontaneous emission* and *stimulated emission*. The emission process involves transition from a higher excited energy level to a lower energy level. *Spontaneous emission* is the phenomenon in which an atom or molecule undergoes a transition from an excited higher energy level to a lower level all by itself without any outside intervention or stimulation and in the process emitting a resonance photon (Figure 2.1(b)). The rate of spontaneous emission process is proportional to the related Einstein coefficient. In case of *stimulated emission* (Figure 2.1(c)), there first exists a photon called stimulating photon having energy equal to the resonance energy ($h\nu$). This photon perturbs another excited species (atom or molecule) and causes it to drop to the lower energy level, in the process emitting a photon of the same frequency, phase, and polarization as that of the stimulating photon. The rate of stimulated emission process is proportional to the population of the higher excited energy level and the related Einstein coefficient. According to rules of quantum mechanics, absorption and stimulated emission are analogous processes and can be treated similarly.

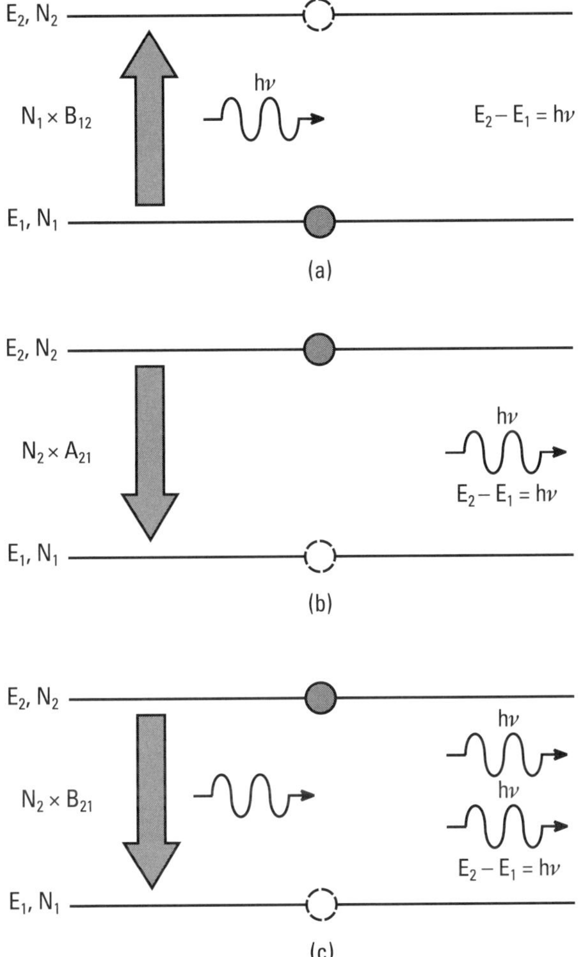

Figure 2.1   (a) Absorption, (b) spontaneous emission, and (c) stimulated emission.

2.1.1.2   Two-, Three-, and Four-Level Lasers

Another important feature that has a bearing on the laser action is the energy level structure of the laser medium. As we will see in the following paragraphs, energy level structure, particularly the energy levels involved in the population inversion process and the laser action, significantly affect the performance of the laser.

In a *two-level laser system* there are only two levels involved in the total process. That is, the atoms or molecules in the lower level, which is also the lower level of the laser transition, are excited to the upper level by the pumping or excitation mechanism. The upper level is also the upper laser level. Once the population inversion is achieved and its extent is above the inversion threshold, the laser action can take place. A two-level system is, however, a theoretical concept only as far as lasers are concerned. No laser ever has been made to work as a two-level system.

In a *three-level laser system*, the lower level of laser transition is the ground state (the lowermost energy level). The atoms or molecules are excited to an upper level higher than the upper level of the laser transition (Figure 2.2). The upper level to which atoms or molecules are excited from the ground state has a relatively much shorter lifetime compared to the lifetime of the upper laser level, which is a metastable level. As a result, the excited species rapidly drop to the metastable level. A relatively much longer lifetime for the metastable level ensures a population inversion between the metastable level and the ground state provided that at least more than half of the atoms or molecules in the ground state have been excited to the uppermost short-lived energy level. The laser action occurs between the metastable level and the ground state. Ruby laser is a classic example of a three-level laser.

It would be an ideal situation if the lower laser level were not the ground state so that it had much fewer atoms or molecules in the thermodynamic equilibrium condition. This would solve the problem encountered in three-level laser systems. Such a desirable situation is possible in four-level laser systems in which the lower laser level is above the ground state as shown in Figure 2.3.

In a *four-level laser system*, the atoms or molecules are excited out of the ground state to an upper highly excited short-lived energy level. Remember that the lower laser level here is not the ground state. In this case, the number of atoms or molecules required to be excited to the upper level would depend on the population of the lower laser level; which is much smaller than the population of the ground state. Also, if the upper level to which the atoms or molecules are

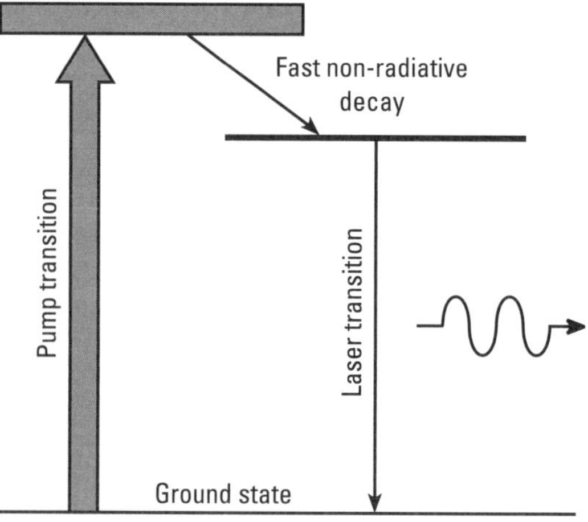

Figure 2.2 Three-level laser system.

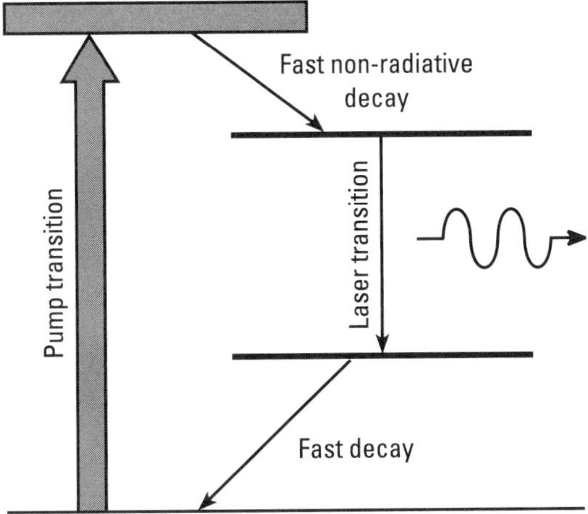

Figure 2.3 Four-level laser system.

initially excited and the lower laser level has a shorter lifetime and the upper laser level (metastable level) a longer lifetime, one can visualize that it would be much easier to achieve and sustain population inversion. Rapid population of the upper laser level and depopulation of lower laser level due to a shorter lifetime of the lower laser level help in sustaining population inversion.

2.1.1.3 Anatomy of a Typical Laser

A typical laser device comprises an active medium, a pumping source, and a resonant cavity (Figure 2.4). The *active medium* in the case of a solid-state laser is a solid-state material comprising lasing species embedded in crystalline or glass host material. In a gas laser, it is a mixture of more than one gas with gases other than the actual lasing species performing certain subtle functions like acting as an intermediate step during transfer of energy from pump source to lasing species (helium (He) in helium-neon lasers and nitrogen in carbon dioxide lasers),

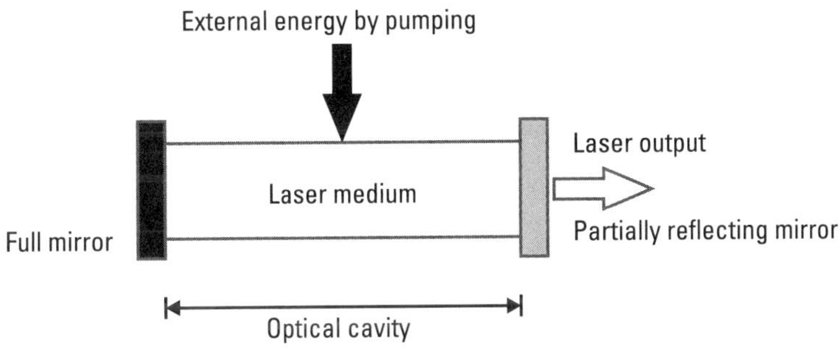

Figure 2.4 Different constituent parts of a laser device.

assisting in heat transfer (helium in carbon dioxide lasers), and depopulating the lower lasing level (helium in helium-neon laser). It is a semiconductor material in the case of semiconductor diode laser.

A *pumping mechanism* is used to create population inversion of the lasing species. Commonly employed pumping mechanisms include optical pumping, electrical pumping, and other mechanisms such as pumping by chemical reactions or electron beams. One aspect that is common to all pumping mechanisms is that the pumping energy/power must be greater than the laser output energy/power. When applied to optical pumping, it is obvious that the optical pump wavelength must be smaller than the laser output wavelength. Optical pumping is employed for those lasers that have a transparent active medium. Solid-state and liquid dye lasers are typical examples. The most commonly used pump sources are flash lamps in the case of pulsed and arc lamps in the case of continuous-wave solid-state lasers. Semiconductor diode laser pumping is another commonly employed optical pumping methodology. Diode pumping with its narrowband output is characterized by relatively much higher efficiency than flash lamps with their broadband output.

Pumping by *electrical discharge* is common in gas lasers. The excited electrons in the gas discharge plasma transfer their energy to the lasing species either directly or indirectly through the atoms or molecules of another element. Diode lasers are also electrically pumped, but not in the same way as the gas lasers. In the case of diode lasers, the electrical current in the forward-biased diode frees electrons to create electron-hole pairs. The electrons and holes recombine to emit photons. While doing so, electrons drop back to the lower state.

Some of the other methods of pumping or creating population inversion that are specific to certain types of lasers include excitation by *combustion reaction* as in gas dynamic CO_2 lasers, *chemical reaction* as in chemical lasers such as hydrogen fluoride (HF) laser, DF laser, and COIL and electron acceleration as in free-electron lasers.

The active laser medium within the closed path bounded by two mirrors, as shown in Figure 2.5, consists of the basic *resonant cavity* provided it meets certain conditions. Also, resonator structures of most practical laser sources would be more complex than the simplistic arrangement of Figure 2.5. As said before, with the help of these mirrors, we can effectively increase the interaction length of the active medium by making the photons emitted by stimulated emission process travel back and forth within the length of the cavity. One of the mirrors in the arrangement is fully reflecting and the other has a small amount of transmission to provide usable laser output. It is clear that if we want the photons emitted as a result of stimulated emission process to continue to add to the strength of those responsible for their emission; it would be necessary for the stimulating and stimulated photons to be in phase. This condition

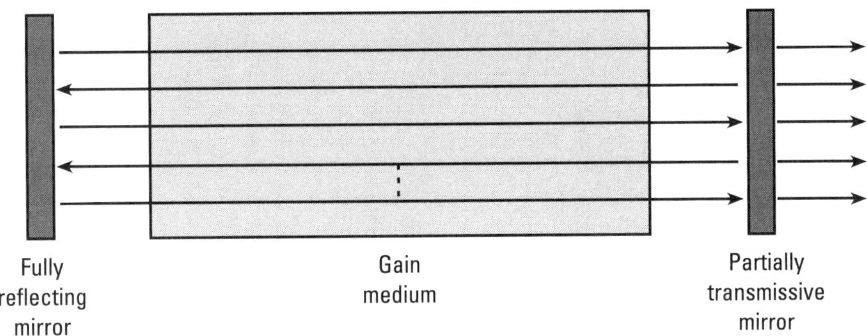

Figure 2.5   Lasing medium bounded by mirrors.

would be satisfied if round-trip length ($= 2L$) equals ($n\lambda$) where $L =$ length of the resonator, $\lambda =$ wavelength, $n =$ integer.

2.2   Laser Characteristics

Laser radiation can be distinguished from the light from conventional sources on the basis of its special characteristics and the effects it is able to produce because of these characteristics:

1. Monochromaticity;
2. Coherence, temporal, and spatial;
3. Directionality.

2.2.1   Monochromaticity

Monochromaticity refers to single frequency or wavelength property of the radiation. Laser radiation is monochromatic and this property has its origin in the stimulated emission process by which laser emits light. While describing the process of stimulated emission, we had said that the stimulated photon has the same frequency, phase, and polarization as those of the stimulating photon. Although all monochromatic radiation is not necessarily coherent, a coherent radiation is necessarily monochromatic.

2.2.2   Coherence

Coherence distinguishes laser radiation from the ordinary light. Light is said to be coherent when different photons (or the waves associated with those photons) have the same phase and this phase relationship is preserved as a function of time (Figure 2.6). There are two types of coherence: *temporal coherence* and *spatial coherence.*

2.2.2.1   Temporal Coherence

While *temporal coherence* is preservation of phase relationship with time, *spatial coherence* is preservation of phase across the width of the beam.

Temporal coherence is measured as *coherence length* or *coherence time.* The two are interrelated as given in (2.1):

$$\text{Coherence length} = c \times \tau_c \tag{2.1}$$

where

τ_c = coherence time

c = speed of light

Coherence length can also be expressed in terms of wavelength/frequency and wavelength/frequency spread as given in (2.2) and (2.3):

$$\text{Coherence length} = \frac{\lambda^2}{2\Delta\lambda} \tag{2.2}$$

$$\text{Coherence length} = \frac{c}{2\Delta\lambda} \tag{2.3}$$

Coherence length is inversely proportional to the frequency spread or line width of the laser. Coherence length for ordinary light is of the order of a fraction of a micron. On the other hand, it could be tens of kilometers for an actively frequency stabilized CO_2 laser.

2.2.2.2   Spatial Coherence

Spatial coherence is about the correlation in phase of different photons transverse to the direction of travel. It is the area in the plane perpendicular to the

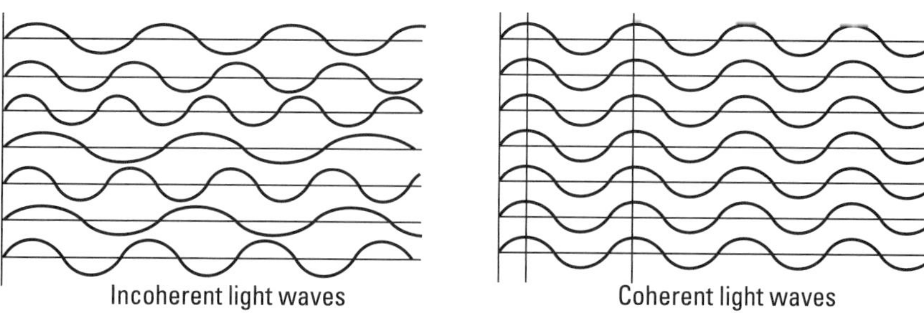

Figure 2.6   Coherence.

direction of travel over which the radiation preserves the coherence. The spatial coherence depends on the transverse mode discrimination property of the laser resonator. Laser radiation operating in the lowest order mode (TEM_{00}) will certainly be more spatially coherent than multimode laser radiation.

2.2.3 Directionality

The *directionality* of laser radiation has its origin in the coherence of the stimulated emission process. All photons emitted as a result of stimulated emission process have the same frequency, phase, direction, and polarization. These photons when emitted carry no information regarding the location of the excited atom or molecule responsible for its emission. It appears as if all photons were emitted from a tiny volume with dimensions that are of the order of a wavelength. If a photon is emitted off-axis, spatial coherence makes it appear as if it were emitted from the axis. Similarly, for a photon that is emitted away from the beam waist on the same axis, temporal coherence makes it appear as if it were emitted from the beam waist.

2.3 Characteristic Parameters

Important laser parameters include wavelength, continuous-wave (CW) power (CW lasers), peak power (pulsed lasers), average power (pulsed lasers), pulse energy (pulsed lasers), repetition rate (pulsed lasers), pulse width (pulsed lasers), duty cycle (pulsed lasers), rise and fall times, irradiance, radiance, beam divergence, spot size, M^2 value, and wall-plug efficiency.

2.3.1 Wavelength

Wavelength is the first and the foremost parameter with which the laser is identified. It is in a way laser-specific. But there are lasers that can possibly emit at more than one wavelength. While Nd-YAG is always associated with 1,064 nm; a helium-neon (He-Ne) laser can emit at 632.8, 543, 1,150, and 3,390 nm. There are some lasers that can be tuned across a band. The vibronic class of solid-state lasers, dye lasers, and free electron lasers belong to this category.

2.3.2 Power

Laser power is the *CW laser power* available from the laser. The typical laser output power level may vary from a fraction of a milliwatt such as in a He-Ne laser or a semiconductor diode laser used in laser pointers to hundreds of kilowatts or even several megawatts in high-power lasers used as directed energy weapons. Other power parameters are the peak power and average power defined with reference to pulsed lasers. *Average power* and *peak power* are defined in the case of pulsed lasers. *Peak power* is average power divided by duty cycle (Figure 2.7). *Duty cycle* is the ratio of pulse width to the time interval between two successive pulses. Average power can also be written as product of pulse energy and the repetition rate.

2.3.3 Pulse Energy

Pulse energy is defined with respect to pulsed lasers. It is in fact the area under the power versus time curve representing the laser pulse. If the laser pulse is considered as a rectangular one with amplitude equal to the peak power, the pulse energy then is the product of peak power and the pulse width. Also, if the laser pulse, which has a Gaussian profile, is approximated as an isosceles triangle as shown in Figure 2.8 and the pulse width is measured as the full width at the points of half of the peak power, the area under the curve turns out to be product of peak power and the pulse width.

2.3.4 Repetition Rate

Repetition rate of a pulsed laser is number of laser pulses produced per second. It is equal to reciprocal of time interval between two successive laser pulses (Figure 2.7).

2.3.5 Pulse Width

Pulse width or *pulse duration* in the case of a pulsed laser is usually measured as full width at half maximum (FWHM) as shown in Figure 2.7. Pulse width is intimately related to bandwidth. The narrower the pulse width, the higher the required bandwidth is. In fact, Heisenberg's uncertainty principle puts a limit on the minimum possible laser pulse width for a given value of available bandwidth. Equation (2.4) gives the minimum pulse width for a given bandwidth.

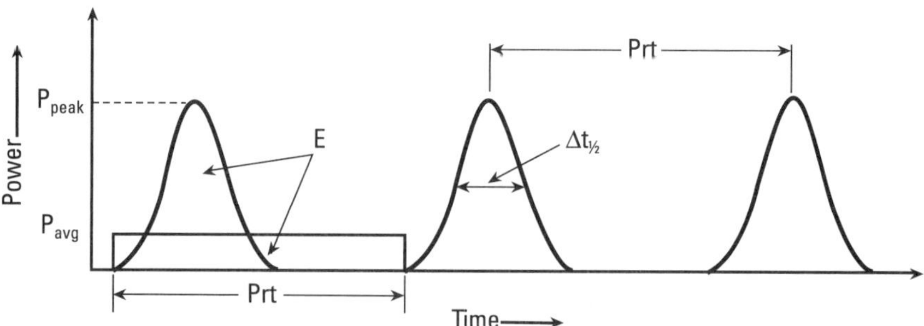

Figure 2.7 Laser pulse parameters.

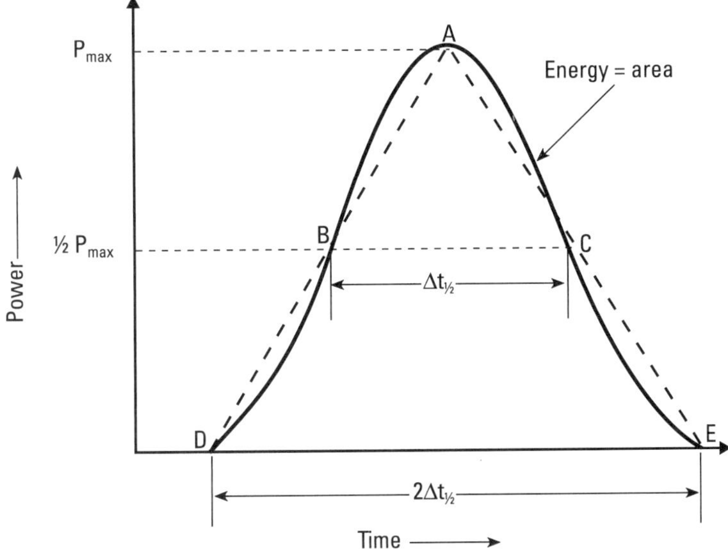

Figure 2.8 Computation of pulse energy of a Gaussian pulse.

$$\text{Minimum pulse width} = 0.441/\text{bandwidth} \qquad (2.4)$$

Pulse width could be a few femtoseconds to hundreds of femtoseconds in the case of some mode-locked lasers, a few nanoseconds to several tens of nanoseconds in the case of Q-switched solid-state lasers and a few tens of microseconds to hundreds of microseconds in the case of free-running lasers. Peak power, average power, pulse width, pulse energy, repetition rate, and duty cycle are interrelated.

2.3.6 Rise and Fall Times

Rise and *fall times* refer to the time duration between 10% and 90% of the peak amplitude of the pulse, respectively, during rising and falling portions of the laser pulse. Pulse rise time becomes particularly important while designing optoelectronic front-end circuit for converting laser radiation pulse into an equivalent electrical signal. The bandwidth of the current-to-voltage converter needs to be commensurate with the rise time specification.

2.3.7 Irradiance

Irradiance also referred to as power density is defined as the power per unit area of the laser radiation falling on the target. It is expressed as watt/m^2. This parameter is particularly important when the laser radiation is used to illuminate a receiving system.

2.3.8 Radiance

Radiance, also referred to as *brightness*, is usually defined with respect to the laser source. It is the power emitted per unit area per unit solid angle. It is expressed as watts/m²-steradian. For small values of angle, planar angle (θ) is related to solid angle (Ω) by [$\Omega = (\pi/4)\theta^2$]. Quite obviously, small exit beam diameter and lower divergence mean higher radiance or brightness.

2.3.9 Beam Divergence

Beam divergence is an indicator of the spread in the laser beam spot as it travels away from the source. It is a function of the wavelength (λ) and size of output optics. If (D) is the diameter of the output optics, then divergence is expressed by (2.5):

$$\theta = \left[(1.27\lambda)/d\right] \tag{2.5}$$

where

 θ is the divergence in radians

 This is the minimum value of divergence the laser can have assuming that it is transmitting the fundamental transverse mode, TEM_{00}. In the presence of higher-order transverse modes, the beam divergence increases more rapidly.

2.3.10 Spot Size

Spot size or beam diameter is defined as the distance across the center of the beam for which the irradiance equals 0.135 ($= 1/e^2$) times the maximum value at the center (Figure 2.9). This implies that if the laser beam were made to fall on a circular aperture of diameter equal to the laser beam diameter as defined above with the center of the beam coinciding with center of aperture, not all laser power is transmitted through the aperture. In fact, fractional transmission through the aperture can be computed from (2.6):

$$T = 1 - \exp\left[-2(r/w)^2\right] \tag{2.6}$$

 $r =$ radius of aperture

 $w =$ spot radius

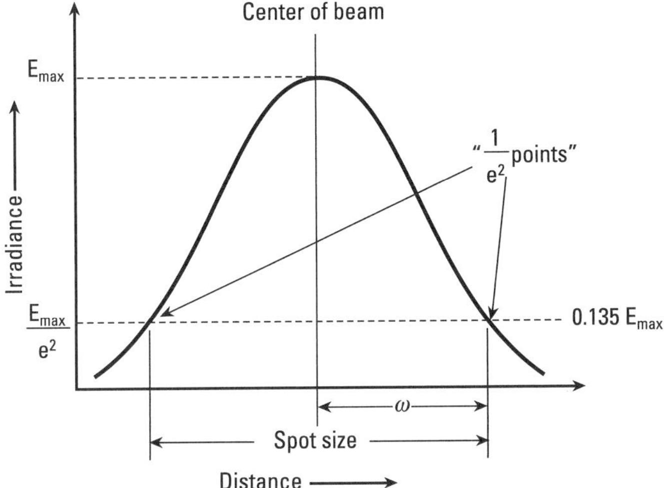

Figure 2.9 Spot size.

2.3.11 M^2 Value

M^2 value is a measure of beam quality. When the laser beam propagates through space, the divergence in the case of an unfocused pure Gaussian beam is given by $(4\lambda/\pi D)$, where (D) is the diameter of the beam waist. In the case of real beams, the divergence is higher due to various factors such as the presence of additional modes and the equation for divergence is usually written as $[(M^2 \times 4\lambda/(\pi D)]$ with the value of (M^2) being greater than one. (M^2) therefore is defined as the ratio of the divergence of the real beam to that of a theoretical diffraction-limited beam of the same waist size with a Gaussian beam profile (TEM_{00} mode).

2.3.12 Wall-Plug Efficiency

Wall-plug efficiency is the overall efficiency of the laser system. It is the ratio of laser power produced (CW power or average power as applicable) to the power drawn from source of input.

2.4 Solid-State Lasers

Lasers have been classified on the basis of various parameters such as nature of active medium (solid-state lasers, semiconductor lasers, and gas lasers), pumping mechanism (optically pumped lasers, gas dynamic lasers, electrically pumped lasers), the nature of laser output in terms of power/energy level or wavelength (visible lasers, IR lasers, CW lasers, pulsed lasers, Q-switched lasers), and so forth. The most commonly used method of classification of lasers seems to be on the basis of the nature of lasing medium. On this basis, solid-state lasers, semiconductor lasers, and gas lasers are the three major categories. In addition, there are a large number of other varieties of lasers that do not fit into any of the above-mentioned broad categories. These include dye lasers, excimer lasers, metal vapor lasers, free-electron lasers, X-ray lasers, chemical lasers, and gas dynamic lasers.

2.4.1 Operational Basics

Like any other laser device, a solid-state laser also has an active medium, a pumping source, and a resonant cavity. The active medium in the case of solid-state lasers is the lasing species embedded into a crystalline or a glass host material. Glasses such as silicate and phosphate glasses, crystalline materials such as YAG, yttrium lithium fluoride (YLF) and yttrium-doped vanadate (YVO_4) are common host materials. Neodymium and chromium are the most widely exploited lasing species. Chromium is used in ruby (chromium-doped aluminum oxide), alexandrite (chromium-doped chrysoberyl, $BeAl_2O_4$), and chromium-doped gadolinium scandium gallium gornet (GSGG). Titanium is used in titanium sapphire (titanium-doped Al_2O_3) lasers. Neodymium is used in Nd-YAG and Nd-glass lasers. Erbium is another lasing species that in YAG and glass hosts becomes the active medium for a rapidly emerging class of solid-state lasers called eye-safe lasers. The word eye-safe comes from the fact that these lasers produce output at 1,540 nm, which poses relatively much less eye hazard compared to neodymium lasers. Neodymium lasers produce an output of around 1,064 nm, which poses serious eye hazards.

This interaction of active species with the host material sometimes assumes interesting proportions in the case of a special class of solid-state lasers called vibrational-electronic (vibronic) lasers where the electronic energy levels of the light-emitting species interact with vibrational levels of the host and get broadened. Instead of being single levels, the lasing levels get transformed to upper and lower lasing bands. This feature makes this class of lasers tunable over a range of wavelengths. Important solid-state lasers in this category include titanium doped *sapphire* (sapphire is the host) and alexandrite, which is chromium-doped *chrysoberyl* (chrysoberyl is the host).

As outlined earlier, the most commonly used pump sources in solid-state lasers are flash lamps in the case of pulsed lasers and arc lamps in the case of continuous-wave solid-state lasers. Semiconductor diode laser pumping is another commonly employed optical pumping methodology. Diode pumping with its narrowband output is characterized by relatively much higher efficiency than flash lamps with their broadband output.

2.4.1.1 Constituents of a Solid-State Laser

Figure 2.10 shows the arrangement of different components of a typical solid-state laser comprising of an active medium (usually in the shape of a rod or slab), pumping source (flash lamp, arc lamp, laser diode) and a resonant structure. A specially contoured enclosure called a cavity houses the laser medium and the pumping source and a set of mirrors placed at either end of the medium transform this cavity into a resonant cavity. In addition, we have all the electronics, most of which go to drive the pumping source. Discussion on the electronics that go along with a solid-state laser is beyond the scope of this text and therefore will not be discussed.

2.4.1.2 Operational Modes

Operational modes here mean different resonator designs leading to different laser output formats. On this basis of different operational modes or more appropriately different operational methodologies, lasers could produce any of the following output formats at the output:

1. CW output;
2. Free-running output;
3. Q-switched output;
4. Cavity-dumped output;
5. Mode-locked output.

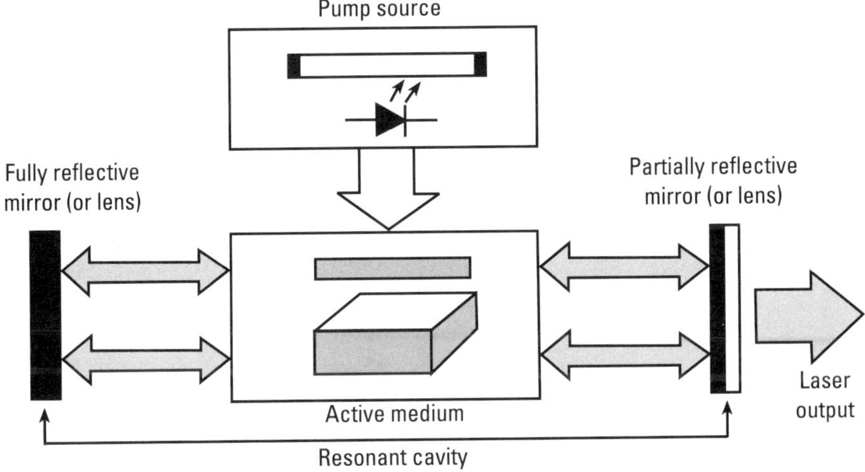

Figure 2.10 Components of a typical solid-state laser.

CW output: In CW output operational mode, which is the simplest mode of operating a laser, the laser produces a continuous output as long as the laser is pumped. Low-power CW semiconductor diode lasers are commonly used as laser-aiming aids on small arms for higher aiming accuracy.

Free-running output: Free-running output is a quasi-CW mode of operation in which the laser operates in the CW mode for a time that is of the order of a few hundreds of microseconds to a few milliseconds. This time period generally equals or is slightly longer than the storage time of the active medium, which in turn is of the same order as the pump input pulse.

Q-switched output: Q-switching is a mechanism of producing short laser pulses with pulse width of the order of a few nanoseconds. The term Q-switching here means switching of the quality factor of the resonator cavity, which is the ratio of energy stored per cycle to the energy lost per cycle, going rapidly from a low value to a high value. Initially, the Q-factor of the resonator cavity is kept at a very low value by introducing some kind of optical switch such as a Pockels cell and a polarizer. When the laser is pumped, it is prohibited from lasing like it does in the case of a CW output. The obvious consequence of this is that the active medium builds up a much higher population inversion density or gain than would have been possible in case of a CW laser. When the inversion density reaches almost its peak value, the Q-factor of the resonator cavity is rapidly switched to the high value leading to a steep fall in the loss value. This manifests itself in the production of a short laser pulse at the output.

Cavity dumped output: The phenomenon of cavity dumping is a slight variation of the Q-switching process with the difference being that in the case of the former, both of the resonator cavity mirrors are 100% reflective. When the active medium is pumped, energy is initially stored in the population inversion as is done in case of Q-switching. During this time, the loss is at a very low value. Since both of the resonator mirrors are 100% reflective, the amplified light remains trapped within the cavity. As peak irradiance is reached, the loss is again switched to a high value thus ejecting the intracavity circulating energy in the form of a pulse.

Mode locked output: Minimum pulse width achievable from the laser is intimately related to its frequency bandwidth. According to Heisenberg's uncertainty principle, minimum achievable pulse width is expressed by (2.7):

$$t_p = \frac{0.441}{B} \tag{2.7}$$

where

t_p = Minimum pulse width

B = Bandwidth

However, it is not practicable to reach this limit with every pulse-forming technique. For instance, in case of Q-switching, minimum achievable pulse width is of the order of 10 ns or so because of the required pulse buildup time. Cavity dumping overcomes this shortcoming to some extent and pulse widths of the order of 1 to 2 ns are achievable. But still, it is nowhere close what is a theoretically achievable value for a given solid-state laser. For instance, the frequency bandwidth of an Nd-YAG is about 150 GHz for a homogeneously broadened line. For a Gaussian pulse, the minimum achievable pulse width would be (0.441/150) ns ≈ 3 ps. Mode locking helps achieve pulse widths approaching the theoretical limit.

The process of *mode locking* forces different longitudinal modes to oscillate with a fixed phase relationship with respect to each other, which produces an ultrashort pulse with well-defined amplitude as a function of time. In the case of an ideal mode-locked laser pulse, the intensities of different longitudinal modes follows a Gaussian distribution and the spectral phases are identically zero.

The repetition rate of the mode-locked pulses is equal to the round-trip transit time of the resonator cavity. That is,

$$PRF = \frac{1}{\Delta T} = \frac{c}{2L} \tag{2.8}$$

It is also possible to combine the processes of Q-switching and mode locking or the processes of cavity dumping and mode locking. Q-switching and mode locking can be achieved simultaneously by introducing Q-switching elements in to the cavity in addition to the mode-locking element. The output in that case has the pulse envelope of a Q-switched pulse comprising of individual short pulses obtained from the mode-locking process.

2.4.2 Types of Solid-State Lasers

In the category of solid-state lasers, we have the ruby laser, the neodymium-doped laser including the Nd-YAG, Nd-YLF, Nd-YVO$_4$, and Nd-glass lasers, and the eye-safe erbium-doped laser. In addition to these solid-state lasers, we also have the tunable vibronic class of lasers including the titanium-sapphire and alexandrite lasers. Neodymium and erbium-doped solid-state lasers that are more relevant to applications for homeland security applications are briefly described in the following sections.

2.4.2.1 Neodymium-Doped Lasers

Neodymium-doped solid-state lasers are the most widely used and exploited type of not only solid-state lasers but also lasers in general. Most laser devices with applications in defense and homeland security are either neodymium-doped lasers or semiconductor diode lasers. Different neodymium-doped lasers differ in the host structure that is doped with neodymium. YAG, yttrium lithium fluoride (YLiF$_4$) and yttrium-doped vanadate (YVO$_4$) are the commonly used crystalline hosts while silicate, phosphate, and fused silica are the popular glass hosts. All neodymium-doped lasers, whether crystalline-host–based or glass-host–based, have an energy level diagram similar in structure to that of the Nd-YAG laser. Interaction of neodymium with the host may lead to a slight change in the output wavelength, to the tune of 1% or so, from one Nd-doped laser to another. For instance, Nd:YAG, Nd:YLF, Nd:YVO$_4$, Nd:glass (silicate), Nd:glass (phosphate), and Nd: glass (fused silica) have wavelengths of 1,064, 1,047/1,053, 1,064, 1,062, 1,054, and 1080 nm, respectively.

Nd:YAG and Nd:YLF lasers can be operated both as pulsed as well as CW lasers. However, due to poorer thermal conductivity, Nd:glass lasers can be operated only as pulsed lasers also at very low repetition rates. Glass as a laser material, however, has the advantage of high-energy storage capability and the ease with which it can be grown in large sizes with the desired optical quality. Rod sizes of 5- to 6-cm diameter and 100-cm length are not uncommon. These are used to build relatively high-energy solid-state lasers with a pulse energy of the order of several kilojoules. Glass lasers also have a broad linewidth, which makes them more adaptable to the mode-locking phenomenon to generate much shorter pulse widths.

Among neodymium-doped lasers, Nd:YAG is the most important and most widely used solid-state laser because of its high gain and good thermal and mechanical properties. In Nd:YAG, trivalent neodymium replaces trivalent yttrium. Some of its noteworthy properties include hardness, stability of structure from very low temperatures up to the melting point, good optical quality, and high thermal conductivity of the YAG host. In addition, the cubic structure of YAG host favors a narrow line width leading to high gain and low lasing threshold. Due to its good thermal and optical properties, Nd:YAG lasers can be used both in CW as well as high repetition rate Q-switched pulsed mode. Average and peak powers in excess of 1 kW and 100 MW, respectively, can be achieved in these lasers. High repetition rate (up to 20 Hz) Q-switched (5- to 20-ns pulse width) Nd:YAG lasers find extensive use in a variety of conventional warfare and homeland security applications for target observation and targeting. Laser range finders and target designators are common examples.

After Nd:YAG lasers, Nd:glass is the other important neodymium-doped solid-state laser. While Nd:YAG is mainly known for its high gain and good thermal and mechanical properties, Nd: glass is capable of being grown in large sizes with diffraction-limited optical quality. As a result, Nd:YAG is the laser of choice in a wide range of applications requiring either CW or high repetition rate pulsed operation. In contrast, the Nd:glass laser is the preferred laser for high-energy and high peak power applications requiring either single-pulse or low repetition rate operation such as laser fusion research. The Nd:YAG laser is definitely far more relevant than the Nd:glass laser when it comes to defense and security-related applications. The Nd:glass laser based military systems of the 1970s and 1980s have largely been replaced by Nd:YAG lasers.

2.4.2.2 Erbium-Doped Lasers

Erbium-doped lasers have some potential medical, military, and homeland security applications because of the two wavelengths they are capable of generating when doped in YAG and glass hosts. These wavelengths are 2,940 nm (erbium:YAG) and 1,540 nm (erbium: glass). Their importance arises from the water absorbent characteristics of these wavelengths. While 2,940 nm holds promise for medical applications in the field of plastic surgery due to its extremely large absorption by water in tissue,1,540 nm is attractive as an eye-safe alternative to neodymium-doped YAG (or glass) based military laser range-finders and laser target designators. Neodymium lasers with emissions around 1,064 nm pose a serious eye hazard. Therefore, when it comes to using laser range finders and laser target designators for training exercises and war games, eye-safe lasers are a much better option. Since erbium-doped lasers producing an eye-safe wavelength of 1,540 nm are more relevant to defense and security applications, these types are discussed further in the following paragraphs.

Erbium-doped glass produces output at 1,540 nm. Three-level behavior of erbium leads to a low laser efficiency. The problem is further worsened by the weak absorption of pump radiation by erbium (Er^{+3}) ions. In order to overcome these shortcomings, ytterbium (Yb^{+3}) and chromium (Cr^{+3}) ions are added. Ytterbium acts as a sensitizing agent. It helps in absorbing pump radiation in the wavelength region (0.9–1 μm) where erbium is more or less transparent. Chromium ions also do a similar job. They help in matching the emission spectrum of flash lamps with the absorption spectrum of ytterbium:erbium:glass.

Q-switched Er:glass lasers find their main application in eye-safe hand-held laser range finders. These are lasers with low repetition rates, typically in the range of 5 to 20 pulses per minute, although in some cases it may be as high as 2 Hz. A large number of manufacturers offer handheld Er:glass laser range finders. LH-40 Eye Safe Laser Finder from M/s Eloptro South Africa, LRB-21K and LRB-25000 Eye Safe Laser Range Finders, both from M/s Newcon Optik, are some examples. Most of the devices in this category have similar performance specifications in terms of operational range, range accuracy, and pulse repetition rate. LRB-25000 has a maximum operational range of 25 km, a range accuracy of ±5m, and a pulse repetition frequency of 0.15 Hz. High repetition rate military lasers of the eye-safe variety currently employ Nd:YAG lasers whose output is wavelength-shifted using an optical parametric oscillator (OPO). Laser range finder type LDM-38 from M/s Carl Zeiss Optronics and laser range finder type G-TOR from M/s SAAB Sweden are examples. Both these laser range finders are configured around OPO-shifted Nd:YAG lasers emitting at 1,570 nm. G-TOR offers pulse repetition rate as high as 25 Hz.

2.5 Fiber Lasers

A *fiber laser* is a type of solid-state laser where the gain medium is rare earth ions doped fiber rather than a rod or slab of host material doped with lasing species. It is an alternative to bulk solid-state lasers in a wide range of industrial and military applications requiring high power levels with high beam quality in a compact and rugged package configuration. Inherent to the fiber laser design and operational regimes are excellent performance characteristics, which include a high level of immunity to misalignment, high beam quality, compactness, and long-term stability. In addition to these characteristics, fiber lasers exhibit outstanding thermo-optical properties arising out of large surface-area-to-volume ratio.

2.5.1 Basic Fiber Lasers

In its simplest form, a fiber laser is comprised of a rare earth element ion-doped fiber as the active medium, a fiber-coupled semiconductor diode laser, or another fiber laser as the pump source and dielectric mirrors or fiber Bragg gratings to form the resonant cavity. The active medium in a fiber laser is a glass fiber doped with rare earth element ions such as neodymium (Nd^{3+}), erbium (Er^{3+}), ytterbium (Yb^{3+}), thulium (Tm^{3+}), holium (Ho^{3+}), or praseodymium (Pr^{3+}). Figure 2.11 shows a simplified arrangement of different components of the basic fiber laser. Both pump and laser radiation is guided through the waveguide structure constituted by the core and the cladding of the single-clad fiber. Dielectric mirrors along with the single-clad fiber constitute the resonant cavity in the arrangement of Figure 2.11. In practical fiber lasers, in most cases, fiber Bragg gratings are used instead. A fiber Bragg grating is a type of distributed

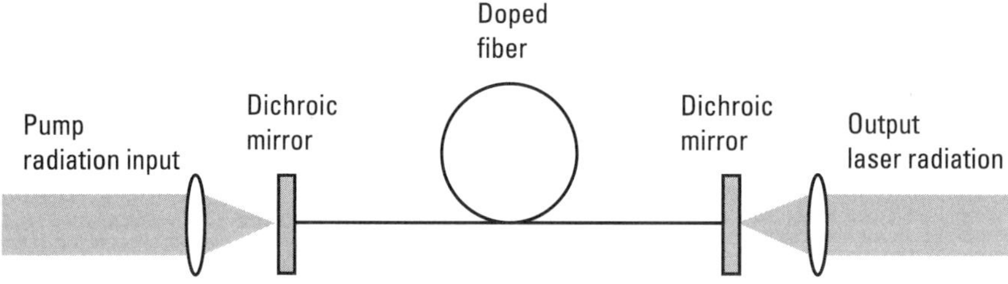

Figure 2.11 Basic fiber laser.

Bragg reflector constructed in a short segment of fiber. The grating is created by a periodic variation of the refractive index of the core and is so designed to reflect a particular wavelength and transmit all other wavelengths. The rare earth ion-doped fiber core guides the light and this necessitates that the pump radiation is spatially coherent.

As the power available from single-mode semiconductor diode lasers is usually limited to only a few watts, such a configuration cannot be used to build relatively higher output power lasers. This limitation is overcome by using a double-clad fiber design. In this case, an active doped core is surrounded by a second waveguide structure constituted by the inner cladding also known as the pump core. Double-clad fiber laser design allows the use of multimode semiconductor diode lasers as the pump source. Pump radiation in this case is launched in the inner cladding. The pump radiation is gradually absorbed over the entire length of the fiber that with laser action is converted into single-mode high-brightness laser radiation. Unlike bulk solid-state lasers where the intensity is limited to Rayleigh length by diffraction, in the case of a double-clad fiber laser, the intensity is maintained over the entire fiber length due to confinement of both pump and laser radiation. Consequently, the gain of the active medium, which is defined by product of light intensity in the gain medium and the interaction length, is significantly higher than what it would be in the case of a bulk solid-state laser. This property gives fiber lasers inherently a high single pass gain and low pump threshold values. Wavelengths emitted by common fiber lasers are in the regions of 1.0 to 1.1 μm from ytterbium-doped lasers, 1.52 to 1.57 μm from erbium-doped lasers, and 1.9 to 2.1 μm from thulium- and holmium-doped lasers.

2.5.2 Applications

High-power fiber lasers with average power levels in the range of tens of watts to multikilowatt levels are finding use in a wide range of industrial and military applications. These lasers are intended for a wide range of industrial applications, which includes cutting, welding, bending, bonding, and sintering, annealing, and surface texturing. Another major application of high-power fiber lasers is in directed energy weapon applications in defense. High power at high beam quality, compactness, ruggedness, reliability, and fiber delivery are the key features that give them an edge over chemical lasers and bulk solid-state lasers when it comes to military applications. Fiber lasers have already been used in a number of directed energy weapon technology demonstrators for a range of applications including counterexplosive devices, counter-RAM, and counter-UAV applications. In the future, directed energy laser weapons for ballistic missile defense and antisatellite applications are not ruled out.

2.6 Gas Lasers

Gas lasers, unlike solid-state and semiconductor diode lasers, have widely varying characteristics including the wavelength range, power levels, and to some extent even the pumping mechanism. The available output powers from gas lasers vary from fraction of a milliwatt in a small helium-neon laser used for optical alignment to a megawatt level in a gigantic high-power chemical laser used as a weapon. The wavelength range also spans almost the entire optical spectrum from ultraviolet to far infrared with thousands of laser wavelengths discovered throughout the region.

2.6.1 Operational Basics

In the following sections we discuss the operational basics of a generic gas laser in terms of the nature of active media, interlevel transitions participating in laser action, and pumping mechanism.

2.6.1.1 Active Media

The active medium in a gas laser is almost invariably a mixture of more than one gas with gases other than the actual lasing species performing certain subtle functions like acting as an intermediate step during transfer of energy from pump source to lasing species (helium in helium-neon lasers and nitrogen in carbon dioxide lasers), assisting in heat transfer (helium in carbon dioxide lasers), and depopulating the lower lasing level (helium in helium-neon laser). The gas mixture is filled in a tube at a pressure that again depends on a number of parameters. Low pressure of the order of a small fraction of atmospheric pressure suitable for having stable discharge for longer periods is mostly the case in CW lasers. In the case of pulsed lasers where the discharge stability is required for a shorter period, a laser gas mixture could be filled at a pressure close to atmospheric pressure and sometimes in excess of one atmosphere. The optimum gas pressure for lasers of a given type also depends on the laser design. The active media of different gas lasers may not be in the same form. For example, in the case of argon-ion and krypton-ion lasers, it is the ionized atoms of rare gases argon and krypton, respectively. In the case of metal vapor lasers such as copper-vapor and gold-vapor lasers, the active medium is hot-metal vapor, but in the case of a helium-cadmium laser, the metal vapor is ionized as well.

2.6.1.2 Interlevel Transitions

Interlevel transitions in the case of most gas lasers are electronic except for carbon dioxide lasers, HF lasers, DF lasers, and carbon monoxide lasers that involve vibrational transitions. Some far infrared lasers producing wavelengths greater than $30\,\mu$m have vibrational or rotational transitions.

2.6.1.3 Pumping Mechanism

Most of the gas lasers are excited by electrical discharge. The active medium is usually excited either by passing an electric discharge current along the length of the tube known as longitudinal excitation as shown in Figure 2.12(a) or by an electric discharge perpendicular to the length of the laser tube known as transverse excitation as shown in Figure 2.12(b). The former is used in relatively low-power CW lasers while the latter is employed in high-power pulsed or CW lasers. In the case of CW lasers, a high DC voltage is initially required to ionize the gas. Once the ionization takes place, the DC voltage is brought to a much lower value needed to sustain the plasma. In the case of a pulsed laser, a hefty capacitor is charged to the required DC voltage and then made to discharge through the laser medium. Some gas lasers like those generating far infrared wavelengths are optically pumped. Another gas laser of shorter wavelength usually pumps these lasers. A carbon dioxide laser may be used to pump such a laser.

2.6.2 Types and Applications

In the following sections, we describe the two most commonly used gas lasers: helium-neon and carbon dioxide lasers.

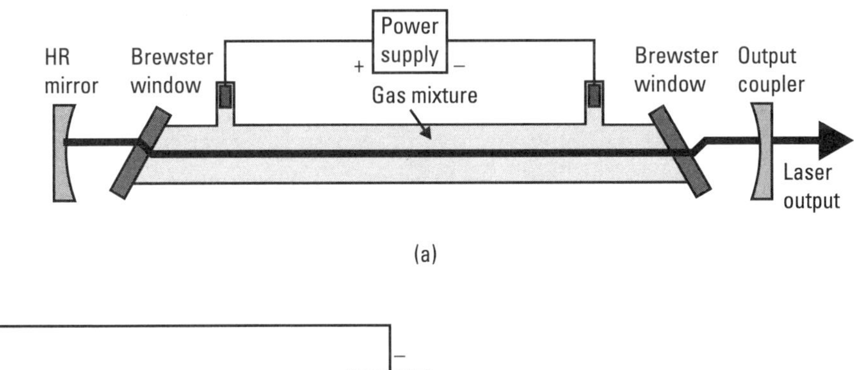

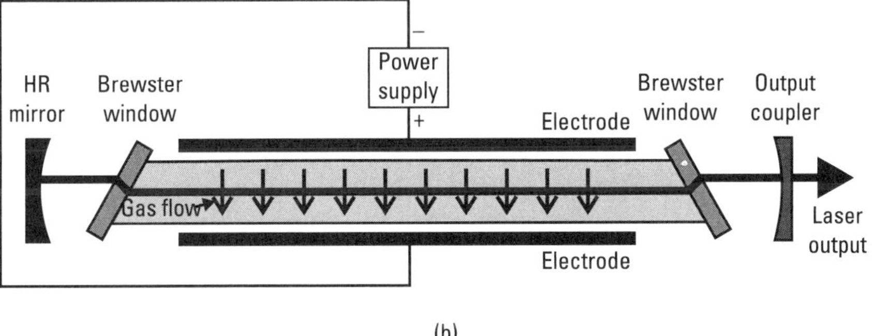

Figure 2.12 Generic gas laser (a) longitudinal excitation and (b) transverse excitation.

2.6.2.1 Helium-Neon Lasers

He-Ne lasers are one of the most commonly used types of gas lasers. Although we are more familiar with the red beam from a He-Ne laser at 632.8 nm, lasing action is also possible at infrared wavelengths of 1.153 and 3.391 μm and visible wavelength of 543.5 nm. In fact, the first successful operation of a CW laser was achieved in a He-Ne laser at 3.391 μm by Javan Benett and Herriott at Bell Labs following the first ever lasing action demonstrated by Maiman in a ruby laser. Incidentally, the He-Ne lasers are usually identified by red output; the lasing transition in He-Ne that that has the highest gain is the one at 3.391 μm. The active medium is the laser gas mixture of helium and neon, which is predominantly helium with only 10% to 20% of neon. The gas mixture is excited by an electrical discharge.

A helium-neon laser can possibly produce laser output at either of the visible (543.5 or 632.8 nm) or infrared (1.153 or 3.391 μm) wavelengths depending on the operating conditions and choice of optics. The standard has become 632.8 nm helium-neon laser wavelength and is also the most widely used one.

Laser power output of a helium-neon laser depends on several parameters; of which the tube length, gas pressure, and discharge bore diameter are the predominant ones. In fact, product of gas pressure (in torr) and bore diameter (in mm) is the figure-of-merit and its optimum value is 3.5 to 4. He-Ne lasers are available both as sealed-off tubes as well as packaged types. The package types are further available as lasers with a separate power supply as well as self-contained lasers for OEM applications. He-Ne lasers are increasingly being replaced by semiconductor diode lasers, particularly in those applications where the high beam quality of a helium-neon laser is not a necessity. Lasers, both visible and infrared, used with squad weapons for target aiming are invariably semiconductor diode lasers. Visible output lasers also used as laser dazzlers for antipersonnel applications are either semiconductor diode lasers or diode pumped solid-state lasers.

2.6.2.2 Carbon Dioxide Lasers

Among the family of gas lasers, carbon dioxide (CO_2) lasers are the most widely used and diversely exploited. This laser finds a myriad of applications in industry, medical diagnosis and treatment, science and technology, and warfare.

The laser medium in this case is a gas mixture of CO_2, He, and nitrogen (N_2). CO_2 is the lasing species. Unlike the gas lasers such as a He-Ne laser, the energy levels responsible for laser action in a CO_2 laser do not correspond to excitation and de-excitation of electrons. Instead they correspond to vibrational levels; namely, asymmetric stretching, symmetric stretching and bending, and rotational levels of CO_2 molecule. Nitrogen participates in the process of creation of population inversion by acting as an intermediate step in the same way as helium does in the case of a He-Ne laser. Helium in the case of a CO_2

laser helps in depopulating the lower laser level. The N_2 molecules transfer the energy to CO_2 molecules. This is the upper lasing level. Laser transitions correspond to the CO_2 molecules dropping from higher energy asymmetric stretching mode to the lower energy symmetric stretching or bending modes. Transition to a lower level corresponding to symmetric stretching produces a 10.6-μm output while its transition to another lower level corresponding to symmetric bending produces a 9.6-μm output. In fact, around both 9.6- and 10.6-μm outputs, there are a large number of closely spaced lines.

Carbon dioxide lasers employ either a sealed tube or flowing gas construction. The gas mixture is excited by an electric discharge. Both DC as well as RF excitation has been successfully and widely used in the case of CO_2 lasers. As the electric discharge breaks down CO_2 molecules to form carbon monoxide and oxygen, a catalyst is added to regenerate CO_2. A flowing gas type of CO_2 lasers may be of longitudinal flow or transversal flow type. In the case of the former, the gas flows along the axis of the cavity while in the case of latter, it flows perpendicular to the cavity axis. In longitudinal flow lasers, the gas pressure is low, the output power is also relatively lower, and lasers operate as CW lasers. Transverse flow lasers are normally used to get higher power outputs. Gas pressures in such lasers can be higher.

Another type of CO_2 laser is the one that is transversely excited and where the gas pressure is about one atmosphere. The CO_2 lasers described in the earlier in this chapter have low gas pressures and usually produce CW output. CW lasers cannot have a high-pressure gas mixture as it is not practical to have a stable continuous discharge at pressures above about one tenth of an atmosphere. However, a pulsed electric discharge is possible at higher pressures and it works very well if the discharge is transverse to the laser axis. That is what is done in a transversely excited atmospheric (TEA) pressure CO_2 laser (TEA laser) (Figure 2.13). A TEA CO_2 laser is a high power pulsed CO_2 laser. The gas pressure is around one atmosphere and the discharge is transverse to the laser axis. TEA CO_2 lasers invariably use a form of preionization to uniformly ionize the space between the electrodes. The primary attraction of a TEA CO_2 laser is in its ability to generate short intense pulses and extraction of high power per unit volume of laser gas mixture. Pulse durations of a few tens of nanoseconds to a few microseconds and pulse energies from a few millijoules to hundreds of joules at repetition rates up to a few hundreds of hertz are achievable. Waveguide CO_2 laser is another type of CO_2 laser that renders compactness. RF-excited sealed-off waveguide CO_2 lasers producing several watts to several tens of watts of CW output hold lot of promise.

Carbon dioxide lasers were extensively used in the early years of industry for cutting, welding, and many other material-processing applications and in medical procedures for soft tissue surgery. As well, in the 1970s and early 1980s, there was an increasing interest toward using sealed-off CO_2 lasers for a

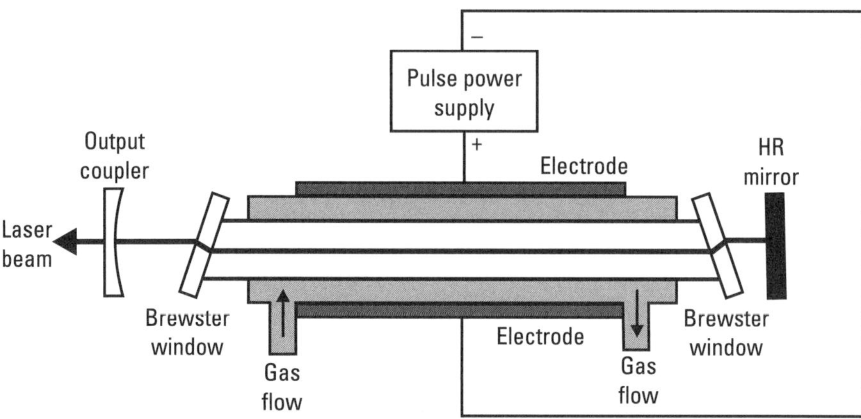

Figure 2.13   TEA CO_2 laser.

variety of military applications, including rangefinders, target designators, laser radars, for laser beamrider missile guidance, laser communications and laser countermeasures, and gas dynamic CO_2 lasers as directed energy weapons. Carbon dioxide lasers have been almost completely replaced by neodymium-doped solid-state and semiconductor diode lasers for most of the above-mentioned applications. High-power solid-state lasers and fiber lasers today are the most potent candidates for use as the high-power laser source in countermeasures and directed energy weapons. Due to their compactness, ruggedness, high beam quality, and freedom from complex logistics and toxicity issues, these lasers have made high-power gas dynamic CO_2 lasers and chemical lasers, including HF/DF lasers and COILs for directed energy weapon applications, obsolete.

2.7   Semiconductor Diode Lasers

Most semiconductor lasers are diode lasers that are pumped by an electrical current. These are also known as injection diode lasers. However, there are also optically pumped semiconductor lasers that are pumped by optical radiation either from a direct-coupled single emitter or fiber-coupled laser diode array and quantum cascade lasers that utilize intraband rather than interband transitions of the conventional semiconductor lasers for laser action. Most of the discussion in this section is centered on semiconductor diode lasers, which typically emit in visible to near infrared bands of electromagnetic spectrum.

2.7.1   Operational Basics

The following sections describe the operational aspects of semiconductor diode lasers. The topics covered include laser action in semiconductor diode lasers, similarities and dissimilarities between laser diodes and light emitting diodes, characteristic parameters of laser diodes, and precautions to be observed while handling laser diodes.

2.7.1.1 Laser Action

The active medium in a semiconductor laser is a semiconductor material. Only direct band gap semiconductor materials are suitable for building diode lasers and LEDs, although some compound semiconductors with indirect band gap can be used to make LEDs. Commonly used semiconductor material compositions for semiconductor diode laser fabrication include gallium arsenide (common laser wavelength 905 nm at room temperature), gallium nitride (common laser wavelength 405 nm), gallium aluminum arsenide (common laser wavelengths 785 and 808 nm), aluminum gallium arsenide (common laser wavelength 1,064 nm), indium gallium arsenide (common laser wavelength 980 nm), indium gallium nitride (common laser wavelengths 405 and 445 nm), gallium antimonide arsenide (common laser wavelengths 1,877, 2,004, 2,330, 2,680, 3,030, and 3,330 nm), indium gallium arsenide phosphide (common laser wavelengths 1,310, 1,480, 1,512, 1,550, and 1,625 nm), and aluminum gallium indium phosphide (common laser wavelengths 635, 657, 670, and 760 nm). The optical gain in this case is usually achieved by a process of stimulated emission at an interband transition triggered by prevailing conditions of high carrier density in the conduction band. In the case of diode lasers, high carrier density in the conduction band is caused by injection current. The emission of radiation is due to recombination of holes and electrons in a forward-biased PN junction diode. Injection diode laser powered by an injected electrical current is the most practical form of diode laser to distinguish it from an optically pumped diode laser.

If the injection current exceeds a certain minimum value, called lasing threshold, the number of electrons and holes available for recombination becomes sufficiently large so as to create a possibility where a spontaneously emitted photon having energy equal to involved recombination energy stimulates an electron-hole pair to recombine to emit a photon of the same frequency, phase, and polarization as that of the stimulating photon. Surrounding the recombination region, also called gain region, by a suitable optical cavity facilitates the process of stimulated emission. The cavity in the case of diode laser is made by cleaving the two ends of the crystal to form perfectly smooth, parallel edges forming a Fabry-Perot resonator. Since semiconductors have a high refractive index, the smooth surfaces offered by cleaved ends reflect about 30% of light back into the material to get sustained laser action in a high gain semiconductor laser material. The stimulated emission produces light amplification as the photons travel back and forth between the two end faces of the cavity, and when the gain due to stimulated emission exceeds the losses due to absorption, imperfect reflections or other circumstances, sustained lasing action is produced.

2.7.1.2 Laser Diode versus Light-Emitting Diode

Light-emitting diodes (LEDs) operate the same way as the laser diodes, with a major difference in the forward-biased current. While the current in the case of an LED is of the order of a few milliamperes, the same in the case of laser diodes emitting a few milliwatts of laser power is of the order of 80 to 100 mA. Figure 2.14 shows optical power versus drive current characteristics of a laser diode. At low levels of drive current, spontaneous emission predominates. In this region, the device operates like an LED. When the drive current is more than the lasing threshold; the light output is predominantly due to stimulated emission and the device functions like a laser. An LED or a laser diode turns on for a forward-bias voltage greater than or equal to the turn-on voltage and the current flowing through the device is nearly zero. In the case of a laser diode, although the device is turned on and a small magnitude of current does start flowing, the lasing action starts only after the forward voltage exceeds the lasing threshold voltage.

2.7.1.3 Characteristic Parameters

Important characteristic parameters of diode lasers relate either to their I-V characteristics or the output beam characteristics. These include threshold current, slope efficiency, and linearity of laser operation in the case of the former and of beam divergence, line width, and beam polarization in the case of the latter. Most of the characteristic parameters are sensitive to changes in tem-

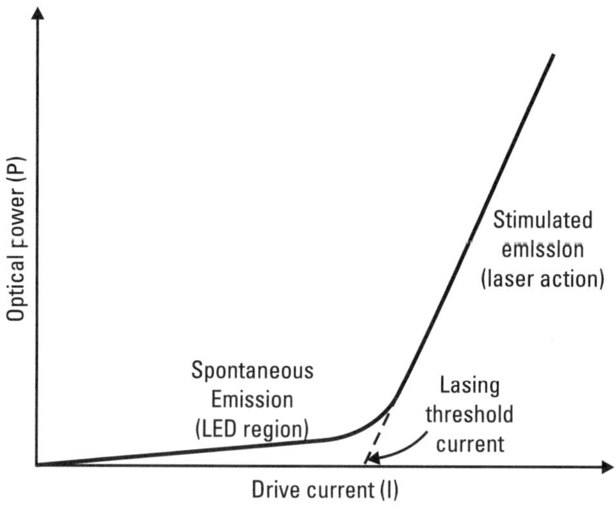

Figure 2.14 Optical power versus drive current characteristics.

perature. Different parameters and their sensitivity to temperature changes are briefly described next.

Threshold current is the minimum forward-biased injection current needed to achieve sustained laser action. When the injection current is below the threshold value, most of the input electrical energy is dissipated as heat and conversion to light output is highly inefficient. Higher threshold current therefore means that more electrical energy must be dissipated as heat energy. Threshold drive current is a strong function of temperature and increases rapidly with temperature. Change in threshold current $I_s(\Delta T)$ due to incremental change in temperature (ΔT) can be computed from (2.9).

$$I_S(\Delta T) = I_S(T) \times \left[e^{(\Delta T/T_0)} - 1 \right] \tag{2.9}$$

where

$I_s(T)$ is the threshold current at absolute temperature (T)

T_0 is a substrate specific characteristic temperature

An increase in temperature also adversely affects the lifetime of the laser as an increase in temperature leads to an increase in threshold current density. Lifetime approximately doubles for every 10-degree decrease in chip temperature. This is taken care of by mounting the chip on a heat sink.

Slope efficiency is determined by the slope of the characteristic I-V curve above the threshold current and is measured in mW/mA (or W/A). In the case of a pigtailed device, slope is reduced by a factor depending on the coupling efficiency of laser power into the fiber. Slope efficiency is strongly dependent on temperature and decreases with increase in temperature.

Divergence in the case of semiconductor diode lasers is primarily influenced by diffraction of light waves as they are coupled out of the chip. Laser diodes characteristically produce a highly diverging laser beam with the exception of surface-emitting diode lasers. Also, due to a rectangular-shaped active light emitting area with strongly differing edge lengths, the divergence in the two orthogonal planes is different. Divergence in the plane parallel to the plane of the active layer is relatively much smaller than that in the plane perpendicular to the active layer. As a result, the laser beam appears as an elliptical spot at some distance from the laser. If required, it is possible to circularize the elliptical beam with the help of a cylindrical lens, which refracts the light in parallel direction.

Line width refers to the frequency or wavelength spread in the laser beam. In the case of gain guided lasers, the envelope of gain profile typically has a 3-dB width of 2 to 3 nm that corresponds to a frequency range of about 1,000 GHz at 800 nm. In the case of an index-guided laser, one spectral line is dominant with the result that the line width is much narrower, typically 10^{-2} nm. This corresponds to a frequency spread of few gigahertz at 800 nm. In the case of distributed feedback (DFB) lasers, which are also index guided, the line width is still smaller, typically of the order of 10^{-4} nm, which corresponds to frequency spread of a few megahertz at 800 nm.

The spectral profile of semiconductor lasers is also strongly affected by temperature changes. Both the gain profile as well as individual lines shift to longer wavelengths with increase in temperature. Typical values of temperature coefficient are as follows:

$$\text{Temperature coefficient (GaAlAs lasers)} = 0.24\,\text{nm/K (envelope variation)}$$
$$= 0.12\,\text{nm/K (individual line variation)}$$
$$\text{Temperature coefficient (InGaAs lasers)} = 0.3\,\text{nm/K (envelope variation)}$$
$$= 0.08\,\text{nm/K (individual line variation)}$$

Beam polarization is yet another important parameter. Diode lasers emit almost linearly polarized light if driven above threshold, which is influenced by polarization dependency of the reflection factor of the emission area of the crystal.

2.7.1.4 Handling Precautions

Diode lasers exhibit high reliability and a long lifetime exceeding 100,000 hours provided certain precautions are taken while handling them and also while designing driver circuits to power them. Diode lasers are particularly sensitive to electrostatic discharge, short duration electric transients such as current spikes, injection current exceeding the prescribed limit, and reverse voltage exceeding the breakdown limit. Damage to the laser diode often manifests itself in the form of reduced output power, shift in the value of threshold current, and its inability to be able to be focused to a sharp spot. The electrostatic discharge caused by human touch is the most common cause of premature failure of a diode laser.

In order to protect the diode lasers from the above failure modes, the driver circuit should be carefully designed and have all the features recommended by the manufacturer of the laser. The driver should be a constant current source with inbuilt features like soft start, protection against transients, interlock control for the connection cable to the laser, and safe adjustable limit for injection current. If the laser is to be operated in the pulsed mode, the injection current should be pulsed between two values above the lasing threshold rather than between the cutoff and lasing mode.

2.7.2 Types

Depending on the structure of various semiconductor materials used to fabricate semiconductor lasers, these can be categorized as part of

1. Homojunction and heterojunction lasers;

2. Quantum-well diode lasers;

3. DFB lasers;

4. Vertical cavity surface emitting lasers (VCSEL);

5. Vertical external cavity surface emitting lasers (VECSEL);

6. External cavity semiconductor diode lasers;

7. Optically pumped semiconductor lasers;

8. Quantum cascade lasers.

9. Lead salt lasers

2.7.2.1 Homojunction and Heterojunction Lasers

In *homojunction lasers*, all layers are of the same semiconductor material. One such type is GaAs/GaAs laser. However, this simple laser diode structure is highly inefficient and therefore could be used only to demonstrate pulsed operation. In *heterojunction lasers*, the active layer and either one or both of the adjacent layers are of different materials. If only one of the adjacent layers is of a different material, it is called a *simple heterojunction* and if both are different, it is called a *double heterojunction*. Some of the popular semiconductor laser types and the corresponding emission wavelength band include the following:

1. AlGaInP/GaAs: heterojunction 620–680 nm

2. $Ga_{0.5}In_{0.5}P$/GaAs: heterojunction 670–680 nm

3. GaAlAs/GaAs: heterojunction 750–870 nm

4. GaAs/GaAs: homojunction 904 nm

5. InGaAsP/InP: heterojunction 1,100–1,650 nm

2.7.2.2 Quantum-Well Lasers

For semiconductor layers that are only a few nanometers thick, the assumption that the material is a continuum is no longer valid and the quantum mechanical properties of atoms and electrons become important. When a thin layer of a semiconductor material with a relatively smaller band gap is sandwiched between two thick layers with larger band gaps, a structure known as a *quantum well* is formed and the electrons passing through the semiconductor are captured in the thin layer. Although these electrons have sufficient energy to be free in the small band-gap quantum-well layer, it is not enough to allow them to enter the large band-gap thicker layers. The thicker layers thus help in confining the captured electrons to the quantum well. If the quantum well is placed in the semiconductor junction, the concentration of electrons as they recombine with holes to emit photons leads to an increase in efficiency and a decrease in lasing threshold. The fact that the quantum-well layer and the outer thicker layers are made from semiconductor materials of different refractive indices further helps the cause by confining the emitted photons to that narrow region.

2.7.2.3 DFB Lasers

Diode lasers have a broad gain-bandwidth curve to accommodate several longitudinal modes. In a conventional diode laser, therefore, due to changes in temperature and other operating conditions, the dominant longitudinal mode may hop from one value to the next adjacent value leading to instability in output power and wavelength. A DFB laser offers a mechanism of ensuring a very narrow output wavelength band. In this laser, a diffraction grating is etched very close to the active layer (Figure 2.15). The grating provides a wavelength selective feedback to the gain region. The feedback causes interference and only the narrow wavelength region for which the interference is constructive is allowed to sustain. The DFB lasers offer single longitudinal and transverse mode operation.

2.7.2.4 Vertical Cavity Surface Emitting Lasers

In the diode laser structures discussed so far, the optical cavity is perpendicular to the direction of current flow. In the case of *vertical cavity surface emitting lasers* (VCSELs), the optical cavity is along the direction of flow of injection current as shown in Figure 2.16. The laser beam in this case emerges from the surface of the wafer rather from its edges. Having mirrors on both sides of the active medium forms the resonant structure. Compared to edge emitting lasers, these lasers produce relatively lower output power levels. However, VCSELs require a very small chip area, typically a few tens of μm^2 with the result that

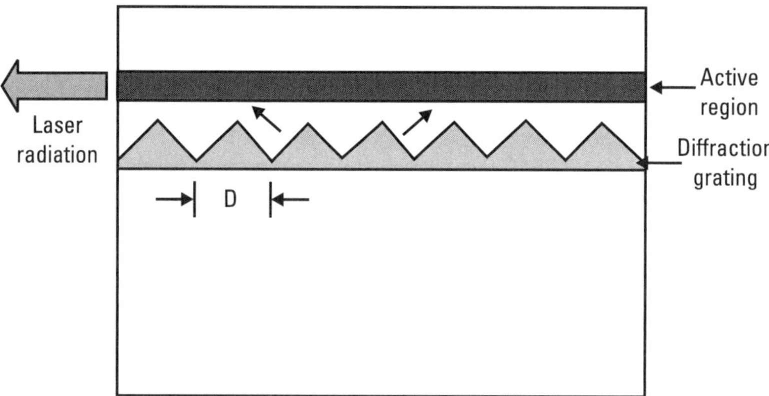

Figure 2.15 DFB laser.

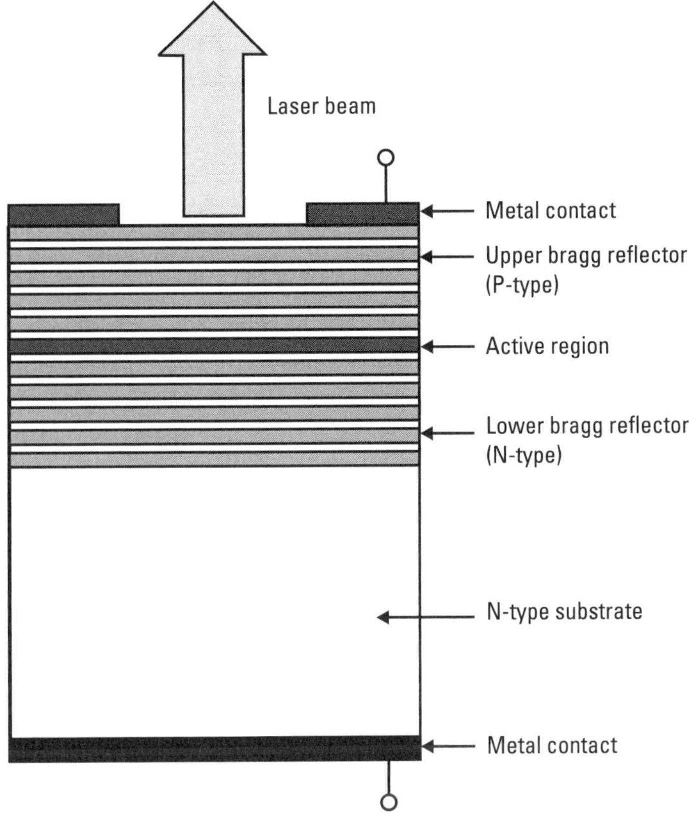

Figure 2.16 VCSEL structure.

a large number of such lasers can be tightly packaged in an array structure on a single chip. These lasers are characterized by a very small threshold current, which could be as low as 1 μA.

2.7.2.5 Vertical External Cavity Surface Emitting Lasers

In the case of a *vertical external cavity surface emitting lasers* (VECSELs), the two mirrors are either grown epitaxially as a part of the diode structure or grown separately and then bonded to the semiconductor chip having the active region. A vertical external cavity surface emitting laser (VECSEL) is a variant of VCSEL where the resonator is completed with a mirror placed external to the diode structure, thus introducing a free-space region in the resonant cavity as shown in Figure 2.17. Compared with other types of semiconductor diode lasers, VECSELs are capable of generating relatively much higher optical powers with high beam quality.

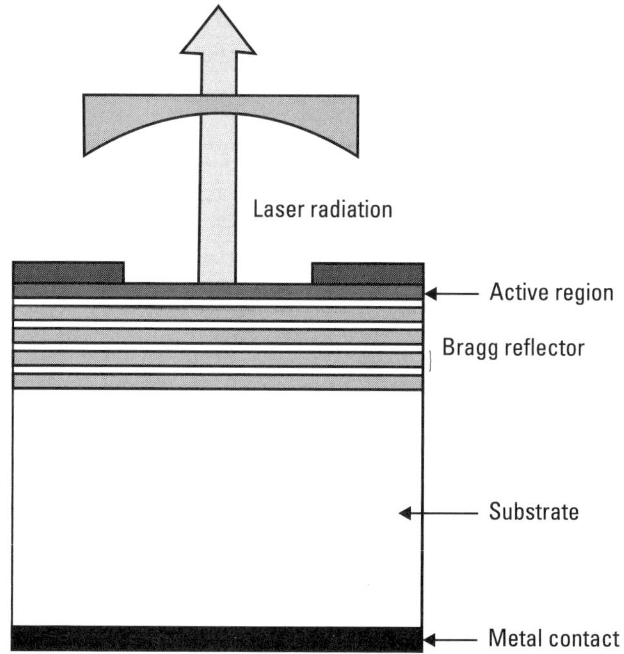

Figure 2.17   VECSEL structure.

2.7.2.6   External Cavity Semiconductor Diode Lasers

External cavity diode lasers are semiconductor diode lasers whose resonators are completed with one or more optical components outside the diode laser chip. In the simplest form, the diverging output from one of the end faces that is antireflection-coated is collimated by an external lens. The collimated beam is made to fall on a partially reflecting mirror that provides optical feedback and also acts as output coupler. An external cavity laser allows for a longer resonator, which in turn leads to lower phase noise and a narrow emission line width resulting from increased damping time of intracavity light. An external resonator also opens up the possibility of introducing suitable intracavity optical elements for wavelength selectivity and tuning, mode locking, and so forth.

2.7.2.7   Optically Pumped Semiconductor Lasers

Optically pumped semiconductor lasers (OPSLs) are simply vertical external cavity semiconductor lasers that are pumped optically. The active region in this case is comprised of alternate layers of a binary semiconductor material and tertiary semiconductor material quantum wells. The emission wavelength depends on the stoichiometry and physical dimensions of the quantum wells.

2.7.2.8 Quantum Cascade Lasers

Quantum cascade lasers (QCLs) are compact high power wavelength agile semiconductor lasers that emit in the midinfrared to far infrared wavelength band. The upper limit of emission wavelength can extend even to terahertz regions. The semiconductor lasers described in earlier sections are all interband devices where emission of laser radiation takes place due to recombination of electrons in the conduction band and holes in the valence band across the band gap of the semiconductor material. The earliest double heterostructure devices had an operating wavelength that was exclusively dependent on the band-gap energy as shown in Figure 2.18(a). In the case of quantum-well structures, as shown in Figure 2.18(b), carriers are confined to energy levels within these wells, opening up the possibility of lower transition energies and thereby extending the operating wavelength. Semiconductor diode lasers are bipolar devices; on the other hand, quantum cascade lasers are unipolar devices and laser emission in this case occurs across inter-subband, also called intraband transitions of electrons in the conduction band, as shown in Figure 2.18(c).

High output power in midinfrared to far infrared spectral region extending to terahertz, tunability and room temperature operation make quantum cascade lasers highly suitable for a wide range of applications in remote sensing of environmental gases and atmospheric pollutants, vehicular cruise control and collision avoidance in poor visibility conditions, medical diagnostics, industrial process control, and homeland security.

2.7.2.9 Lead Salt Semiconductor Lasers

Lead salt semiconductor lasers are semiconductor diode lasers generating wavelength tunable pulsed and CW output in the midinfrared spectral band that is particularly useful in high-resolution absorption spectroscopy used for detection and identification of trace gases with high sensitivity and selectivity. A lead salt semiconductor laser is a P-N junction diode laser that consists of a single crystal of lead telluride (PbTe), lead selenide (PbSe), lead sulfide (PbS) or their alloys with themselves or with strontium selenide (SnSe), strontium telluride (SnTe), cadmium sulfide (CdS), and other materials. The structure of the laser cavity is similar to the conventional semiconductor diode laser employing a Fabry-Perot resonator cavity comprising two parallel end faces. The injection current populates the near-empty conduction band and laser radiation is emitted by the process of stimulated emission across the band gap. The band-gap energy in the case of lead salt semiconductor lasers is very small, in the range of 0.25 to 0.30 eV. These lasers require cryogenic cooling for population inversion and laser action.

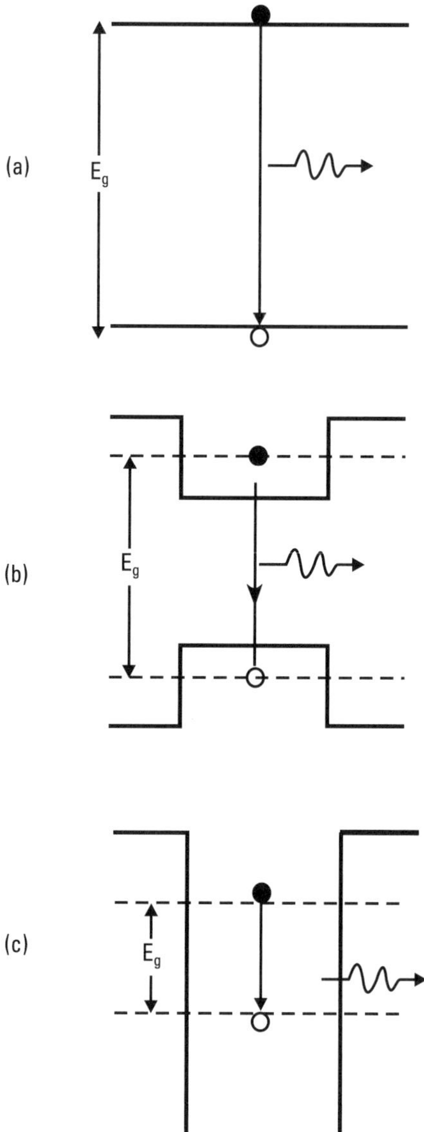

Figure 2.18   Interband and intraband laser transitions in semiconductor lasers.

2.7.3 Applications

Diode lasers find extensive use in commercial, industrial, medical, military, and scientific applications. Different applications exploit one or more of the inherent characteristics of the laser beam, which include directed energy due to low divergence, coherence, and monochromaticity (or narrow bandwidth of emission spectrum).

The application areas that primarily make use of directed energy property of the laser light include laser printing, bar code reading, image scanning, optical data recording, laser surgery, target illumination and designation, laser-based ignition of combustion reaction, and laser-based ignition of explosives. Infrared laser diodes are also used as target illuminators in night vision devices to enhance their performance. Applications that exploit the coherence property of laser light include distance measurement with interferometer setup, proximity sensors, ladar sensors, laser range finders, holography, and coherent communications. Applications that make use of monochromaticity mainly include telecommunications, spectroscopy, and a large number of biomedical diagnostic and therapeutic applications. Spectroscopy helps in the study of materials by measuring wavelengths absorbed and emitted by them. One security-related application of laser spectroscopy is in detection and identification of chemical, biological, and explosive agents.

2.8 Photosensors

Photosensors constitute the heart of a variety of systems ranging from simple gadgets like light meters to the most complex of military systems like precision guided munitions, laser range finders, target trackers, remote sensing systems, navigation sensors, sniper and explosive detectors, fiber-optic and laser based communication systems, night vision devices, and lidar and spectroscopic sensors. Understanding photosensors and related sensor systems therefore is essential to understanding the design and operation of a large number of defense systems. While individual photosensors such as PIN photodiodes and avalanche photodiodes find applications in military optoelectronic systems including laser range finders and target designators, lidar sensors, navigation sensors, and many more, sensor arrays such as CMOS, CCD, and APD arrays are at the core of imaging sensor systems including ladar sensors, night vision devices, and laser and imaging infrared seekers.

2.8.1 Types of Photosensors

Photosensors are classified into two major categories: photoelectric sensors and thermal sensors. Photoelectric sensors are further classified into two types: devices that depend on an external photo effect for their operation and devices that make use of some kind of internal photo effect. These devices together include photoemissive sensors, photoconductors, and junction-type photo sensors such as photodiodes and phototransistors.

Photoemissive sensors are based on an external photo effect. Photoconductors and junction-type photo sensors use an internal photo effect. Common photoemissive sensors include nonimaging sensors such as vacuum photo cells and photomultiplier tubes and imaging sensors such as image intensifier tubes. Photoconductors are bulk semiconductor devices whose resistance decreases with an increase in incident light intensity. They are also known as photoresistors, light-dependent resistors, and photo cells. Junction-type photosensors are further classified into amplifying and nonamplifying types. The amplifying type of junction photosensors include phototransistors, photo thyristors, and photo FETs. Nonamplifying types of junction photosensors include photo diodes, solar cells, CMOS sensors, and CCDs. CMOS and CCDs are imaging sensors. In the category of thermal sensors, we have thermocouple- (or thermopile-) type sensors, bolometric sensors, and pyroelectric sensors. Thermal sensors absorb incident radiation and operate on the resulting temperature rise whereas photoelectric sensors are based on quantum effect. Thermal sensors are relatively sluggish in their response to incident radiation than photoelectric sensors; however, thermal sensors offer a much wider operational wavelength band than photoelectric sensors.

2.8.2 Characteristic Parameters

Major characteristic parameters used to characterize the performance of photosensors include responsivity, noise equivalent power (NEP), sensitivity usually measured as detectivity and dee-star, quantum efficiency, response time, and noise

Responsivity is defined as the ratio of electrical output to radiant light input determined in the linear region of the response. It is measured in amperes per watt (A/W) or V/W if the photosensor produces a voltage output rather than a current output. Responsivity is a function of wavelength of incident radiation and band-gap energy. *Spectral response* is a related parameter. It is the curve that shows variation of responsivity as a function of wavelength. Most photoelectric sensors have narrow spectral response whereas most thermal sensors have wide spectral response. Responsivity can be expressed by (2.10):

$$NEP = \frac{I_N}{R_v} \tag{2.10}$$

where

I_N = total noise current (A)

R_V = responsivity in (A/W)

NEP is the input power to a sensor that generates an output signal current equal to the total internal noise current of the device. In other words, it is the minimum detectable radiation level of the sensor. The noise power and thus the noise-equivalent power depend on the assumed detection bandwidth.

If one were to use the full detection bandwidth of the device to compute NEP, then the NEP would not allow a fair comparison of sensors with different bandwidths. Therefore, it is a common practice to assume a bandwidth of 1 Hz, which is usually far below the detection bandwidth. NEP is usually specified in units of W/√Hz rather than watts.

Detectivity of a sensor is the reciprocal of its NEP. Detectivity, like NEP, depends on noise bandwidth and sensor area. To eliminate these factors, a normalized figure of detectivity referred to as dee-star is used. It is defined as the detectivity normalized to an area of 1 cm² and a noise bandwidth of 1 Hz. Dee-star (D*) is expressed by (2.11):

$$D^* = D\sqrt{A\Delta f} \tag{2.11}$$

where

D^* = dee-star ($W^{-1}cm\ Hz^{1/2}$)

D = detectivity (W^{-1})

A = sensor area (cm^2)

Δf = bandwidth (Hz)

Quantum efficiency is the ratio of the number of photoelectrons released to the number of photons of incident light absorbed. It is the percentage of input radiation power converted into photo current.

Response time is expressed as a rise/fall time parameter in photoelectric sensors and as a time constant parameter in thermal sensors. Rise and fall times are the time durations required by the output to change from 10% to 90% and 90% to 10% of the final response, respectively. It determines the highest signal frequency to which a sensor can respond. Time constant is defined as the time required by the output to reach to 63% of the final response from zero initial value.

Noise is the most critical factor in designing sensitive radiation detection systems. Noise in these systems is generated in photosensors, radiation sources, and postdetection circuitry. Photosensor noise mainly is comprised of Johnson noise, shot noise, generation-recombination noise, and flicker noise. Johnson noise, also known as Nyquist noise or thermal noise, is caused by the thermal motion of charged particles in a resistive element. The RMS value of the noise voltage depends on the resistance value, temperature, and the system bandwidth. It is expressed by (2.12):

$$V_{RMS} = \sqrt{4KRT\Delta f} \tag{2.12}$$

where

V_{RMS} = RMS noise voltage (V)

R = resistance value in ohms

K = Boltzmann constant (1.38×10^{-23} J/K)

T = absolute temperature (K)

Δf = system bandwidth (Hz)

Shot noise in a photosensor is caused by the discrete nature of the photoelectrons generated. It is related to the statistical fluctuation of both the dark current and the photo current. It depends on the average current through the photosensor and system bandwidth. It is the dominant source of noise in the case of photodiodes operating in photoconductive mode. It is expressed by (2.13):

$$I_{SRMS} = \sqrt{2eI_{av}\Delta f}\qquad(2.13)$$

where

I_{SRMS} = RMS shot noise current (A)

I_{AV} = average current through the photo sensor (A)

e = charge of an electron (= 1.60×10^{-19} C)

Δf = detection bandwidth (Hz)

Generation-recombination noise is caused by the fluctuation in current generation and the recombination rates in a photosensor. This type of noise is predominant in photoconductive sensors operating at infrared wavelengths. It is expressed by (2.14):

$$I_{GRMS} = 2eG\sqrt{\eta EA\Delta f}\qquad(2.14)$$

where

I_{GRMS} = RMS generation-recombination noise current (A)

e = charge of an electron (1.60×10^{-19} C)

Δf = detection bandwidth (Hz)

E = radiant intensity (W/cm^2)

A = sensor receiving area (cm^2)

G = photoconductive gain

η = quantum efficiency

Flicker noise or 1/f noise occurs in all conductors where the conducting medium is not a metal and exists in all semiconductor devices that require bias current for their operation. Its amplitude is inversely proportional to the frequency. Flicker noise is usually predominant at frequencies below 100 Hz.

2.8.3 Photoconductors

Photoconductors are bulk semiconductor devices whose resistance decreases with increasing incident light intensity. The resistance change in a photoconductor is of the order of 6 decades, ranging from a few tens of megaohms under dark conditions to a few tens or hundreds of ohms under bright light conditions. Other features include wide dynamic response, spectral coverage from ultraviolet to far infrared, and low cost. However, they are sluggish devices having a response time of the order of hundreds of milliseconds.

2.8.4 Photodiodes

Photodiodes are junction-type semiconductor devices that generate current or voltage when the P-N junction in the semiconductor is illuminated by light of sufficient energy. The spectral response of the photodiode is a function of the band-gap energy of the material used in its construction. The upper cutoff wavelength of a photodiode is given by (2.15):

$$\lambda_c = \frac{1240}{E_g} \tag{2.15}$$

Photodiodes are mostly constructed using silicon, germanium, indium gallium arsenide, lead sulfide, and mercury cadmium telluride. Figure 2.19 shows the spectral characteristics of these photodiodes. Depending on their construction, there are several types of photodiodes. These include PN photodiodes, PIN photodiodes, Schottky-type photodiodes, and avalanche photodiodes (APDs).

A *P-N photodiode* is comprised of a P-N junction. When light with sufficient energy strikes the photodiode, the electrons are pulled up into the conduction band, leaving behind holes in the valence band. These electron-hole pairs occur throughout the p-layer, depletion layer, and N-layer materials. When the photodiode is reverse-biased, the photo-induced electrons will move down the potential hill from the P side to the N side. Similarly, the photo-induced holes will add to the current flow by moving across the junction to the P side. Shorter wavelengths are absorbed at the surface while the longer wavelengths penetrate deep into the diode. P-N photodiodes are used for precision photometry applications like medical instrumentation, analytical instruments, semiconductor tools, and industrial measurement systems.

In *PIN photodiodes*, an extrahigh resistance intrinsic layer is added between the P and N layers. This has the effect of reducing the transit or diffusion time of the photo-induced electron-hole pairs that in turn results in improved response time. PIN photodiodes feature low capacitance, thereby offering high

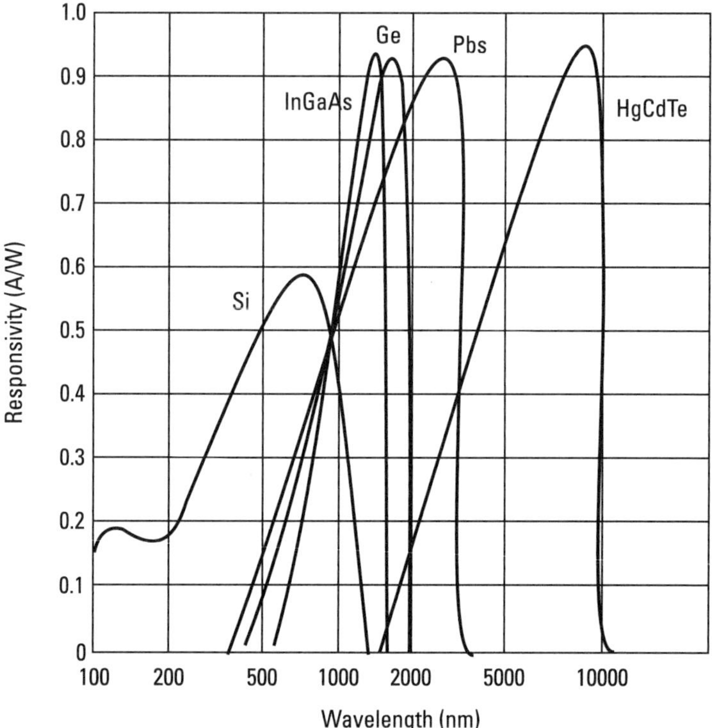

Figure 2.19   Spectral characteristics of photodiodes.

bandwidth that makes them suitable for high-speed photometry as well as optical communication applications. Single-element PIN photodiodes, quadrant photodiodes, and one- and two-dimensional arrays of photodiodes find extensive applications in a variety of sensor systems for military applications. Prominent among these include laser warning sensor suites on armored fighting vehicles and airborne platforms, laser seekers in laser-guided munitions, laser receivers in laser communication systems, and focal plane arrays in ladar sensors. Photodiodes are available in a large number of package styles including customized ones. In *Schottky-type photodiodes*, a thin gold coating is sputtered on to the N-material to form a Schottky effect P-N junction. Schottky photodiodes have enhanced UV response.

APDs are high-speed, high-sensitivity photodiodes utilizing an internal gain mechanism that functions by applying a relatively higher reverse-bias voltage than that is applied in the case of PIN photodiodes. APDs have fast response times similar to that of PIN photodiodes. As well, responsivity figures for silicon PIN photodiodes are in the range of 0.4 to 0.6 A/W whereas for APDs they are between 40 to 80A/W, around 100 times more than that of PIN photodiodes. Moreover, they offer excellent signal-to-noise ratio of the order of that offered by photomultiplier tubes. Hence, they are used in a variety of applications requiring high sensitivity such as long-distance optical communication and optical distance measurement. Military laser range finders based on the time-of-flight principle have their receivers' front end invariably employing a silicon or indium gallium arsenide APD depending on whether it is an Nd:YAG laser range finder or an eye-safe laser range finder. APDs, just like PIN photodiodes, are also available as both single detectors as well as linear or two-dimensional arrays in similar package configurations.

Photodiodes can be operated in two modes: the *photovoltaic* mode and *photoconductive* mode. In photovoltaic mode of operation, no bias voltage is applied and due to the incident light, a forward voltage is produced across the photodiode. In photoconductive operational mode, a reverse-bias voltage is applied across the photodiode. This widens the depletion region, resulting in a higher speed of response. As a rule of thumb, all applications requiring bandwidth less than 10 kHz can use photodiodes in photovoltaic mode. For all other applications, photodiodes are operated in photoconductive mode. Moreover, the linearity of a photodiode is also improved when it is operated in the photoconductive mode. However, there is an increase in the noise current of the photodiode when it is operated in the photoconductive mode. This is due to the reverse saturation current referred to as the dark current flowing through the photodiode. The value of dark current is typically in the range of 1 to 10 nA at a specified reverse-bias voltage. When the photodiode is operated in the photovoltaic mode, the value of dark current is zero.

2.8.5 Image Sensors

Imaging sensors are an important constituent of a wide range of defense systems and are predicted to play a growing role in the coming years. Some of the well-established military systems using imaging sensors include thermal imaging cameras used in law enforcement, marine safety, and other surveillance related applications; military satellites, space vehicles, and imaging missile seekers; directed energy weapon systems for target detection and tracking; electronic and electro-optic countermeasure systems for target detection and night vision devices for low-light-level imaging applications; sensor systems for space science imaging to capture detailed images across ultraviolet, visible, and near infrared spectra, ladar sensors for 3-D imaging of targets; and hyperspectral imaging sensor systems that look at many spectra of light in closely spaced bandwidths to allow not only the presence of a target but also its features such as the material it is made of and the kind of paint it has. Two of the most common types of imaging sensors with potential military applications include CCDs and CMOS sensors. Ladar sensors that employ a 2-D array of APDs are another important imaging sensor finding application in precision-guided weapon seeker systems.

Both CCD and CMOS sensors use a 2-D array of thousands to millions of discrete pixels. The amount of light falling on each of the pixels generates free electrons with the number of electrons and hence the quantum of charge depending on the intensity of impinging photons. The two types of sensors differ in the mode in which this charge is converted into voltage and subsequently read out of the chip for further processing.

Figure 2.20 shows a typical CCD sensor with a single point readout. The basic CCD sensor can only determine number of photons collected by each pixel and therefore it carries no information about the wavelength or color of those photons. As a result, the CCD sensor is capable of recording the image only in monochrome. In order to record image in full color, a filter array is bonded to the sensor substrate. One such common color filter array is the Bayer filter. Bayer's color filter array (CFA) is comprised of an arrangement of red, green, and blue filters to capture color information.

In a CCD sensor, all of the pixels can be devoted to light capture, and the output's uniformity, which is a key factor in high image quality of CCD sensors. On the other hand, in a CMOS sensor, each pixel has its own charge-to-voltage convertor, amplifier, and a pixel select switch (Figure 2.21). This is called active pixel sensor architecture in contrast to passive pixel sensor architecture used in

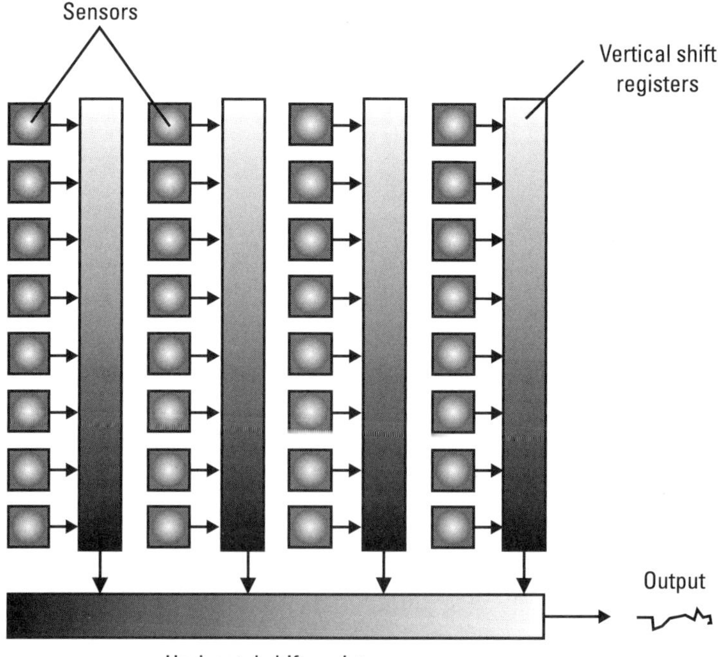

Figure 2.20 CCD sensor.

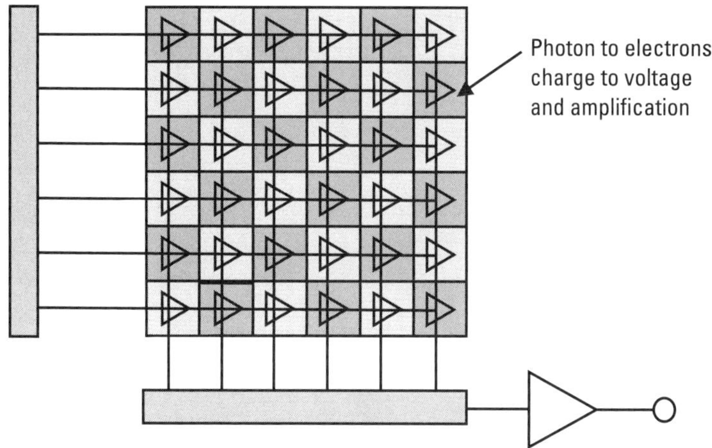

Figure 2.21 CMOS sensor.

a CCD sensor. As well, the sensor often also includes on-chip amplifiers, noise correction, and analog-to-digital conversion circuits and other circuits critical to pixel sensor operation. The chip in this case outputs digital bits. Inclusion of these functions reduces the area available for light capture. Also, with each pixel doing its own conversion, uniformity and consequently image quality is lower. While the readout mechanism of a CCD sensor is serial, it is massively parallel in the case of a CMOS sensor allowing high total bandwidth for high speed.

2.8.6 Photoemissive Sensors

Photoemissive sensors rely on external photo effects wherein the photo-generated electrons travel beyond the physical boundaries of the material. Some of the commonly used photoemissive photosensors include vacuum photodiodes, photomultiplier tubes, and image intensifier tubes. Of the three types, photomultiplier tubes (PMTs) and image intensifier tubes are important photoemissive sensors. Photomultiplier tubes are extremely sensitive photosensors operating in the ultraviolet, visible, and near-infrared spectrum. PMTs have internal gain of the order of 10^8 and can even detect a single photon of light. They are constructed from a glass vacuum tube that houses a photocathode, several dynodes, and an anode. When the incident photons strike the photocathode, electrons are produced as a result of the photoelectric effect. These electrons accelerate toward the anode and in the process electron multiplication taken place due to secondary emission process from the dynodes. Salient features of photomultiplier tubes include low noise, high frequency response, and large active area. By virtue of these features, PMTs are used in nuclear and particle physics, astronomy, medical imaging, and motion picture film scanning.

2.8.7 Thermal Sensors

Thermal sensors absorb radiation, which produces a temperature change that in turn causes a change in the physical or the electrical property of the sensor. In other words, thermal sensors respond to change in their bulk temperature caused by the incident radiation. Thermocouple, thermopile, bolometer, and pyroelectric sensors belong to the category of thermal sensors. Thermal sensors lack the sensitivity of photoelectric sensors and are generally slow in response, but have a wide spectral response. Most of these sensors are passive devices, requiring no bias.

2.8.7.1 Thermocouple and Thermopile Sensors

Thermocouple sensors are based on the Seebeck effect (i.e., the temperature change at the junction of two dissimilar metals generates an EMF proportional to the temperature change). Commonly used thermocouple materials are bismuth-antimony, iron-constantan, and copper-constantan. Their temperature coefficients are 100 μV/°C, 54 μV/°C, and 39 μV/°C, respectively. To compensate for the changes in the ambient temperature, thermocouples generally have two junctions: the measuring junction and the reference junction.

The responsivity of a single thermocouple is very low and therefore to increase the responsivity several junctions are connected in series to form a *thermopile*. Thermopiles are a series combination of around 20 to 200 thermocouples. Spectral response of thermocouples and thermopiles extends into the far infrared band up to 40 μm. They are suitable for making measurements over a large temperature range up to 1,800 K. However, thermocouples are less suitable for applications where smaller temperature differences need to be measured with great accuracy such as 0°C to 100°C measurement with 0.1°C accuracy. For such applications, thermistors and RTDs are more suitable.

2.8.7.2 Bolometer

A *bolometer* is the most popular type of thermal sensor. The sensing element in a bolometer is a resistor with a high temperature coefficient. A bolometer is different from photoconductor; in a photoconductor a direct photon-electron interaction causes a change in the conductivity of the material whereas in a bolometer the increased temperature and the temperature coefficient of the element causes the resistance change. Bolometers can be further categorized as metal bolometers, thermistor bolometers, and low-temperature germanium bolometers.

A metal bolometer uses metals such as bismuth, nickel, or platinum with temperature coefficient in the range of 0.3%/°C to 0.5%/°C. Thermistor bolometers are the most popular types and they find applications in burglar alarms, smoke sensors, and other similar devices. The sensor in this case is a thermistor, an element made of manganese, cobalt, and nickel oxide. They have high temperature coefficients up to 5%/°C. The temperature coefficient varies with temperature as $1/T^2$. They are classified as a negative temperature coefficient (NTC) and positive temperature coefficient (PTC) thermistors depending on whether their temperature coefficient of resistance is negative or positive.

2.8.7.3 Pyroelectric Sensors

Pyroelectric sensors are characterized by spontaneous electric polarization that is altered by temperature changes as light illuminates these sensors. Pyroelectric sensors are low-cost, high-sensitivity devices that are stable against temperature variations and electromagnetic interference. Pyroelectric sensors only respond to modulating light radiation and there will be no output for a CW incident radiation. Pyroelectric sensors operate in two modes: the voltage mode and the current mode. In the voltage mode, the voltage generated across the entire pyroelectric crystal is detected. In the current mode of operation, current flowing on and off the electrode on the exposed face of the crystal is detected. Voltage mode is more commonly used than current mode.

Selected Bibliography

Endo, M., and R. F. Walter, *Gas Lasers*, Boca Raton, FL: CRC Press, 2006.

Hecht, J., *Understanding Lasers: An Entry Level Guide*, Third Edition, Piscataway, NJ: IEEE Press, 2011.

Hertsens, T., *Overview of Laser Diode Characteristics: Application Note # 5*, Bozeman, MT: ILX Lightwave Photonic Test and Measurement, 2000.

Injeyan, H., and G. D. Goodno, *High Power Laser Handbook*, New York: McGraw-Hill, 2011.

Junji, O., *Semiconductor Lasers*, Heidelberg, Germany: Springer, 2017.

Kapon, E., *Semiconductor Lasers-I: Fundamentals*, San Diego: Academic Press, 1999.

Kasap, S. O., *Optoelectronics & Photonics: Principles and Practices,* Prentice Hall, 2012.

Koechner, W., *Optoelectronics & Photonics: Principles and Practices*, Sixth Edition, Springer, 2006.

Koechner, W., and M. Bass, *Solid State Lasers: A Graduate Text,* New York: Springer, 2003.

Sennaroglu, A., *Solid State Lasers and Applications,* Boca Raton, FL: CRC Press, 2006.

Silfvast, W. T., *Laser Fundamentals*, Second Edition, Cambridge, UK: Cambridge University Press, 2012.

Svelto, O., *Principles of Lasers*, New York: Springer, 2009.

Vinter, B., and E. Rosencher, *Optoelectronics*, Cambridge, UK: Cambridge University Press, 2002.

Waynant, R., and M. Ediger, *Electro-Optics Handbook,* New York: McGraw-Hill Professional, 2000.

Webb, C. E., and J. D. C. Jones, *Handbook of Laser Technology and Applications: Volume II,* London: Institute of Physics Publishing, 2003.

Webb, C. E., and J. D. C. Jones, *Handbook of Laser Technology and Applications: Volume I,* CRC Press, 2003.

3

Less-Lethal Laser Weapons

This chapter comprehensively covers less-lethal laser weapons designed for use in antipersonnel applications. The chapter begins with an overview of less-lethal nonoptical technologies and then moves on to discuss at length operational fundamentals of laser dazzlers, choice of operational parameters, deployment scenarios including industrial security, counterinsurgency and antiterrorist operations, riot and mob control, protection of critical assets from unwanted aerial threats, and protection of sea vessels from asymmetric threats. Safety issues associated are discussed next followed by an outline on representative laser dazzler systems including handheld and weapon-mounted laser dazzlers, vehicle-mounted laser dazzlers for riot and mob control, ground-based laser dazzlers against aerial threats, shipborne laser dazzlers against asymmetric threats, laser dazzlers on airborne platforms, and spaceborne laser dazzlers.

3.1 Less-Lethal Laser Weapons

Development of less-lethal weapons, facilitated by ever-increasing interest shown by political class, armed forces and law enforcement agencies, and large-scale investment in research and development of nonlethal technologies by both government and private agencies, has witnessed a massive growth in the last 10 to 15 years. The fact that most of these technologies have dual-use military/civilian applications has only helped the cause further. This has led to deployment of a large number of these less-lethal devices in a variety of application scenarios, which includes riot and mob control, counterinsurgency and antiterrorist operations, protection of critical assets in industry, defense, communications, food and drinking-water resources, science and technology, and space assets. Less-lethal weapons are found to be particularly useful in operations such as those when security agencies must counter violent lawbreakers; when combatants and noncombatants are mixed together unintentionally or deliberately; when collateral damage needs to be minimized if it cannot be completely avoided; when there is demand for having alternatives to lethal weapons for military peacekeeping operations; and when there is ever-increasing pressure on security agencies from both the public and government sectors to make the operations bloodless and humane. A number of nonlethal or less-lethal technologies belonging to the broad categories of kinetic energy, directed energy, electrical, acoustic, chemical, biochemical, biological, and barriers and entanglements have been developed. In the following sections, we briefly describe each of these technologies.

3.1.1   Review of Less-Lethal Technologies

Based on the technology behind the functioning of less-lethal weapons, these are grouped together in the following major categories:

1. Kinetic energy;
2. Directed energy;
3. Electrical;
4. Acoustic;
5. Chemical;
6. Biological;
7. Biochemical;
8. Barriers and entanglements.

3.1.1.1   Less-Lethal Kinetic Energy Weapons

Kinetic energy weapons achieve the desired effect by transferring kinetic energy from the weapon to the targeted person or material object. Common kinetic energy weapons include kinetic impact projectiles such as rubber and plastic bullets, beanbag rounds, pellet rounds, foam rounds, and sponge rounds, and water cannons. *Rubber* and *plastic bullets* are solid, spherical, or cylindrical projectiles of variable sizes and made of rubber, plastic, polyvinyl chloride (PVC), or a composite material including metal. These are fired as single shots or in groups of multiple projectiles. *Beanbag rounds,* also known as flexible batons, are comprised of synthetic cloth bags filled with small metal pellets that are fit into a cartridge. They expand as they travel to create a wide surface area impact. *Sponge rounds* are projectiles with a softer tip or nose. This limits penetration of the projectile into the skin. Sponge rounds include foam rounds with a hard foam nose or attenuated energy projectiles with a hollow nose. *Pellet rounds* are cartridges filled with small lead, steel, or plastic/rubber pellets that spread out when fired. Figure 3.1 shows an assortment of kinetic impact projectiles.

A new class of less-lethal ammunitions is the *blunt impact projectile* (BIP). It also depends on the kinetic energy imparted to the targeted person or object for its operation. It is larger than a rubber bullet and comes with a silicon head that expands and collapses on impact. The impact is strong enough to incapacitate an aggressive and noncompliant subject without the ill effects of regular bullets. One such blunt impact projectile is a 40-mm BIP from Defense Technology, Safariland Group. It employs a collapsible gel nose technology that enables kinetic energy to be applied over a larger surface area of the body, targeting the lower torso or extremities. These areas provide sufficient pain stimulus while greatly reducing serious or life-threatening injuries. Figure 3.2(a, b) shows

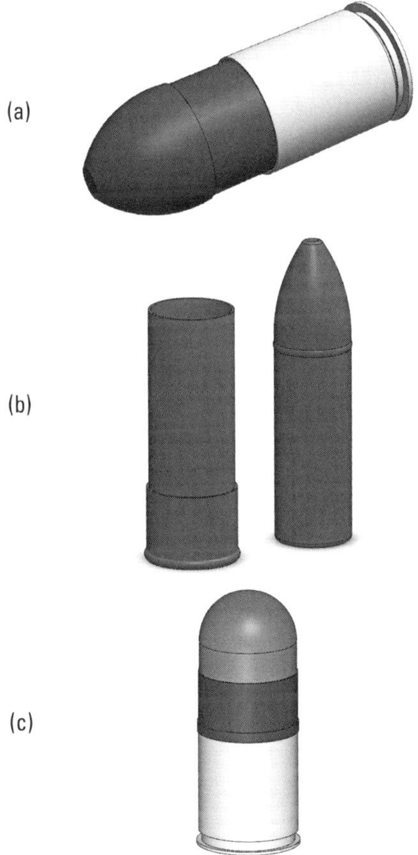

Figure 3.1 Kinetic impact projectiles.

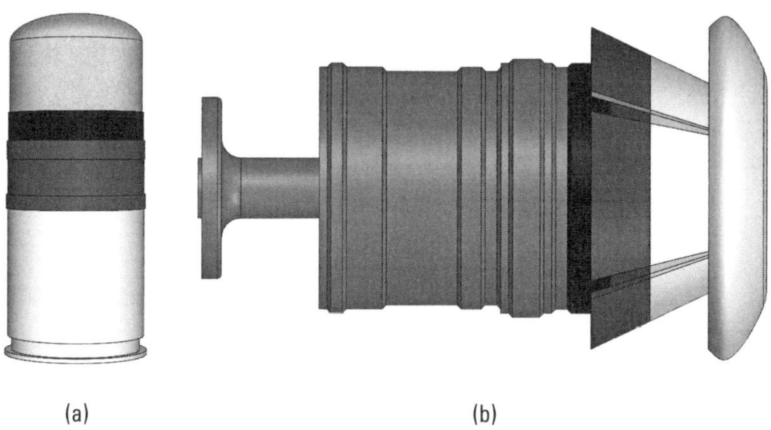

Figure 3.2 Blunt impact projectile (a) before impact, and (b) after impact.

a blunt impact projectile using collapsible gel nose technology before and after impact.

Kinetic impact projectiles are intentionally designed to inflict pain and incapacitate an individual without penetrating the body. This is achieved by the larger shape and slower speed of these projectiles, which limits their ability to penetrate the skin or cause deep blunt trauma injury. These projectiles are deployed from a wide range of launchers and guns. The launchers may be either specially designed, such as those used for crowd control operations, or be an add-on to rifles used for live ammunition. Depending on the type of bullet and launcher, single or multiple projectiles may be fired at once. Kinetic impact projectile weapons are both antipersonnel and antimaterial weapons.

Water cannons are high-pressure jets that shoot out a high-pressure stream of water. Water cannons have proven to be very effective for riot control. These may be marked with dye for easy identification of rioters, be electrified, or have an added chemical irritant for enhancing efficacy. These are either mobile vehicle-mounted systems, backpack systems, or fixed in-place systems. Figure 3.3 shows a Tanqueta riot control vehicle equipped with a water cannon. The water cannon has a storage capacity of 11,500 L and is effective up to 50m of range. Recent developments in the technology of water cannons have led to the development of a sophisticated water cannon system by Israel that fires bullets of water that are nothing but very small quantities of water released at high pressure.

All kinetic impact less-lethal weapons are primarily short-range weapons since due to their slower speed and irregular shape, accuracy deteriorates for longer ranges. This increases the probability of hitting vulnerable parts of the body or even unintended targets, which limits their use to close combat engagements.

Figure 3.3 Tanqueta riot control vehicle with water cannon. (Source: Wikimedia Commons.)

Safety considerations and the control of blunt and penetrative trauma remains a serious concern. While blunt injuries cause internal damage without breaking the skin barrier, penetrative injuries pierce the skin or soft tissue. Kinetic energy weapons, projectiles in particular, can cause serious and sometimes irreversible damage to the eyes, musculoskeletal system, cardiorespiratory system, brain, head and neck, organs of the abdominal region, including kidney, spleen, and liver, and skin and soft tissue. Eyes are highly vulnerable to irreversible damage due to direct trauma to the eye from kinetic energy projectiles. It is highly likely that it causes total blindness due to rupture of the eyeball as well as trauma to nearby structures. These projectiles can even penetrate through the eye socket and enter the brain, causing brain injury. Injuries to the musculoskeletal system including muscles and bones may cause sprains, bruises, and fractures. Deeper injuries can even lead to permanent damage to the neurovascular structures, sometimes leading to what is known as compartment syndrome. Compartment syndrome is a painful condition that occurs when pressure within the muscles builds to dangerous levels causing decrease in blood flow, thereby preventing nourishment and oxygen from reaching nerve and muscle cells. Kinetic energy projectiles can cause serious injuries to cardiorespiratory system including lungs and heart. Penetration into the chest may lead to fatal injuries such as bleeding, collapsed lung, and heart attack. Blunt trauma to the *brain* can cause concussions, intracranial hemorrhage, and skull fractures. The projectile may penetrate the skull or enter the brain tissue causing hemorrhage or injury to the spinal cord. The delicate structures of the face and neck including bones of the face and skull, the spinal cord, and the blood vessels in the neck are particularly vulnerable to traumatic injury as they are all close to the skin surface. Blunt and penetrative injuries can cause serious damage to the organs of the abdominal region. Blunt injuries may cause bleeding in liver, kidney, and spleen. Penetrative injuries can also cause bleeding, perforations, and urogenital injuries. Kinetic energy projectiles can cause bruising and contusions of the skin and soft tissue, as well as deep cuts and tears in skin or flesh. Some of these may even cause muscle or nerve damage.

3.1.1.2   Less-Lethal Directed Energy Weapons

A DEW system primarily uses directed energy in the form of a concentrated beam of electromagnetic energy in the targeted direction to cause intended damage to the enemy's equipment, facilities, and personnel. Intended damage could be lethal or nonlethal. Based on the wavelength of the directed electromagnetic energy, these are classified as high-power microwave (HPM) weapons, MMW weapons, and laser weapons. While HPM weapons are primarily antimaterial weapons, MMW weapons are mainly used for antipersonnel applications. Laser weapons can be either antimaterial or antipersonnel depending on the laser power or energy used.

HPM weapons are designed to disrupt, degrade, or destroy electronics of the target by radiating electromagnetic energy in the microwave frequency band. These weapons generate an intense blast of microwave energy strong enough to overload electrical circuitry, inducing currents large enough to temporarily disrupt electronic systems or permanently damage integrated circuits, causing them to fail minutes, days, or even weeks later. The microwave blast may even melt the circuitry in some cases. Humans exposed to the blast of the HPM weapon remain unharmed and might not even know they have been hit. HPM weapon systems with their speed-of-light delivery, all-weather capability to destroy an adversary's electronics systems, area coverage of multiple targets, minimum collateral damage, simplified tracking and beam pointing, and a deep magazine address most of the requirements of military commanders of the day.

Frequencies around 100 GHz, which falls into the MMW band, have been exploited to build nonlethal weapons due to the significant effects this frequency has on human beings. The *active denial system* (ADS) is one such nonlethal MMW-directed energy system that can be used in a counterpersonnel role against hostile human targets at distances beyond the effective range of small arms. Active denial technology is a breakthrough nonlethal technology that uses millimeter-wave electromagnetic energy for area denial, perimeter security, and riot control applications from relatively longer ranges without the application of any lethal force. With a range of about 500m and a beam that is a little less than 2m across, and ADS is capable of dispersing riots from a safe distance. It has been developed by Raytheon for the U.S. Air Force Research Labs and Department of Defense Joint Non-Lethal Weapons Directorate. It operates as follows. Microwave energy at 95 GHz generated by a gyrotron is focused and directed at the targeted personnel by a directional planar antenna.

The microwave energy almost instantly produces an intolerable heating sensation within seconds, forcing the subject to flee from the scene to avoid exposure to the millimeter-wave energy beam. The millimeter-wave energy in fact heats up the water molecules just under the subject's skin, heating them enough to cause extraordinary pain. The pain sensation immediately ceases when the individual moves out of the beam or when the beam is turned off. The millimeter-wave beam does not cause any injury because of the low energy levels used and the shallow penetration depth of about 0.5 mm at the operating wavelength. In addition, the pain induced as a consequence of the subject's natural defense mechanism serves as a warning to help protect him or her from any injury. A simple control console on the ADS enables the operator to view the scene and precisely aim the millimeter-wave beam to affect only the intended target.

The ADS has been comprehensively evaluated on a large number of volunteers to establish its nonlethality. Two variants of the ADS integrated on land vehicles include an ADS mounted on a wheeled vehicular platform (Figure 3.4) and another integrated on a high-mobility multipurpose wheeled vehicle (HMMWV) platform more commonly known as a Humvee. These platforms would be equipped with the required power source to drive the weapon system. A midrange version of the active denial system called *Silent Guardian* with an operational range of about 250m has also been developed by Raytheon. Silent Guardian was primarily developed for use by law enforcement agencies. The system is easily transportable on standard military tactical vehicles and can also be integrated into combat vehicles. It is also being considered for operation from airborne platforms such as the AC-130 gunship. The airborne version will use a more powerful and lightweight version of its land-based counterpart. The

Figure 3.4 Operational version of an active denial system. (Wikimedia Commons.)

United States reportedly deployed an active denial system as a crowd control weapon in Afghanistan but it was withdrawn in 2010 without ever seeing it used due to serious questions raised by critics about the ethics of using a pain beam to break up riots.

According to reports, Russia and China are developing their own versions of active denial systems. As reported by the Interfax News Agency, the Russian military is testing its own beam weapon under development at the 12th Central Military Research and Development Institute. The device reportedly has a range of about 270m and is intended to be used for riot control applications. The Chinese system called Poly WB-1 pain beam weapon also uses millimeter-wave energy to cause intolerable pain in human subjects up to a kilometer away from the source.

Directed-energy laser systems are generally categorized as nonlethal or more appropriately less-lethal directed-energy laser weapons, such as laser dazzlers used in antipersonnel applications for antiterrorist and counterinsurgency operations and lethal high-energy laser weapon systems aimed at inflicting structural damage to the intended targets at tactical and strategic ranges. High-energy lasers used in antisensor applications as a part of electro-optic countermeasure equipment also come under less-lethal directed-energy laser systems. Another important application of high-energy directed-energy laser systems when operated at relatively lower power levels is in safe neutralization of unexploded ordnances with minimized collateral damage. These systems generally configured around kilowatt-class bulk solid-state lasers or high-power fiber lasers have amply demonstrated their efficacy in disposal of unexploded ordnances including improvised explosive devices from a safe distance. The laser dazzler is specifically designed for applications where subject vision impairment is to be achieved at a distance ranging from a few tens of meters to several kilometers in bright ambient conditions. They are particularly effective in situations where use of lethal force is not preferred; examples being limiting escalation and temporarily disabling facilities and equipment. Laser-based less-lethal weapons such as laser dazzlers can be used for counterinsurgency, antiterrorism, countersniper, self-defense, and crowd control and infrastructure protection applications. Laser dazzlers are emerging internationally as a new nonlethal alternative to lethal force for law enforcement, homeland security, border patrol, coastal protection, ship defense from asymmetric threats, aircraft defense against shoulder-fired missiles, infrastructure protection, and a host of other low-intensity conflict scenarios. Different types of laser dazzlers ranging from short-range handheld devices for close combat operations to medium-range portable devices for crowd control and long-range platform-mounted systems for protection of critical assets along with their important performance parameters and safety aspects are discussed at length in subsequent sections of this chapter.

3.1.1.3 Less-Lethal Electrical Weapons

Most electrical weapons work by sending a high-voltage, low-current electrical discharge through the body of the targeted person. The discharge voltage typically varies from 1 megavolt to a few tens of megavolts. The discharge current is limited to less than 5 milliamperes. The electrical shock interferes with the communication between the brain and the muscles, causing involuntary muscle contractions and impairment of motor function. There are electrical weapons such as stun guns, stun belts, stun batons, electrified shields and nets, and electrified water cannons that require direct contact with the body to produce the intended effect. There are electrical weapons such as Tasers that fire projectiles and administer electrical shock through thin flexible wires. Stun action depends on the mode of application. In the case of direct contact, the overwhelming factor is the creation of pain and hence compliance. In the case where the electrodes are fired toward the target as projectiles, neuromuscular stimulation occurs over a larger area. In this case, in addition to pain, the device incapacitates the target by stimulating his or her motor nerves and muscles as well as sensory neurons. Different devices have varying effects depending on the frequency of stimulation and the shape of the electrical pulse. We also have stun grenades that produce a blinding flash of light accompanied by a loud bang to temporarily disorient the adversary.

Stun guns are often combined with flashlights to provide twin options of flash blinding or incapacitating the opponent using electrical shock. Figure 3.5 shows one such device. The JOLT police tactical stun flashlight uses megavolts of electrical discharge voltage and combines it with an ultrabright flashlight that produces brightness of up to 200 lumens. The flashlight may be used to illuminate an entire area or use it to temporarily blind or switch on the stun feature and apply a high-voltage electric shock to incapacitate the assailant. Stun guns generally have an inbuilt safety feature that disables the device in case it is snatched away from the owner by an assailant. The safety mechanism comprises

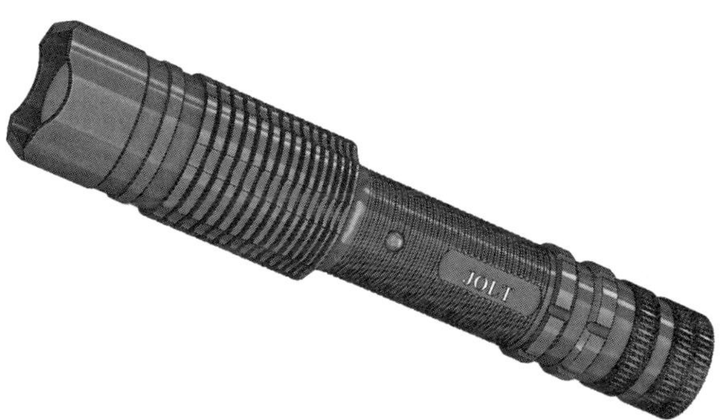

Figure 3.5 JOLT police tactical stun flashlight.

a safety lanyard around the owner's waist and a pin that plugs into the unit. In normal operation, the pin is plugged into the device. Any attempt to snatch the device away from the owner pulls the pin out of the device, thereby rendering it inoperable.

A *stun baton* is similar in operation to a stun gun except for its longer reach that provides added personal protection. It is the preferred electroshock weapon in situations where it is desired to have some distance between the user and the person you are attempting to stun. Stun batons usually come with a wrist strap attached to them that prevents the baton to be snatched out of the user's hands. Some stun batons have metal strips running down the sides of the baton. These metal strips are also charged, which prevents the assailant from grabbing it as he or she is being stunned. Also, like stun guns, most stun batons are combined with an ultrabright flashlight to provide the option of flash-blinding the assailant. Figure 3.6 shows a representative stun baton flashlight called the Stun Master. It is 30 cm long, uses A 12-mV discharge voltage and 4.5 mA of discharge current, and has side strips that prevent grabbing. It is equipped with a disable pin lanyard safety.

A *stun belt* is a remotely operated electroshock device that is fastened around the subject's waist, leg, or arm and is equipped with features that disallow the subject from unfastening or removing it. The stun belt comprises a battery and a control pack. The control pack is commanded to give electric shock to the subject by sending a remote control signal. The REACT belt is one such device. Each shock delivers 50 kV for 8 seconds at a current of 3 to 4 mA, pulsed at 17 to 22 pulses per second. The primary application of remotely controlled stun belts is by law enforcement for restraint of in-custody subjects in various situations, including courtroom appearances and transportation. *Electrified shields* are stun shields with electrodes embedded into the face. These have been adopted as the ultimate safeguard when it comes to riot control, handling disorderly crowds, civil disturbances, prison disturbances, and forced cell entries. Typical electrical shock specifications include discharge voltage in the range of 50 to 200 kV, a discharge current of 5 to 8 mA, and a pulse rate of 17

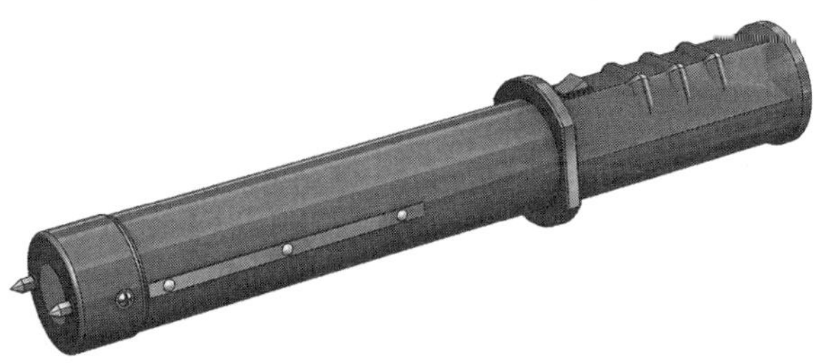

Figure 3.6 The Stun Master stun baton with flashlight.

to 22 pulses per second. *Electrified nets* are used to fence a certain area to check unwanted intrusions.

A *Taser gun* is a less-lethal conducted electric weapon used for self-defense. It can instantly incapacitate an assailant from a distance of typically 4 to 6m. As outlined earlier, a Taser gun incapacitates by its ability to cause neuromuscular incapacitation by mimicking the body's natural electrical impulses and disrupting them and not just providing pain compliance as is the case with non-Taser stun guns. When the trigger on the Taser gun is pulled, it releases two high-voltage dartlike electrodes that stay connected to the unit by conductive wire. The electrodes are on a pointed tip to penetrate clothing and are barbed to prevent removal once in place. Most Taser guns have a built-in laser for aiming and an ultrabright white LED to allow use in low- to no-light situations. Figure 3.7 shows one such device, the X26 Taser gun from Axon Enterprise, Inc.

3.1.1.4 Less-Lethal Acoustic Weapons

Use of *acoustics technology* to build less-lethal weapons has begun to demonstrate maturity. A number of prototypes have been built in the last couple of decades using different aspects of acoustic technology. Some of these devices have even been reportedly used by the U.S. Marines in Iraq and Afghanistan. Acoustic weapons, employing audible sound, infrasound, or ultrasound represent one emerging nonlethal technology that is beginning to mature. Less-lethal acoustic weapons fall into three broad categories. These are *acoustic-optical devices* such as stun grenades that produce a flash bang or a loud sound with an ultrabright flash of light, and are antipersonnel weapons that can be used to cause disorientation. The second category is that of *acoustic generators* that project sound in audible, ultrasonic, or infrasonic frequency ranges at a decibel level generally exceeding 85 dB. Sound level exceeding 85 dB for extended periods can cause permanent hearing loss (Table 3.1). These devices cause pain, discomfort, nausea, and disorientation depending on frequency range, decibel level, and duration of exposure. Both fixed and portable or handheld acoustic generators exist. Like acoustic-optical devices, these are also antipersonnel weapons. The third category is that of *vortex generators* that project a vortex of air at high speed, also

Figure 3.7 X26 Taser gun. (Courtesy of Wikimedia Commons.)

Table 3.1
Noise Level and Possible Sources and Effects

Noise Level (dB)	Sources	Effects
>140	—	Painful and dangerous
140	Fireworks Gunshots Car stereos at full volume	Painful and highly uncomfortable
130	Jackhammers Ambulances	Highly uncomfortable
120	Jet planes during takeoff	Dangerous over 30 seconds
110	Concerts Car horns Sporting events	Very loud, dangerous over 30 minutes
100	Snowmobiles 2MP3 players at full volume	Very loud, dangerous over 30 minutes
90	Lawn mowers Power tools Blenders Hair dryers	Very loud, dangerous over 30 minutes
80	Alarm clocks	Loud
70	Traffic	Loud
60	Normal conversation Dishwashers	Moderate
50	Moderate rainfall	Moderate
40	Quiet library	Soft
30	Whisper	Soft
20	Leaves rustling	Faint

called an acoustic projectile. Such vortices may also be used as a carrier of other substances such as chemical agents and are used in antipersonnel applications.

A *stun grenade*, also known as a flash bang, is a small explosive device. When activated, it emits an ultrabright flash of light and a very loud bang. The flash and bang are, respectively, intense and loud enough to temporarily blind and deafen those in the vicinity. These are antipersonnel weapons and are commonly used by military and special police forces for riot/mob control and during surgical raids. Unlike a fragmentation grenade, the casing of a stun grenade is constructed to remain intact during detonation. This ensures that most of its explosive force is confined and shrapnel injuries are avoided. The casing is made to have large circular cutouts to allow the light and sound of the explosion through. Though the grenades are specifically designed to be nonlethal, they are not without hazards. They can still ignite secondary explosions if set off in close proximity to a flammable material such as a gasoline tank. The intensity of the blast can cause heart attacks in elderly people and those with existing heart problems. There are different grenade varieties specifically designed for various application scenarios. Some grenades typically emit vast plumes of irritating smoke in order to discourage crowds from marching forward. There are grenades such as the M84 stun grenade (Figure 3.8); when activated, the M84 grenade emits an ultrabright blinding flash of light of more than 10 million lumens within 1.5m of initiation accompanied by a loud bang of 170 to180 dB sufficient to cause immediate flash blindness, deafness, tinnitus, and inner-ear disturbance. There are other types like the XM99 that are designed for riot control and combine the flash-bang effect with small rubber projectiles. The XM99 Blunt Trauma Grenade uses a pyrotechnic charge to discharge 32-caliber rubber balls to achieve crowd control through audio, visual, and physical stimuli.

Acoustic generators, as outlined earlier, project discomforting and disorienting high-intensity sound exceeding the 85-dB level in audio, ultrasonic, or infrasonic bands over long ranges. One such device is the *Long Range Acoustic Device* (LRAD) designed by the LRAD Corporation to deliver audible warning messages over long distances. The device produces a sound that can be directed in a 30-degree-wide beam. The device is available in different variants that are capable of producing varying degrees of sound over different operational ranges. Both portable handheld and platform-mounted LRAD devices are available. The LRAD 500X, generally used by police departments for example, is designed to communicate at up to 2,000m during ideal conditions. In a typical outdoor environment, the device can be heard from 650m. The LRAD 500X is also capable of short bursts of directed sound that cause severe headaches in anyone within a 300m range. The military-grade LRAD 2000X can transmit voice commands at a maximum of 162 dB up to a range in excess of 8 km. The device can produce dangerously high sound levels at distances of less than 10m, which could lead to permanent hearing loss. Figure 3.9 shows the LRAD system in use onboard a U.S. naval vessel during a small boat attack drill. A prototype handheld system based on the same technology called the *directed stick radiator* has also been demonstrated. It fires high-intensity pulses

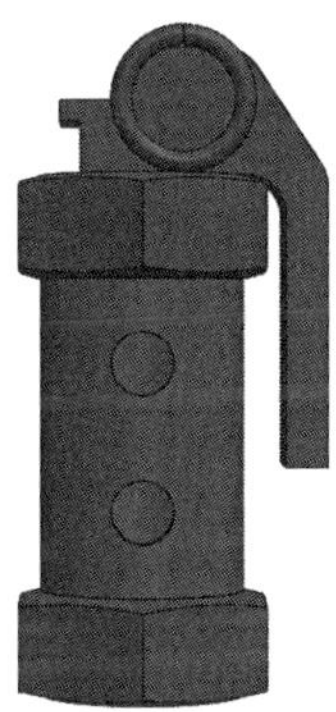

Figure 3.8 M84 stun grenade.

Figure 3.9   LRAD system. (Courtesy of Wikimedia Commons.)

of sound called sonic bullets at 120- to 150-dB sound levels. Acoustic weapons employing infrasonic frequencies have been reported to be capable of causing nausea, disorientation, and bowel spasms. A mobile *infrapulse generator* generating low-frequency shock waves to resonate with body organs and cause physical damage is also under development. Acoustic weapons of the LRAD type have reportedly been used in the Iraq and Afghanistan wars. Like many other less-lethal weapons, acoustic weapons also have safety issues.

3.1.1.5   Less-Lethal Chemical Weapons

Chemical agents used as less-lethal weapons mainly include riot control agents, malodorants, antitraction materials, obscurants, foams, antimaterial chemicals, and defoliants/herbicides. *Riot control agents* are compounds that cause temporary incapacitation by irritation of the eyes and the upper respiratory tract. They are often called irritants, irritating agents, and harassing agents popularly known as tear gas. These include synthetic chemicals 2-chlorobenzalmalononitrile (CS), chloroacetophenone (CN), and dibenzoxazepine (CR), as well as oleoresin capsicum (OC), known as pepper spray, which is biological in origin. *Pepper spray,* which is available in a variety of package styles, is widely used in policing, riot control, crowd control, and by individuals for self-defense. Small pepper spray devices designed for personal use by women, senior citizens, security guards, highway motorists, and also for protection against charging animals are common. The spray causes burning and severe irritation in the eyes and the facial tissues of the attackers, immobilizing them completely. Pelargonic acid vanillylamide (PAVA) is a synthetic version of OC. It contains a 0.3% solution of PAVA in a solvent of aqueous ethanol. Due to its higher potency than a natural product, it has become more popular for use in law enforcement. Nitrogen is used as the propellant. There are a variety of shells, grenades, and spray devices commonly used for delivering riot control agents. *Malodorants* are foul-smelling chemical compounds that are considered to be very effective for controlling crowds, clearing facilities, and area denial. *Antitraction materials* are lubricating polymers spread on ground or other surfaces to prevent access by people or vehicles. These have both antipersonnel and antimaterial applications. Antitraction materials are used for area denial to attackers. They may be used to deny access through doors, hallways, and windows. A vehicle-mounted system could apply large quantities of antitraction material to streets, intersections, and open areas as needed. Commonly employed antitraction materials include oil-slick-in-a-can, a combination of drilling mud additive, flocculent, and water that renders surfaces as slippery as wet ice. Rigid or sticky foams as antitraction materials are used as a barrier. These are not for direct use against humans because of risk of blocking airways. Aqueous foams are also used as personnel barriers. Chemical irritants could be added to enhance efficacy. These are available as spray devices.

Antimaterial chemicals such as combustion modifiers, fuel contaminants, supercorrosives, embrittling agents, superadhesives, and depolymerization agents have been proposed for use against structures and vehicles. These are deployed either directly or as a spray device or delivered through a projectile containing the chemical substance.

Obscurants are smokes used to obscure vision. Dyes are used for underwater applications. Smoke is an aerosol comprising many small particles suspended in air. An aerosol is a colloid of fine solid particles or liquid droplets in air or in another gas. There are natural aerosols such as fog, forest exudates, and geyser steam, or anthropogenic aerosols such as haze, dust, particulate air pollutants, and smoke. The particles in an aerosol have a diameter that is mostly smaller than 1 μm or so. Larger particles with a significant settling speed make the mixture a suspension. These particles scatter or absorb the light, thus reducing visibility to a level depending on smoke density. When the density of smoke material aerosol particles exceeds a certain minimum threshold value, visibility becomes nearly zero. There are three basic types of screening smokes: hexachloroethane (HC) smoke, phosphorous smoke, and fog oil smoke. Most smokes are not hazardous in concentrations used for obscuration purposes if the exposure is not too long. Sufficiently long exposure can be hazardous to health.

Toxic chemical/biochemical agents are distinct from riot control agents. Toxic chemical or biochemical agents acting on neuroreceptors in the central nervous system to cause sedation, disorientation, hallucination, mood changes, unconsciousness, and death, and are delivered as an aerosol. These incapacitants are considered illegal under the Chemical Weapons Convention (CWC) and the Biological and Toxin Weapons Convention (BTWC) and are not discussed here.

3.1.2   Advantages of Laser Dazzlers as Less-Lethal Weapons

The primary advantages of any laser weapon, less-lethal or lethal, include speed-of-light delivery, near-zero collateral damage, reusability, deep magazine, multiple target engagement and rapid retargeting capability, immunity to electromagnetic interference, and no influence of gravity.

Laser weapons, antipersonnel, and antimaterial engage targets at the speed of light with essentially no time of flight compared to conventional kinetic energy weapons that require a finite travel time. When compared to kinetic energy weapons, it would reduce the time to hit the target from seconds and minutes, depending on the range, to microseconds and milliseconds. For example, a rubber bullet with a velocity of 60 meters per second would take a couple of seconds to hit a target after it is fired. In contrast, the blinding laser beam from laser dazzler would take only a fraction of a microsecond to hit the targeted person after it is activated. The biggest advantage of laser weapons in general and laser dazzlers in particular is their *near-zero collateral damage*. This is attributed to the pinpoint accuracy of laser weapons.

While kinetic energy weapons, including both lethal and less-lethal, get destroyed during the mission, laser weapons are reusable. Every time a kinetic impact projectile is fired, it is lost in the process regardless of whether the intended objective is achieved or not. Laser weapons, including less-lethal devices such as laser dazzlers, are reusable. A laser dazzler shoots out a dazzling laser beam as long as there is electrical energy to power the unit.

The multiple target engagement and rapid retargeting feature of laser weapons is attributed to their being powered by rechargeable energy, which is electrical in the case of laser dazzlers. Shifting from one target to another in the case of multiple target engagement involves only repointing the laser beam.

Laser pointing is practically without any inertia and a light bullet has no mass and therefore is not influenced by gravity. Laser weapons are far more accurate and precise than their kinetic energy counterparts. The accuracy of less-lethal kinetic energy weapons, such as rubber and plastic bullets, deteriorates with range due to their irregular shape and slower speed. Laser beams do not have any of these issues. A laser-based directed-energy weapon has practically unlimited magazines. The total number of shots a laser can fire is only limited either by the amount of chemical fuel, as is the case in chemical lasers, or electrical power for solid-state and fiber lasers.

3.1.3   Potential Applications

Laser-based less-lethal weapons such as laser dazzlers can be used for counterinsurgency, antiterrorism (Figure 3.10), countersniper, self-defense, crowd control, infrastructure protection, patrolling operations, coastal protection, protection of naval vessels against asymmetric threats (Figure 3.11), aircraft defense

Figure 3.10 Laser dazzler in a close combat operation.

Figure 3.11 Laser dazzler aboard a naval vessel. (Courtesy of Wikimedia Commons.)

against MANPADS, and a host of other low-intensity conflict scenarios. Laser dazzlers are also being considered for warning the crews of commercial airliners or military aircraft who violate a no-fly zone intentionally or unintentionally. Such systems in a networked configuration of multiple laser dazzler stations and radars could be effectively used for protection of critical infrastructure or assets. Before the aircraft cockpit is flooded with dazzling laser light, much lower power levels are employed to send a warning signal to a crew to find out their intent. This helps in discriminating rogue aircraft from those that might have gone astray unintentionally.

3.2 Deployment Scenarios

Laser dazzlers have established themselves as a potent less-lethal alternative to the use of lethal force for a wide range of applications as outlined earlier. Laser dazzlers offer potent solutions to many applications in the defense and civilian sectors, including industrial security and protection of critical infrastructure, counterinsurgency and antiterrorist operations, deployment by law enforcement agencies at checkpoints and roadblocks, crowd and mob control, protection of sea vessels against asymmetric threats, protection of aerial platforms against MANPADS, and thwarting suicide planes and aircraft from violating no-fly zones and protected airspace. Each of these deployment modes is briefly described in the following sections.

3.2.1 Industrial Security and Protection of Critical Infrastructure

Industrial security involves a variety of functions that includes access management and control, day/night video surveillance using CCD cameras, night vision devices, and thermal imaging cameras, and perimeter protection and intrusion detection using a host of available laser and nonlaser technologies. The use of biometric security devices are commonplace in the areas of access management and control. These devices, including fingerprint readers, retinal eye scanners, hand geometry readers, and X-ray baggage scanners offer a simple and effective solution for security personnel to scan the interior of bags and other items along with metal detectors used for security screening and provide an efficient way to securely allow the maximum number of people through while still pinpointing hidden metal objects. Sensitive areas within large industrial establishments and in places like airports, railway stations, power stations, and nuclear installations may even employ advanced instrumentation such as terahertz imaging for detection of explosives and concealed weapons. Video surveillance is another vital component of security setup of industrial establishments and critical infrastructure. An integrated setup of CCD cameras, night vision devices, and thermal imaging cameras may be used to provide comprehensive day/night surveillance. There are a host of technologies, including electronic intrusion detection systems and fiber-optic cable-based perimeter intrusion detection systems, which can be used for buried or fence-mounted sensing fiber applications, line-of-sight laser fences, lidar-sensor–based laser walls, and microwave-sensor–based detection systems designed to trace walking, running, or crawling human targets and are used mainly in an outdoor environment. It is mainly in the areas relating to perimeter protection and intrusion detection where laser dazzlers can potentially be an important asset.

There can be several possible deployment modes of laser dazzlers when used as a part of perimeter protection and intrusion detection instrumentation setup. One of these modes could be the use of short- to medium-range handheld laser dazzlers by security guards entrusted with the responsibility of manning entry and exit points and also patrolling the perimeter. In an industrial setup these security guards are generally equipped with torchlike flashlights that are primarily used for nighttime patrolling and for operations in twilight conditions. The flashlight can only confirm an intrusion or identify an intruder if there were an attempt to breach the perimeter, but an ultrabright flashlight can deter the miscreants away from the protected site. Flashlights employing an array of superbright LEDs and producing hundreds to thousands of lumens of white light are commercially available. Figure 3.12 shows one such 250-lumen flashlight. Multicolor flashlights employing arrays of superbright LEDs are also available commercially. Examples of these flashlights include the Pelican 2370 flashlight with 358 lumens of high brightness output and 190m of beam range that produce selectable white, red, and green outputs, the Coast TX-10 multi-color flashlight with white, red, green, and blue outputs at 73 lumens, and the Coleman CPX-4.5 multicolor flashlight with a 155-lumen output and 90m range with white/red output options.

A laser dazzler can act as a force multiplier in such a scenario. It can effectively drive intruders away. As well, a laser dazzler can cause complete confusion and disorientation of the intruders through flash-blinding, thereby allowing for their easy capture. Dazzlers can also be particularly effective against armed intrusions. While security could be equipped with flashlights as well as short-range laser dazzlers, it would be a good idea to combine the two functions into one device with a selector switch to use either the flashlight or dazzling green laser light. Such a device could use an array of ultrabright LEDs and laser diodes emitting at 532 nm. Figure 3.13 shows the concept of such a device, which would use six ultrabright LEDs and one laser diode. One could possibly use one or two laser diodes, each producing about 50 mW of continuous-wave power and about six to eight LEDs. Note that laser diodes and LEDs are available in more or less similar package configurations. The actual number of LEDs and laser diodes will of course depend on operational range and size of the bright spot of white light or laser light at the target. Both LEDs and laser diodes are driven by constant current driver electronics, which is powered by a recharge-able battery pack.

Another possible deployment mode can be in the integration of a laser dazzler with a perimeter intrusion system. Multiple laser dazzler devices can be

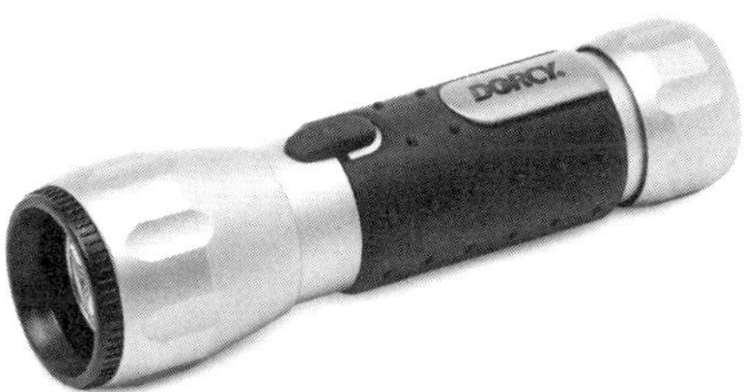

Figure 3.12　DORCY high-power LED flashlight. (Courtesy of Wikimedia Commons.)

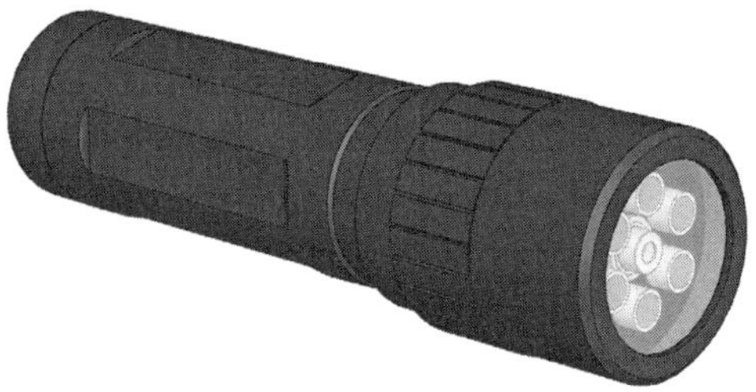

Figure 3.13   Flashlight-laser dazzler combo concept.

used to cover the full 360 degrees of a perimeter. On detection of an intrusion and the location, the relevant laser dazzler could be automatically activated to flash-blind the intruder/s. The alarm would alert the security guards who could capture the intruders without the use of lethal force. If the laser dazzlers are integrated with a lidar-sensor–based laser wall, the miscreants could be warned with a relatively lower-power laser beam when they are some distance away from the fence or wall. This gives them a loud message that they are under observation, which in normal circumstances should be enough to deter them. However, any attempt to breach the fence could be countered with a high-power dazzling beam. Dazzlers could also be programmed to switch to high-power output if the intruders come within a certain specified distance from the fence.

3.2.2   Counterinsurgency and Antiterrorist Operations

Although not the same, the terms *insurgency* and *terrorism* are often used interchangeably. An *insurgency* is a form of rebellion specifically aimed against the government with a direct political objective. Insurgencies sustain usually with the support of some or majority of the population of the affected area because of their disagreement with government's methods or policies. Use of violence or terror is not a prerequisite of insurgencies, although some insurgencies do use violence. *Terrorism* is characterized by use of unauthorized extreme violence to achieve a political or personal aim, and does not only target a specific group. Their targets can be anyone, including government officials, armed and paramilitary forces, or innocent civilians. Unlike insurgencies, which targets government forces directly, terrorism targets noncombatants to influence the views and perceptions of a government.

When it comes to counterinsurgency and counterterrorism operations, a fundamental principle not to forget is that the combatants and military objectives must be distinguished from noncombatants, civilians, and civilian objects. Only combatants and military objectives can be legitimately targeted. There needs to be a balance between military necessity, humanity, and proportionality. This brings to the fore the relevance of less-lethal weapons in such operations.

Laser dazzlers, high-energy millimeter-wave devices, acoustic hailing devices, flash-bang munitions, blunt impact munitions, and human electromuscular incapacitation munitions are some commonly employed less-lethal devices that have been briefly described earlier in this chapter. In this book we will limit the discussion to the use of laser dazzlers in counterinsurgency and counterterrorist operations. Two of the most frequently encountered scenarios relevant to the use of laser dazzlers include dispersal of violent and unruly mobs and close-quarter engagements.

A close-quarter engagement usually refers to relatively small-sized units of military, paramilitary, or police forces engaging an adversary with personal weapons at a short range. Most engagements that take place in built-up areas are close-quarter engagements. These engagements are often between small groups of combatants within the confines of a single room. There is an element of uncertainty in such combat actions generally accompanied by instinctive actions taken by both sides. Both parties in a close-quarter engagement aim at completing all offensive action before the adversary being engaged is able to react. Another element of such an engagement is the intermixing of combatants with noncombatants in the same building, often in the same room. Therefore, actions taken during close-quarter operations are meant to achieve the objective by minimizing friendly losses, avoiding collateral damage, and conserving ammunition and demolitions for subsequent operations. These principles are equally valid for crowd/mob control operations.

The technology of the short-range laser dazzler and crowd control laser dazzler is more or less similar except that the latter system uses a greater power that can produce the desired power density in a larger beam spot. Short-range laser dazzlers with operational ranges up to 100m and producing around 200 to 300 mW of laser power are used in close-quarter engagements. The GLARE RECOIL from B.E. Meyers with its inbuilt range-finder–based safety is one such short-range weapon-mounted laser dazzler suitable for close-quarter engagement scenarios. Laser dazzlers used in crowd control applications have operational ranges up to 500m and typically employ 1 to 10W of laser power. The required power depends on the laser beam spot diameter. In crowd control laser dazzlers, some means of scanning the laser beam spot may also be necessary. In the case of a scanning laser, a relatively smaller laser spot diameter may be scanned at a rate faster than the persistence of vision across a much wider cross section to create the effect of a much larger spot diameter. Note that in the absence of scanning, one would require tens of kilowatts of laser power to produce the dazzling power density of 0.1 to 1 mW/cm^2 across the cross section of a typical crowd. Scanning allows the use of relatively much lower laser power to simultaneously irradiate the crowd spread over a large cross section. Several optical schemes have been used for scanning the laser beam. One of these uses a Risley prism pair comprised of a pair of identical prisms rotating about a common optical axis to steer the beam. Another is to use a large concave mirror, onto which the focused laser beam is quickly played across and reflected at the targets. Both methods allow the laser weapon to cover a wide conelike area in a fraction of a second. There are several vehicle-mounted laser dazzler systems particularly suited to management of unruly crowds and violent mobs.

3.2.3   Deployment at Checkpoints and Roadblocks

A laser dazzler can be a potent hail and warning device for a variety of applications including close combat operations and for use by security agencies at checkpoints and roadblocks. These checkpoints could be a border crossing, a base, a protected zone in a city under siege, or something else. The requirement is to force the errant drivers to a halt when they tend to ignore warning signs while approaching checkpoints and roadblocks. Use of lethal force is not a preferred choice in such situations as the intent of the driver cannot be known with certainty. The driver could be a careless driver or a suicide bomber. Therefore, safety of the security personnel, and equally important, that of the targeted individual is extremely important. Short-range laser dazzlers—both handheld and weapon-mounted with operational ranges of 100 to 300m and employing 200 to 300 mW of laser power—were reportedly used during the wars in Iraq and Afghanistan. A large number of incidents of injuries inflicted on both security personnel and civilians have also been reported. This has primarily occurred because the laser dazzling devices of that time did not have enough safety measures built into them. These devices had a NOHD that was unacceptably large for such an operation, which significantly increased the probability of causing an eye injury if the targeted individual came within the NOHD. It also increased the probability of injury to security personnel due to the dazzler's radiation scattered from the atmospheric constituents.

The GLARE family of handheld and weapon-mountable laser dazzlers, including GLARE-MOUT, GLARE-MOUT PLUS, GLARE LA-9/P, and GLARE RECOIL from B.E. Meyers have reportedly been used at traffic control points, vehicle checkpoints, entry control points, and for convoy protection. The United States military reportedly encountered problems in manning checkpoints during the Iraq war. Probably due to language and cultural differences, the Iraqis would not even heed to warning shots or slow down while approaching a checkpoint. Reportedly, there were incidents of unarmed locals being shot by U.S. soldiers for failing to stop at checkpoints. They had resorted to using laser dazzlers mounted on M4 rifles that would cause temporary blindness when shone on those approaching checkpoints and who had initially failed to follow verbal commands to stop.

The GLARE LA-9/P is equipped with a safety feature that rapidly determines if an unintended observer is within the NOHD and immediately shuts off the glaring output to prevent accidental eye injury. The laser beam is instantly resumed once the bystander has moved out of the nominal ocular hazard distance. The GLARE RECOIL integrates laser range finding into the system's safety controls. The Green Laser Optical Warner (GLOW) from Thales, U.K. is a gun-mounted laser warner that uses an intense green light to warn approaching personnel that they are under observation and should not approach further. GLOW has inbuilt safety features that ensure that the output laser power remains within certain specified limits, and features a choice of a narrow beam to be used for longer range and a wide beam to be used in close-quarter engagements in the urban scenario. GLOW is in service with several countries including the United Kingdom.

3.2.4 Ship Defense

A laser dazzler is an effective nonlethal and nonviolent alternative to provide protection to ships and other naval platforms against a variety of threats from aerial platforms and asymmetric threats such as those from sea pirates and small suicide boats. Figure 3.14 shows the image of a laser dazzler in operation on a naval vessel. The laser dazzler suitable for the purpose is generally mounted on a stabilized IP-controlled platform and has output laser power in the range of a few hundreds of milliwatts to a few watts depending on desired operational range and laser beam spot size. As well, the laser dazzler is integrated with multiple camera sensors that include a day/night color camera, thermal camera, and SWIR camera for extended performance capabilities. The integrated suite might also have optional video tracking and radar slew-to-cue features. Availability of a nonlethal laser dazzler allows the user to choose between suppres-

Figure 3.14 Laser dazzler onboard a naval vessel.

sion/deterrence, hail/warning, and illumination operations. The first reported deployment of a medium-power laser dazzler was by the British during the Falklands War of 1982. They were reportedly fitted to various Royal Navy warships in the South Atlantic. These laser dazzlers were designed to dazzle low-flying Argentinian pilots as they attacked the Royal Navy ships. As per a persistent rumor in defense circles, the loss of several Argentine fighters was attributed to use of laser dazzlers against pilots, although reportedly this nonlethal laser weapon was never used in action.

3.2.5   Protection of Critical Ground Assets from Suspect Aerial Platforms

An integrated set-up of multiple laser dazzler stations and radars has the potential of protecting critical infrastructure from suspect aircraft and other airborne platforms that would tend to violate no-fly zone. In a system like this, radar provides the initial cue about the rogue or suspect aircraft when it is still more than 100 km away from the actual asset to be protected. The radar keeps a continuous vigil on the suspect aircraft until it comes within the tracking range of the electro-optic tracker, which is usually an integral part of the laser dazzler station. The electro-optic tracker station takes a cue from the radar and tracks the target with much higher accuracy needed for the dazzling action. Figure 3.15 illustrates the concept.

Exposure to a strong laser light source results in flash blindness and afterimages. In flash blindness, exposure to a very bright light source completely deprives the pilot of vision for a period of time ranging from a few seconds to a few minutes (Figure 3.16). The laser illumination fills the flight deck with a bright light, making it difficult for the pilot to concentrate on the flight instruments, which adversely affects the pilot's intended actions. Before the aircraft cockpit is flooded with dazzling laser light, much lower power levels are employed to send a kind of warning signal to the crew to know their intent. This helps in discriminating rogue aircraft from those that might have gone astray unintentionally.

3.2.6   Protection of Aerial Platforms against MANPADS

MANPADS are surface-to-air missiles used as antiaircraft weapons. They can be fired by an individual or by a small team of people. These weapon systems have an operational range of 4 to 5 km, which makes low-flying aircraft and attack helicopters as well as commercial airliners during the take-off and landing phases particularly vulnerable to attack. MANPADS are often described as shoulder-fired antiaircraft missiles. There are three types of MANPADS: those employing command line-of-sight guidance, laser beam rider guidance, and passive infrared homing guidance. Command line-of-sight (CLOS) MANPADS are guided to their targets through the use of a remote control. Laser

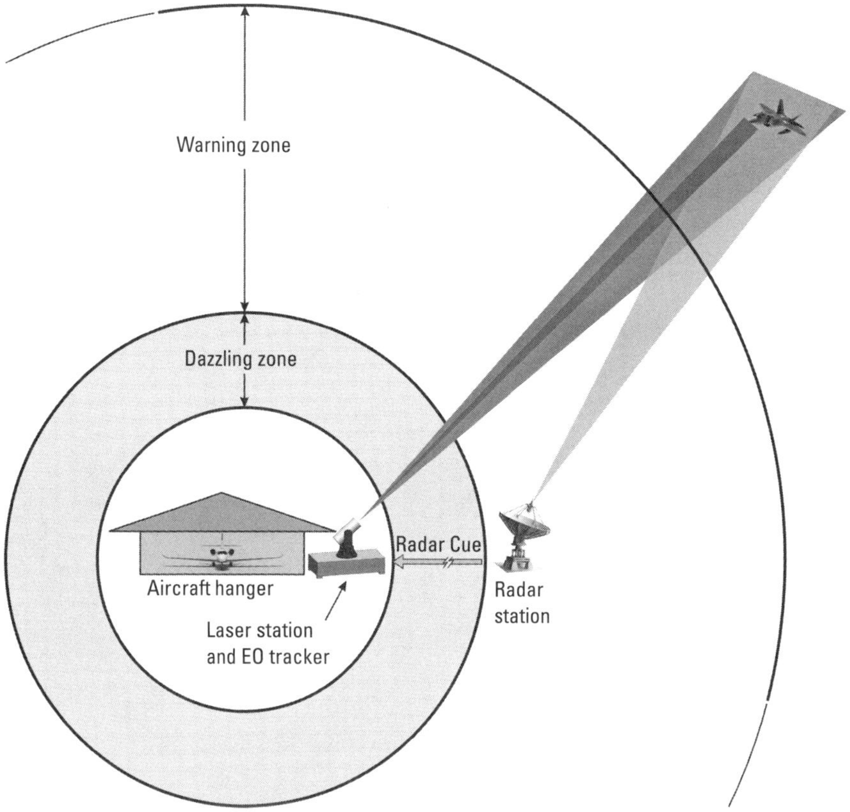

Figure 3.15 Protection of critical ground assets from suspect aerial platforms.

Figure 3.16 Pilot exposed to blinding laser radiation.

beam rider MANPADS follow a laser projected onto the target. The most common MANPADS, however, use passive infrared homing guidance. They lock on to the heat energy emanating from an aircraft's engine. Strela and Igla weapon systems from the Soviet era and Stinger from the United States are some representative examples.

To understand the relevance of use of laser dazzlers to force an approaching missile to fly away from its intended path, it is important to understand the functioning of MANPAD guidance methodologies. In the case of semiactive command to line-of-sight guidance, which is the most common form of CLOS used with MANPADS, the target tracking is manual carried out by the operator or a member of the operating team. In the case of laser beam rider guidance, the operator ensures that the guiding laser beam is always on the target during the entire period of the weapon's flight. In the case of passive infrared homing, the most common form of MANPADS, the missile is launched by the operator in the rough direction of the target. The infrared seeker in the missile takes care of the rest. It is a kind of fire and forget missile. As is evident, the operator plays a key role in all three types of guidance mechanisms related to MANPADS. It is more so in the case of infrared homing MANPADS because the operator must have a clear vision of the target at the time of launching the missile. A laser dazzler can be very effective in temporarily flash-blinding an operator engaged in this crucial task during the launch phase, as shown in Figure 3.17. This may force the operator to launch the missile in the wrong direction. In the case of laser beam rider and SACLOS guidance methodologies, dazzling of the operator forces the missile to deviate from its intended path.

Figure 3.17   Laser dazzler for protection of aircraft against shoulder-launched missiles.

3.3 Potential Laser Sources

The most widely exploited wavelength for flash-blinding or dazzling of personnel is the green wavelength at 532 nm. The blue-green wavelength region is the chosen wavelength band for dazzling because the human eye is most sensitive to the green wavelength and also because the green wavelength has ability to interact with human eyes in daylight and reduced light conditions. Further, 532 nm is the preferred wavelength in the blue-green wavelength band as it can be conveniently generated by frequency doubling of 1,064-nm wavelength generated by the well-established neodymium-doped solid-state lasers. Contemporary laser dazzlers almost invariably use 532 nm as the laser wavelength. A red wavelength around 635 nm has also been used in some dazzlers in the past. The Saber-203 laser dazzler is a representative example. However, it is no longer the preferred wavelength as the human eye responds to green laser five times more than a red laser of the same power. Laser dazzlers using a combination of red, green, and blue (RGB) wavelengths are an emerging concept. RGB laser dazzlers, as they are called, have the potential of being more effective compared to green laser dazzlers particularly if the targeted personnel are using protective eyewear. As well, RGB dazzlers are reported to cause nausea in exposed personnel, which makes these devices particularly effective in a sea environment. One possible application could be against sea pirates. Different types of laser sources have been used to build laser dazzlers for various application scenarios. These include diode-pumped solid-state lasers, semiconductor diode lasers, and fiber lasers. Each of these is briefly described in the following sections.

3.3.1 Diode-Pumped Solid-State Lasers

Diode-pumped solid-state lasers are the preferred choice when the desired operating wavelength is 532 nm and output laser power level is of the order of a few hundreds of milliwatts, which is the power produced by most short- and medium-range laser dazzlers. This will be evident when we discuss some representative portable handheld laser dazzlers along with their features and specifications in the latter part of the chapter. The desired wavelength is generated by frequency doubling the output of a neodymium-doped solid-state laser. As a result of frequency doubling, the fundamental wavelength of 1,064 nm gets halved to 532 nm. Neodymium-doped yttrium vanadate (Nd-YVO$_4$) and Nd-YAG are the solid-state laser materials used for this purpose, with the former being more commonly used. Neodymium-doped yttrium vanadate as a laser material is important because of several properties that make it particularly attractive for laser diode pumping. These include a large stimulated emission cross section and a strong broadband absorption around 808 nm. Frequency doubling is achieved by using a nonlinear crystal. Potassium titanyl phosphate

($KTiOPO_4$), abbreviated as KTP is the commonly used nonlinear material for frequency doubling of neodymium-doped solid-state lasers. KTP crystals coupled with Nd-YAG and $Nd-YVO_4$ crystals are commonly found in green laser modules. Figure 3.18 shows the inside of a frequency-doubled diode-pumped $Nd-YVO_4$ laser head highlighting different components. State-of-the-art laser heads are very small. Two such laser heads are the OEM industry-grade diode-pumped solid-state green laser heads from Frankfurt Laser Company, which offer optical output in 1 to 5 and 50 to 100 mW of power and respectively known as MicroGreen (Figure 3.19(a)) and MiniGreen lasers (Figure 3.19(b)). MicroGreen and MiniGreen laser heads measure 5.6 mm × 9 mm and 9 mm × 13 mm, respectively.

The laser diode in the laser head is driven by a driver circuit, which is a constant current source with inbuilt protection features. The OEM laser modules used for building laser dazzlers have the laser head and the driver electronics packaged as a single unit. Figure 3.20(a) shows the schematic arrangement of such a module along with images of some representative OEM DPSSL modules (Figure 3.20(b)). The source of input electrical power, which is a battery, is usually external to the laser module.

Frequency-tripled solid-state laser modules can also be used to generate an ultraviolet wavelength of 355 nm (= 1,064/3 nm). The third harmonic frequency is generated by adding a second harmonic frequency to the fundamental frequency. A common methodology is to use two LBO (lithium triborate, LiB_3O_5) crystals, or an LBO and a BBO [beta barium borate, BaB_2O_4, or $(BaBO_2)_2$] crystal. The first crystal is phase-matched for the second-harmonic generation and the second for sum frequency generation. Although the process would be efficient when using pulses from a Q-switched or mode-locked laser, it is also possible in continuous-wave operation for example with intracavity frequency doubling and resonant sum frequency generation. This wavelength

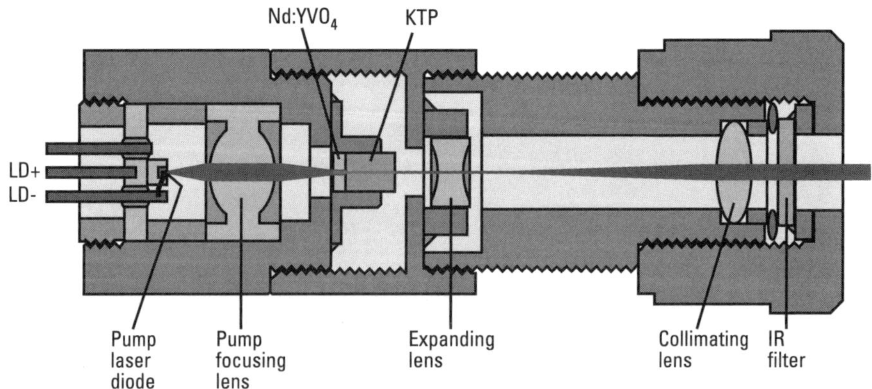

Figure 3.18 Frequency-doubled Nd-YVO4 laser head.

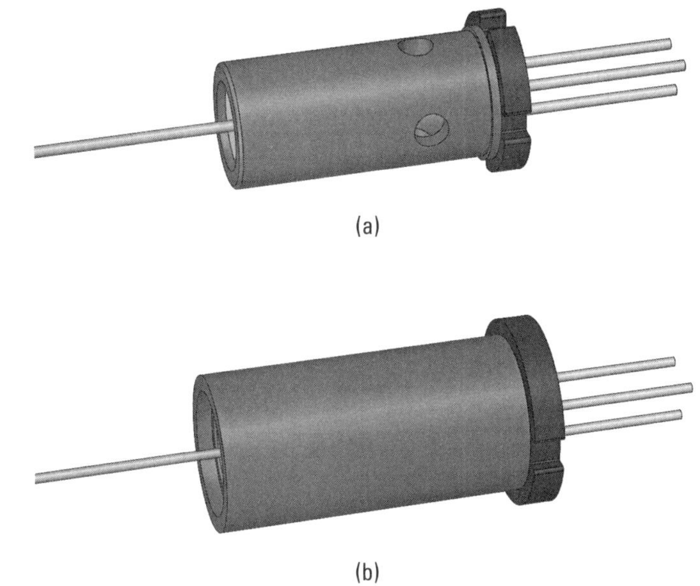

(a)

(b)

Figure 3.19 MicroGreen and MiniGreen laser heads from Frankfurt Laser Company.

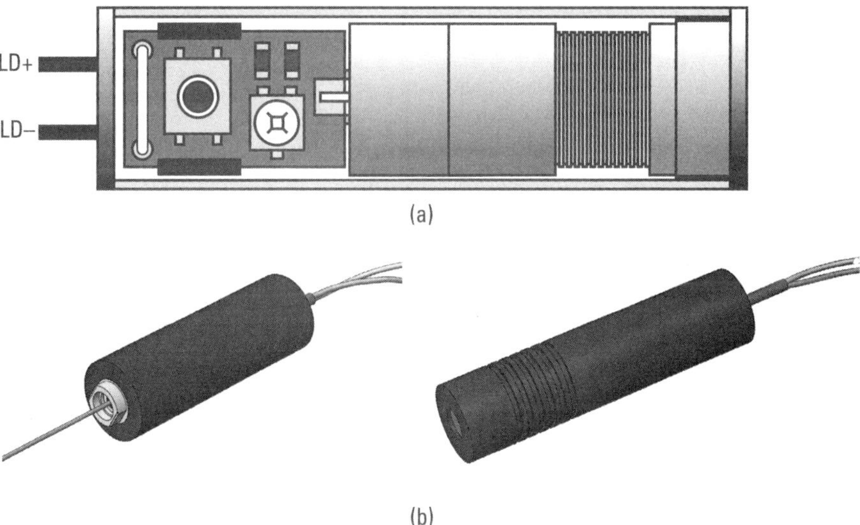

(a)

(b)

Figure 3.20 OEM green laser module.

could possibly be used for RGB dazzlers, although a blue wavelength generated by semiconductor diode lasers is more common. A blue wavelength of 440 nm could also be generated by frequency tripling of the output from a diode-pumped Nd-YAG laser operating at 1.3-μm fundamental frequency using two LBO crystals.

3.3.2 Semiconductor Diode Lasers

After diode-pumped solid-state lasers, different types of semiconductor diode lasers are the next most commonly used laser sources for building laser dazzlers. Aluminum gallium indium phosphide/gallium arsenide (AlGaInP/GaAs) laser diodes are used for emitting wavelengths in the red region at 635, 650, and 670 nm. Blue and blue-purple wavelengths (405, 450, and 470 nm) are generated by Indium Gallium Nitride/Gallium Nitride (InGaN/GaN) laser diodes. Indium gallium nitride laser diodes have been used to generate green wavelengths in the region of 510 to 530 nm and beyond. The PL 520 laser diode from OS-RAM Opto Semiconductors is capable of generating 50 mW of laser power in the wavelength band of 515 to 530 nm. The PLP 520 laser diode from the same company generates up to 120 mW of power at 520 nm. Laser diodes generating up to 1,000 mW of CW power in green are also available. As an example, the laser diode FVLD-520-1000M from Frankfurt Laser Company is a multimode laser diode with 1,000 mW of CW output power at 520 nm. Green laser diodes emitting at wavelengths beyond 532 nm are also being reported. A team from Sumitomo and the Advanced Materials Laboratory of Sony (Atsugi, Japan) has reported InGaN laser diodes emitting more than 100 mW CW power at wavelengths beyond 532 nm and CW emission of unspecified power at 536.6 nm. Red laser diodes with CW output power approaching 5,000 mW are commercially available. Blue laser diodes with CW output power in excess of 100 mW are also commercially available. With these laser power levels conveniently available at red, green, and blue wavelengths, semiconductor diode lasers have begun to be used for building short-range laser dazzlers using semiconductor diode lasers. The SABER-203 laser dazzler uses a 250 mW laser diode emitting in the red wavelength. Most short-to-medium–range green laser dazzlers use laser power of 200 to 500 mW. Given the state of the art in green laser diode power levels, more than one semiconductor diode laser may be used in the array form to achieve these power levels. Presently, these power levels are conveniently available from frequency-doubled diode-pumped solid-state lasers.

One of the biggest advantages of using semiconductor diode lasers is their relatively much higher wall-plug efficiency as compared to diode-pumped solid-state lasers. In the case of semiconductor diode lasers, the wall-plug efficiency is same as the electrical-to-optical conversion efficiency, which is in the range of 50% to 60%. Efficiency figures approaching 70% to 80% have also been reported. In the case of diode-pumped solid-state lasers, the wall-plug efficiency is due to the combined effect of electrical-to-optical efficiency of the pump diode, optical-to-optical efficiency of the solid-state laser crystal such as Nd-YVO$_4$, and conversion efficiency of nonlinear crystals such as KTP. The three figures, respectively, are 50% to 60%, 50% to 60%, and 80%. This gives an overall efficiency figure of about 20%. The efficiency of diode-pumped solid-state laser (DPSSL) generating the blue wavelength is still lower.

3.3.3 Fiber Lasers

Fiber lasers due to their immunity to misalignment, rugged construction, compactness, high beam quality, outstanding thermo-optical properties, and long-term stability are an alternative to bulk solid-state lasers, such as diode-pumped solid-state lasers, in a large number of industrial and military applications. CW power level of a few tens of watts to a few hundreds of watts can be conveniently obtained in the wavelength band of 1,050 to 1,090 nm using ytterbium-doped fiber lasers. Second harmonic generation using fiber laser as a pump source can lead to a fiber laser emitting in green. These lasers may be used as the dazzling source in long-range systems such as those used for protection of ground-based critical assets against attack from rogue aerial platforms and also those violating a no-fly zone.

YLM and YLR Series diode pumped CW Ytterbium fiber lasers from IPG Photonics are capable of generating CW output power in the range of 1 to 100W in the 0.98- to 1.1-μm wavelength band. Second harmonic generation could be used to generate output in green. The 100W laser YLM-100 is air-cooled, measures just $270 \times 255 \times 275$ mm, and weighs less than 7 kg.

3.4 Operational Parameters

Important operational parameters of laser dazzlers include laser wavelength, laser power, spot size at target, and NOHD. Each of these parameters is described in the following sections, with particular reference to their relevance to different application scenarios.

3.4.1 Operating Wavelength

The blue-green region is the chosen wavelength band as the human eye is most sensitive to green light. Relative spectral response (Figure 3.21) is used in determining lumens of a light source and thus how bright the light will appear in terms of lux. For a red light to appear as bright as a green light, the red light must emit more radiation than the green light because humans are not as sensitive to red light. Peak response of the human eye occurs at 550 nm. As is evident from the curve shown in Figure 3.21, for a red laser at 650 nm and a green laser at 532 nm to appear equally bright to the human eye, the red laser would need to have about seven times higher power. The most commonly

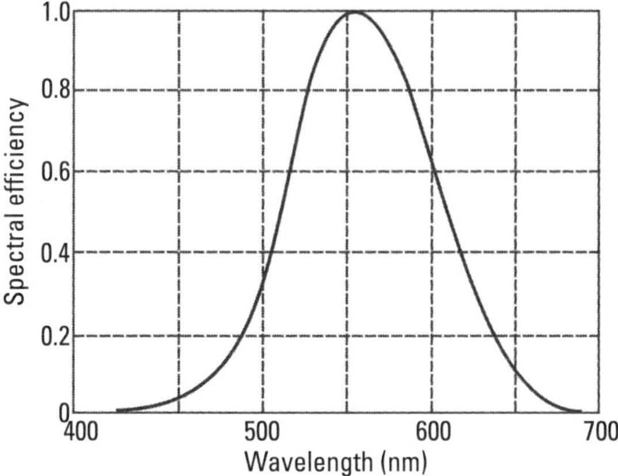

Figure 3.21   Spectral efficiency of the human eye.

employed wavelength for the purpose is 532 nm as it can be generated using well-established frequency doubled neodymium-doped solid-state lasers such as Nd-YAG and Nd-YVO$_4$ lasers. Semiconductor diode lasers are also used. The red wavelength in the 635- to 670-nm band has also been used in the past. The Saber-203 laser dazzler is an example. Current laser dazzlers almost invariably use the green wavelength. However, red and blue or blue/purple wavelengths along with the green wavelength are being tested as a combo to build RGB dazzlers for enhanced effect.

3.4.2 Laser Power

Laser power together with laser spot size at the target determines the power density available at the target plane. Laser power is chosen to produce the desired power density at the target plane. The power density is less than the MPE specification for the chosen wavelength and exposure time as per ANSI standards for eye safety. MPE expressed as power density is dependent on wavelength and exposure time. At 532 nm, maximum permissible power density equals 2.5 mW /cm^2 for a 0.25-second exposure and 1.0 mW/cm^2 for an exposure time of 10 seconds. Power density of 0.5 mW/cm^2 is considered as a safe value for dazzling operation. For a known value of maximum operational range, required value of laser beam divergence can be computed to produce desired spot size. Laser power is then chosen to produce the desired value of power density for a given value of spot size. Short- and medium-range laser dazzlers with operational range up to 2,000m produce laser power typically in the range of 100 to 500 mW. For example, the GLARE MOUT laser dazzler from B.E. Meyers with a maximum operational range of 1,500m employs a laser power of 125 mW. The GLARE MOUT PLUS from the same manufacturer employs a laser power of 200 mW.

A CHP laser dazzler with an operational range of 200m uses a 500 mW laser module. The Guardian and Defender laser dazzlers from Laser Energetics use 250 and 500 mW laser modules, respectively. The two laser dazzler variants have maximum operational ranges of 300m and 2400m, respectively. Note that the laser power required to achieve a certain maximum operational range also depends on the desired value of the laser beam spot size at target plane, which in turn depends on the application scenario.

3.4.3 Spot Size

As mentioned above, the desired value of spot size depends on the application scenario. Most laser dazzlers designed for use in close combat engagements offer narrow and wide beam options; the former for long-range operation and latter for shorter ranges. This is because for a fixed value of laser beam divergence, the power density at the target plane increases with a decrease in source-to-target distance. The beam divergence and hence the spot size is changed with target distance to ensure that the MPE limit is never exceeded throughout the operational range of the laser dazzler. It also helps in achieving the desired NOHD. A distance sensor may be integrated with the laser dazzler to switch the spot size from one value to another. In some cases, the spot size may be continuously adjusted as the target distance changes. Feedback from the distance sensor may also be used to switch off the dazzler in case the target comes within a certain specified distance. The GLARE LA-9/P laser dazzler from B.E. Meyers is one such device that employs eye-safe technology to shut off the laser dazzler immediately after the target comes within NOHD. The GLARE RECOIL from the same company also integrates a laser range finder with the device to ensure nonoccurrence of any accidental eye injury.

3.4.4 NOHD

NOHD is the minimum distance from the source at which the power density in the case of CW lasers or energy density in the case of pulsed lasers becomes equal to the MPE on the cornea and skin. A laser beam can be considered dangerous if the targeted person were closer from the source than the NOHD. Beyond the NOHD, the MPE value is not exceeded. The NOHD parameter depends on laser power, laser spot size, and laser beam divergence. The laser beam divergence and laser spot size in turn depend on beam shaping and beam directing optics. As mentioned earlier, most state-of-the-art laser dazzlers have inbuilt laser safety features. The zone beyond NOHD and up to twice the NOHD is known as the sensitive zone exposure distance (SZED). In this zone, the laser beam is bright enough to cause temporary impairment of vision. Assuming MPE for eye exposure to be 0.5 mW/cm^2, beyond this distance the beam irradiance is less than the SZED irradiance limit of $100\,\mu$W/cm^2. Beyond the SZED and up to about ten times the NOHD is the *critical zone exposure distance* (CZED). In this zone, the laser beam is bright enough to cause distraction interfering with critical task performance. Beyond this distance, the laser beam irradiance is less than the CZED irradiance limit of 5 μW/cm^2. Beyond CZED is the *laser free exposure zone* (LFEZ). Beyond LFED, the laser beam irradiance is less than the LFED irradiance limit of 50 nW/cm^2. Different zones are illustrated in Figure 3.22.

In order to ensure that there is no eye hazard even in case of someone accidently coming within the NOHD, we must have some means of either switching off the laser or increasing the laser beam divergence at the NOHD limit. In the context of use of a laser dazzler, switching off the laser would mean abandoning the operation. In the second option of increasing the laser beam divergence, the laser remains on and increased beam divergence ensures that the beam is bright enough to cause distraction or disorientation without causing any eye injury. This is illustrated in Figure 3.23. In both cases, some kind of distance monitoring is required to provide feedback to control laser on/off operation or its beam divergence. By controlling the laser beam divergence, we are effectively reducing the NOHD. As outlined earlier, most state-of-the-art laser dazzlers are equipped with such a laser safety measure.

NOHD in the case of commercially available short- and medium-range laser dazzlers varies from a few meters to a few tens of meters. For example, the CHP laser dazzler from LE Technologies that can engage targets up to 200m reportedly has a NOHD of 25m. The GLARE MOUT with a daylight effective range of 300 to 1,500m and the GLARE MOUT PLUS with an effective range of 450 to 2000m, both from B.E. Meyers, have a NOHD of 18m. Dazer laser variants Guardian and Defender from Laser Energetics with effective operational ranges of 1 to 300m and 1 to 2,400m, respectively, have a NOHD of 1m. The GLOW from Thales has an effective range up to 350m and a NOHD of less than 10m.

3.5   Laser Safety

Laser radiation is potentially hazardous to human eyes above a certain threshold intensity and also to the skin above a relatively much higher intensity threshold. In the context of use of less-lethal laser weapons such as laser dazzlers for homeland security, we will confine our discussion to laser eye safety. Laser safety is an important concern for both the users of laser devices as well as for those who would be potential recipients. The latter is relevant to the use of less-lethal laser weapons. It is therefore important that the users of lasers and laser-based devices take necessary precautions to avoid accidental injury while using these devices. In the context of less-lethal laser weapons, the weapon should be designed to

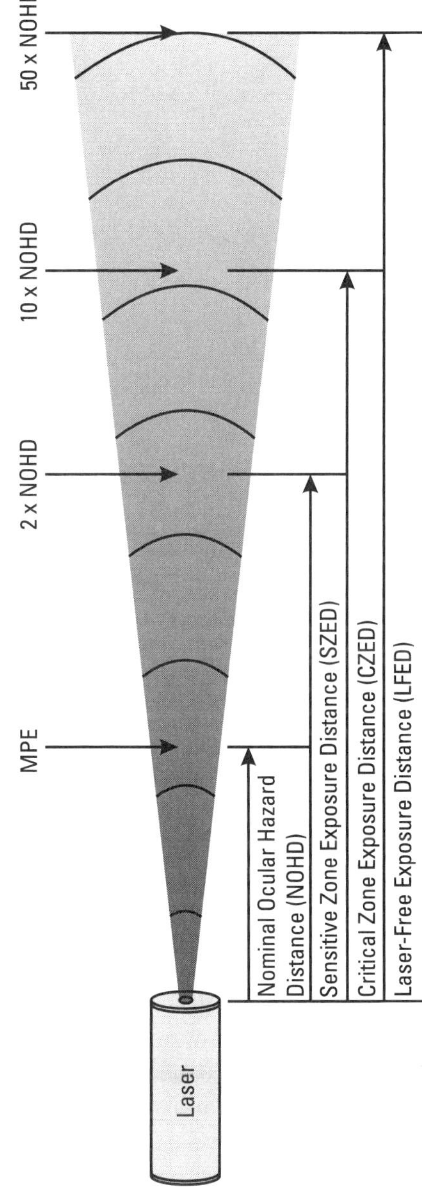

Figure 3.22 NOHD, SZED, CZED, and LFED.

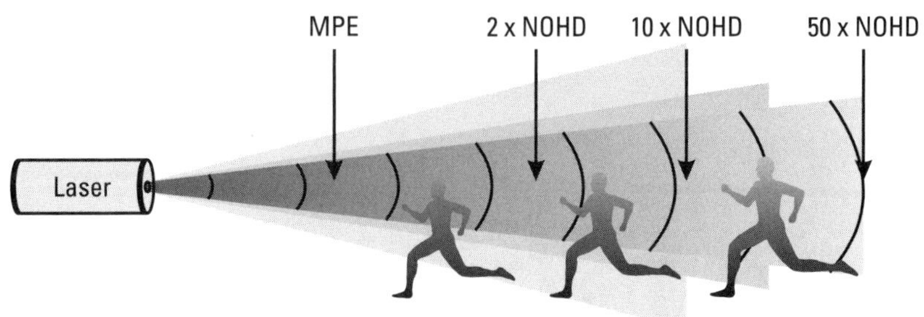

Figure 3.23 Controlling beam divergence to eye safety.

obviate any possibility of causing an accidental injury to the operator. Similarly, the weapon should have inbuilt design and safety features to eliminate the possibility of any permanent eye damage throughout the operational range of the weapon. The following sections briefly describe important aspects of laser safety including laser damage and damage mechanism, laser safety classification, maximum permissible exposure, and protocol-IV for use of laser-blinding weapons.

3.5.1 Laser Damage

Due to laser radiation's low divergence and resultant high intensity even after traveling long distances, lasers can be potentially hazardous to human eyes at moderate power/energy levels and to skin at relatively higher power/energy levels. Human eyes are likely to be permanently damaged when exposed to laser radiation even at moderate power level due to the focusing action of the eye lens that concentrates the impinging laser energy into an extremely small spot on the retina. The damage to the photoreceptor cells of the retina is caused by the transient rise in temperature at the focal spot.

Damage can result from both thermal and photochemical effects. Thermal damage occurs when tissues are heated to the point of denaturation of proteins. In the case of photochemical damage, laser radiation triggers chemical reactions in the tissues. Different laser wavelengths produce different pathological effects. While photochemical damage mainly occurs with ultrashort wavelength lasers emitting in the ultraviolet and blue regions of the electromagnetic spectrum, visible and infrared wavelengths are harmful due to thermal damage. Figure 3.24 illustrates the response of the human eye to different wavelength regions of the electromagnetic spectrum. Table 3.2 lists different wavelength regions and corresponding pathological effects.

3.5.2 Laser Safety Classification

Based on how hazardous a given laser is to the designers, operators, users, and targeted personnel, it is assigned a safety class. The classification is based on

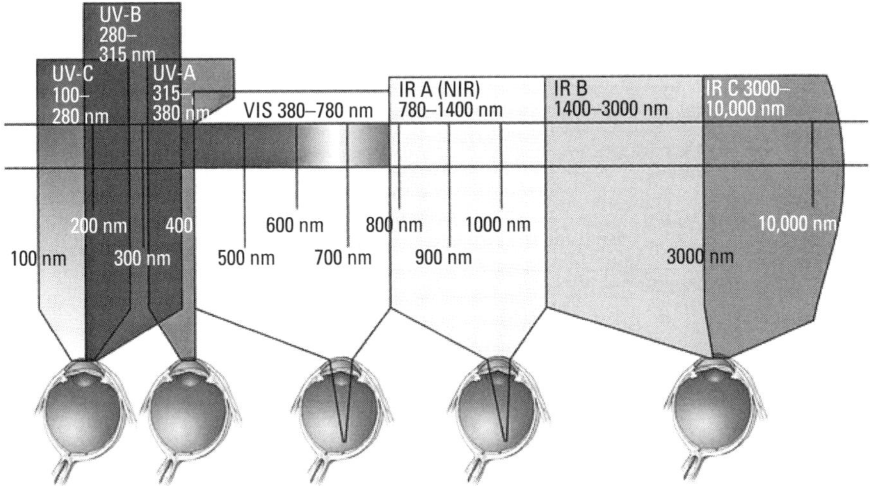

Figure 3.24 The human eye's response to different wavelength regions.

Table 3.2
Pathological Effects of Different Laser Wavelengths

Wavelength Range (nm)	Pathological Effect
180–315	Photokeratitis: Inflammation of cornea equivalent to sunburn
315–400	Photochemical cataract
400–780	Photochemical damage to retina, retinal burn
780–1,400	Cataract, retinal burn
1,400–3,000	Aqueous flare, cataract, corneal burn
3,000–1,000,000	Corneal burn

the concept of accessible emission limit (AEL) defined for each of the laser classes. There are two classification systems. The first one is the American system, which was in use until 2002. The second one is the revised system, which is a part of the IEC 60825 and has been included in the ANSI Z136.1 laser safety standard of the United States since 2007. The former designated safety class using Roman numerals in the United States and in Arabic numerals in the European Union. The revised system uses Arabic numerals in all jurisdictions. In both the old and revised classifications, lasers are classified into four main classes and a few subclasses in terms of maximum output power for different wavelength ranges. Different laser safety classes in the old and revised systems are briefly described in Tables 3.3 and 3.4, respectively.

Table 3.3
Laser Safety Classification: Old System

Safety Class	Description
I	A class-I laser is safe and there is no possibility of eye damage. This can be either due to low output power with no risk of eye damage after hours of exposure or due to the laser being contained inside an enclosure such as in the case of a compact disk player or laser printer.
II	This class only refers to lasers emitting in the visible spectrum with output power up to 1.0 mW. The lasers are safe due to the blink action of the eye unless deliberately staring into the beam for an extended period of time. Most laser pointers belong to this category.
IIa	For continuous exposure for a period greater than 1,000s, lasers at the low power end of the class II category may produce retinal burn.
IIIa	Lasers with a power level greater than 1.0 mW and less than 5.0 mW and a power density less than 2.5 mW/cm^2 belong to class IIIa. These lasers are dangerous when used in combination with optical instruments and also dangerous to the naked eye for direct viewing of more than 2.0 minutes.
IIIb	Lasers with a power level of 5.0–500 mW belong to class IIIb. Direct viewing of these lasers may cause damage to the eye. A diffuse reflection is generally not hazardous but both direct viewing and specular reflection are equally dangerous. Class IIIb lasers, which are toward the high power end, may present a fire hazard or cause skin burn.
IV	Lasers belonging to class IV have power levels greater than 500 mW. Class IV lasers can cause permanent eye damage or skin burn even when used without optical instrumentation. Diffuse reflection from these lasers can also be hazardous to eyes or skin within the nominal ocular hazard zone. Many industrial, medical, scientific, and military lasers are of the class IV category.

3.5.3 MPE

MPE is the highest power or energy density of the light source measured in W/cm^2 or J/cm^2, respectively, that is considered safe and has negligible probability of causing any damage to the eyes. MPE is usually taken as 10% of power or energy density that has 50% probability of causing damage under worst-case conditions. The MPE values of power/energy density as a function of wavelength and exposure duration for eye and skin exposure are derived from the ANSI Z136.1 standard. Laser dazzlers mainly use green wavelength and in some cases red and blue wavelengths. MPE figures for continuous-wave and nanosecond-pulsed lasers in these wavelength bands are as follows:

1. For continuous-wave lasers in the wavelength band of 400 to 700 nm covering both green and red lasers, the MPE is 1.0 mW/cm^2 for exposure duration in the range of $10 - 3 \times 10^4$ seconds.

2. For continuous-wave lasers in the wavelength band of 400 to 450 nm covering blue lasers, the MPE is 0.1 mW/cm^2 for exposure duration in the range of $100 - 3 \times 10^4$ seconds.

Table 3.4
Laser Safety Classification: Revised System

Safety Class	Description
1	The accessible laser radiation is not dangerous under reasonable conditions of use. A class I laser is safe under all conditions of normal use. This implies that while viewing a class I laser with the naked eye or with typical magnifying optics such as a telescope or a microscope, the MPE limit cannot be exceeded.
1M	The accessible laser radiation is not hazardous provided that no optical instruments are used which may, for example, focus the radiation. The classification is applicable to lasers with power levels greater than the limit specified for class I lasers provided the laser energy entering the pupil of the eye does not exceed the limits of class I lasers due to large divergence of the laser.
2	The accessible laser radiation is limited to the visible spectral range (400–700 nm) and to 1.0 mW.
2M	A class 2M laser is also safe because of the reflex action of the eye, with the additional restriction that it is not viewed through optical instruments. As with class 1M lasers, the classification is also applicable to lasers with power levels greater than 1.0 mW provided the beam divergence is large enough to prevent the laser energy passing through pupil of the eye to exceed the limits of class 2 lasers.
3R	The accessible laser radiation may be dangerous for the eye but can have at the most five times the permissible optical power of class 2 lasers emitting in the visible spectrum and class 1 lasers for other wavelengths. The MPE can be exceeded with class 2M lasers with a low risk of injury.
3B	The accessible laser radiation may be dangerous for the eye, and under particular conditions, also for the skin. Diffuse radiation scattered from diffused targets is normally harmless. The accessible radiation limit is 500 mW for continuous-wave lasers emitting in the visible spectrum and 30 mW for pulsed lasers. In the case of direct viewing of class 3B lasers, protective eyewear is required.
4	The accessible radiation of a class 4 laser is very dangerous for the eye and for the skin. Light from diffuse reflections and indirect viewing may be hazardous for the eye. Class 4 lasers must be equipped with a key switch and have an inbuilt safety interlock. Most industrial, medical, scientific, and military lasers belong to this category.

3. For pulsed lasers in the wavelength band of 400 to 700 nm covering red, green, and blue lasers, the MPE is 0.5 μJ/cm^2 and $1.8 \times t^{0.75} \times 10^{-3}$J/cm^2 for exposure durations of 1 ns to 18 μs and 18 μs to 10s, respectively.

3.5.4 Protocol IV for Blinding Laser Weapons

Protocol IV for blinding laser weapons, from the *Treaties and Other International Acts Series 09-721.2*, prohibiting the use and transfer of blinding laser weapons was adopted on October 13, 1995 during the Vienna Convention. The four articles of the protocol are reproduced as follows:

Article 1: It is prohibited to employ laser weapons specifically designed, as their sole combat function or as one of their combat functions, to cause permanent blindness to unenhanced vision, that is, to the naked eye or to the eye with corrective eyesight devices. The High Contracting Parties shall not transfer such weapons to any State or non-State entity.

Article 2: In the employment of laser systems, the High Contracting Parties shall take all feasible precautions to avoid the incidence of permanent blindness to unenhanced vision. Such precautions shall include training of their armed forces and other practical measures.

Article 3: Blinding as an incidental or collateral effect of the legitimate military employment of laser systems, including laser systems used against optical equipment, is not covered by the prohibition of this protocol.

Article 4: For the purpose of this protocol "permanent blindness" means irreversible and uncorrectable loss of vision which is seriously disabling with no prospect of recovery. Serious disability is equivalent to visual acuity of less than 20/200 Snellen measured using both eyes.

Article 1 of the protocol underlines the importance of less-lethal laser weapons such as laser dazzlers to be designed to have minimum possible NOHD and inbuilt safety features to ensure that the device is safe throughout its operational range. The safety feature should preferably shut off the laser beam if a targeted individual accidentally comes within the NOHD. Another interesting corollary of article 1 of the protocol is that although it is prohibited to cause permanent injury to unaided eyes or eyes using corrective eyesight devices, it is not a violation of the protocol if permanent blindness were caused to an eye using optical instrumentation such as optical sights.

3.6   Representative Systems

In the following sections we describe some representative laser dazzlers, including short- and medium-range devices for close combat operations, vehicle-mounted systems for crowd/mob control, long-range laser dazzlers for protection against aerial threats, and spaceborne laser dazzlers.

3.6.1   Handheld and Weapon Mounted Laser Dazzlers

Many companies offer handheld and weapon-mountable laser dazzlers. Some of the better-known devices include GLARE MOUT, GLARE MOUT PLUS, GLARE GBD-IIIC, GLARE LA-9/P, and GLARE RECOIL from B.E. Meyers, United States, Dazer laser available in two variants: DEFENDER and GUARDIAN from Laser Energetics, United States, Saber-203 developed by a U.S. military research lab, Chinese JD-3 laser dazzler, Threat Assessment Laser Illuminator (TALI) from Wicked Lasers, United States, Medusa and Hydra laser dazzlers from Passive Force, United Arab Emirates, Green Light Optical Warner (GLOW) from Thales Group, United Kingdom, and Compact High Power (CHP) laser dazzler from L.E. Systems, United States.

The *GLARE MOUT* is a nonlethal visual disruption laser with an effective range of 20m to 2 km. The device is ideally suited for small arms integration as well as mobile crew-served applications. It is configured around a 125-mW green laser emitting at 532 nm. It has a nominal ocular hazard distance of 18m. GLARE MOUT reportedly saved numerous lives of both soldiers and noncombatants in Iraq and Afghanistan. *GLARE MOUT Plus* offers more than a 60% increase in power over the standard GLARE MOUT. The *GLARE GBD-IIIC* is a long-range variant of GLARE MOUT ideally suiting ship-to-ship signaling or airborne over-watch. With twice the power and a more concentrated beam than the GLARE MOUT, the device has an operational range of 72 to 4,000 meters with the lower limit being the NOHD for an unaided eye. The *GLARE LA-9/P*, also built around a 250-mW green laser, is yet another long-range visual deterrent laser device for hail and warning applications and is intended to be effective out to a range of 0.3 to 4 km for ship-to-ship signaling or airborne over-watch applications. The GLARE LA-9/P has an additional feature of automatically shutting off the device if the subject target were within the nominal ocular hazard distance. The *GLARE RECOIL* (Figure 3.25), also configured around a 250-mW laser, has a safety feature that integrates the laser range finder into the system's safety controls.

The *Dazer Laser* is yet another very popular device. It comes in two variants: GUARDIAN, which has a range from 1 to 300m (model-dependent), and the DEFENDER, which has a range from 1 to 2,400m (model-dependent). Both variants of the Dazer Laser temporarily impair the vision of the target adversary and succeed in eliminating the threat's ability to see, engage, or effectively target the user. Both variants of the Dazer are designed to be eye-safe at all ranges beyond 1m.

The *Saber 203* Grenade Shell Laser Intruder Countermeasure System is a type of laser dazzler that uses a 250-mW red laser diode mounted in a hard

Figure 3.25 GLARE RECOIL.

plastic capsule in the shape of a standard 40-mm grenade. It is suitable for being loaded into a M203 grenade launcher. It has an effective range of 300m. The Saber 203 dazzlers were used in Somalia in 1995 during Operation United Shield.

The Chinese *JD-3 laser dazzler* is another established system. The JD-3 laser dazzler is reported to be mounted on the Chinese type 98 main battle tank and is coupled with a laser radiation detector. It automatically aims for the enemy's illuminating laser designator, attempting to overwhelm its optical systems or blind the operator.

The *Photonic Disruptor*, classified as a TALI, is yet another nonlethal high-power green laser developed by Wicked Lasers, United States, in cooperation with Xtreme Alternative Defense Systems. This tactical laser is equipped with a versatile focus-adjustable collimating lens to compensate for range and power intensity when used to either incapacitate an attacker in close range or safely identify threats from a distance. TALI-series devices are configured around a 100-mW, 532-nm laser producing a laser beam with 1.5 to 7.5 milliradian adjustable beam divergence.

The *MEDUSA laser gun* is a completely self-contained high-power green laser device configured around a 5,000-mW green laser module that can be effectively used on both static and moving targets. The *HYDRA laser dazzler* is a more compact unit configured around a 1,000-mW green laser module. Like the MEDUSA laser gun, it can also be effectively used on static and moving targets.

The *GLOW* is a gun-mounted laser dazzler. It can also be operated remotely or used as a handheld device with a pistol grip or stock. The GLOW laser dazzler provides a selection of two different beam divergence values suitable for operation in close quarter engagements and longer operational ranges. In both modes the beam can be pulsed to give a higher level of warning. The maximum operational ranges in the two cases are 50m (wide) and 350m (narrow), respectively. Figure 3.26 shows the device.

A CHP laser dazzler emits a 500-mW flashing green dazzling laser beam. The CHP with its higher power creates a credible glare effect in a larger spot size for use on moving vehicles or individuals. This feature is particularly important for protection of entry control points and convoys, at distance, in bright ambient conditions.

3.6.2 Vehicle- and Platform-Mounted Laser Dazzlers

A number of static installation and vehicle-mounted laser dazzler systems are also developed for various application scenarios including crowd/mob control and also those involving long operational ranges, such as protection of critical ground assets from aerial threats and protection of naval platforms against

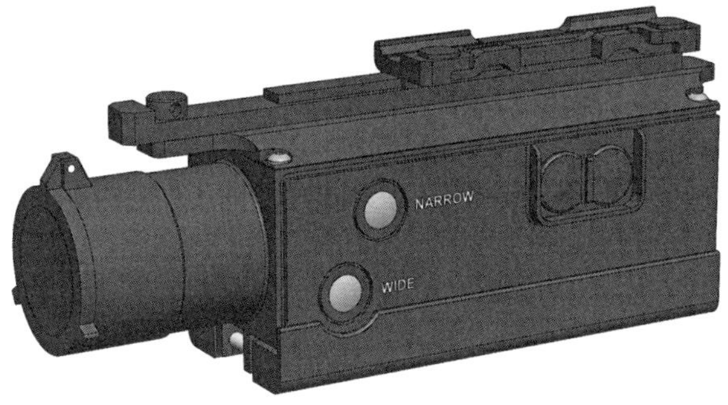

Figure 3.26 Green light optical warner.

asymmetric threats. One such dazzler system is BeamQ Lasers' *Vehicle Long Range Laser Dazzler* configured around a 3,000 mW green laser and integrated with an infrared night vision device. It has an operational range of 3,000 to 5,000m.

The *SEALASE-II* from Passive Force LLC is a high-power nonlethal laser dazzler mounted on a stabilized IP-controllable pan-and-tilt platform. The system is configured around a 5,000-mW green laser and can be integrated with multiple camera sensors. The integrated camera sensors can be day/night color, thermal, or ultralow-light CMOS cameras for extended performance capabilities. Optical sensor for visual tracking and the optional feature of radar based target tracking and slew-to-cue ability allow building a long-range system for protection of critical assets against aerial threats. The system allows the user to choose between suppression/deterrence, hail/warning, and illumination operations.

The *STORM* system is not a dedicated laser dazzler system. It is a family of systems that integrate a range of nonlethal technologies into one packaged system that can be vehicle-mounted or static positioned and operated by a user-friendly controller operating system with a video monitor, thereby offers to the user a multicapability system that can be used from inside a vehicle or from a control room. The system integrates several nonlethal technologies, including use of light, laser, and sound deterrents. The STORM system is designed to address multiple application requirements involving friendly personnel directly facing threats.

3.7 Emerging Trends

Laser dazzlers have shown a lot of promise and potential for a variety of low-intensity conflict applications, including close-quarter battle scenarios, border patrol and coastal surveillance, controlling unruly and violent crowds and mobs, countering asymmetric threats particularly encountered in a naval environment, and protection of critical infrastructure from terrorist attacks. Different variants of these devices are being developed to suit different application requirements. While handheld portable devices are being offered to meet the short-range requirements of paramilitary and law enforcement agencies, tripod-mounted and vehicle-mounted versions equipped with electro-optic sighting support, target acquisition, and tracking devices are also being developed to meet the requirements of emerging scenarios.

While short- and medium-range handheld and weapon-integrated versions of the laser dazzler are in widespread use for antiterrorist and counterinsurgency applications, vehicle-mounted systems designed for operations against unruly and violent crowds are beginning to appear on the scene. Shipborne medium- to long-range laser dazzlers mounted on stabilized platforms are also catching the attention of security forces to defeat asymmetric threats. Another interesting development in the recent past has been a keen interest in long-range laser dazzler systems mounted on Gimbal platforms and equipped with an electro-optic tracker and integrated with a network of radars to provide 24/7 protection to strategic assets from airspace violation by aerial platforms.

RGB laser dazzlers that emit a randomly flickering, intense flashing pattern of red, green, and blue lights at a frequency of a few kilohertz are also reportedly under development although not much information is currently available about the existence of any commercial devices. These devices generally use an array of laser diodes and/or diode-pumped solid-state laser modules to generate the desired wavelengths. Arrays of red, green, and blue LEDs could also be used instead. Studies have revealed that RGB laser dazzlers can be more effective than single wavelength (red or green) dazzlers in a sea environment while countering asymmetric threats such as those arising from suicide boats and sea pirates.

In the not-too-distant future, one would see deployment of laser dazzlers with global coverage as briefly mentioned in Chapter 1. These systems are proposed to use remotely controlled membrane reflectors to receive the dazzling laser beam from the source station and guide it to the intended target location.

Selected Bibliography

Davison, N., *Non-Lethal Weapons,* Palgrave Macmillan, 2009.

Hecht, J., *Understanding Lasers: An Entry Level Guide,* Third Edition, Piscataway, NJ: Wiley-IEEE Press, 2008.

McAulay, A. D., *Military Laser Technology for Defence,* Hoboken, NJ: Wiley-Interscience, 2011.

Maini, A. K., *Handbook of Defence Electronics and Optronics: Fundamentals, Technologies and Systems*, Hoboken, NJ: John Wiley-Blackwell, 2018.

Maini, A. K., *Lasers and Optoelectronics: Fundamentals, Devices and Applications*, Chichester, UK: Wiley-Blackwell, 2013.

Perram, G.P., *An Introduction to Laser Weapon Systems*, Directed Energy Professional Society, 2009.

Titterton, D. H., *Military Laser Technology and Systems,* Norwood, MA: Artech House, 2015.

Waynant, R., and M. Ediger (eds.), *Electro-Optics Handbook*, New York: McGraw-Hill, 2000.

4

Directed-Energy Lasers

In the overview of the use of optoelectronic technologies and systems presented in Chapter 1, the role of relatively low-power directed-energy laser systems in homeland security and low-intensity conflict scenarios was briefly discussed. This chapter describes in further detail use of directed-energy laser devices and systems for military applications with an emphasis on those of relevance to use by paramilitary forces and security agencies. Major topics included for discussion include laser countermeasures used for incapacitation of optoelectronic sensors used by adversarial forces on small arms, land-based platforms and unmanned aerial vehicles, kilowatt-class laser-based systems for neutralization of unexploded ordnances, and neutralization of terrorist hideouts. Another major application of low-power directed-energy lasers is in less-lethal antipersonnel laser weapons, which were comprehensively discussed in Chapter 3.

4.1 Directed-Energy Lasers

Directed-energy laser weapons use the directed-energy property of lasers to do the intended damage. The damage could be temporary impairment of vision of the adversary personnel, as is the case with laser dazzlers; disabling or incapacitation of optoelectronic sensors mounted on small arms and vehicular platforms and saturation and neutralization of laser seekers for protection against laser-guided attacks in a laser countermeasure application; and disabling of optoelectronic sensors onboard unmanned aerial vehicles and low-earth orbit satellites using ground-based lasers. Directed-energy lasers have certain specific properties and consequent operational advantages that have led to emergence of these potential applications. However, these advantages come with certain limitations. The operational advantages and the limitations are briefly discussed in the following sections.

4.1.1 Advantages and Limitations

The primary advantages of low-power directed-energy laser systems as well as high-power directed-energy laser weapons include speed-of-light delivery, near-zero collateral damage, multiple target engagement and rapid retargeting capability, immunity to electromagnetic interference, and no influence of gravity. Deep magazine and low cost per shot are the other advantages, which were also discussed in Section 3.1.2.

Limitations of use and efficacy of directed-energy laser systems include their *line-of-sight dependence*, requirement of finite dwell time, problems due to atmospheric attenuation and turbulence, and ineffectiveness against hardened structures. Laser weapons require direct line-of-sight to engage a target. Their effectiveness is reduced or neutralized by the presence of any object or structure in front of the target that cannot be burned through. Directed-energy laser systems require a certain *minimum dwell time* depending on the application to deposit sufficient energy for target destruction. The effectiveness of the laser weapon is adversely affected by the *atmospheric conditions*. The laser beam suffers attenuation due to absorption and scattering by airborne particles and gas molecules and deterioration of beam quality in the form of deformation of the laser beam wavefront and increase in the laser beam spot size at the target.

4.1.2 Potential Applications

Potential application areas of directed-energy laser systems are broadly categorized as short- and medium-range tactical missions and long-range strategic missions. Applications belonging to the tactical class are more relevant to homeland security and low-intensity conflict scenarios. Some of the important application areas of tactical class systems include

1. Laser countermeasures used for neutralization of optoelectronic sensors mounted on small arms, land vehicles, unmanned aerial vehicles, laser-guided munitions, and low-earth orbit satellites on surveillance missions;

2. Standoff neutralization of ordnances such as mines, unexploded ordnances, and IEDs;

3. Ground-based defense against RAMs;

4. Ground-based capability to destroy UAVs of the adversary, airborne defense of aircraft against MANPADS such as shoulder-fired surface-to-air missiles;

5. Ship defense against maneuvering cruise missiles and tactical ballistic missiles.

Except for anti-RAM, anti-MANPADS, antiballistic missiles, and anti-UAV high-power laser weapons, other applications including saturation or neutralization of optoelectronic sensors and surveillance cameras, laser countermeasures, and neutralization of unexploded ordnances have found wide acceptance with paramilitary forces and security agencies fighting insurgencies and terrorism.

4.2 Laser Countermeasures

Laser countermeasures constitutes a subset of electro-optic countermeasures employed against systems operating in the optical spectrum of electromagnetic radiation from ultraviolet to infrared. Examples include sighting, observation and surveillance devices, laser designators and range finders, laser radar, and optoelectronic sensors such as laser seekers and infrared seekers, respectively, used in laser-guided munitions and infrared-guided weapons. The relevance of electro-optic countermeasures in general and laser countermeasures of different types stems from the fact that whenever a defense technology has matured to a high level leading to its widespread usage, it has triggered huge R&D investment in the corresponding countermeasures technologies and systems. It happened earlier in the case of radar and other related military systems operating in the RF spectrum of electromagnetic radiation; it has happened subsequently in the case of lasers and related devices operating in the optical spectrum.

4.2.1 Relevance

The need to develop and deploy countermeasure lasers that can offer effective countermeasures against similar systems deployed by the armed forces of an adversary is much more pertinent and relevant in the present-day scenario. No military platform, be it land-based, aerial, or shipborne, is currently free from the risk of being exposed to laser radiation. The activities of these platforms are under constant surveillance by various kinds of electro-optic devices and optoelectronic sensors. Rendering these devices and sensors ineffective in the battlefield therefore makes a huge difference to the battlefield competence of a nation. Deployment of eletro-optic countermeasures (EOCM) devices and systems designed to incapacitate or neutralize the more conventional laser devices and systems would act as a force multiplier as a platform incapacitated in the enemy camp is a platform added to your own. This goes a long way in enhancing the survivability quotient of the armed forces equipped with such a capability.

4.2.2 Passive and Active Countermeasures

Electro-optic countermeasures are broadly classified as passive and active countermeasures. *Passive countermeasures* for platform protection include use of armor, camouflage, fortification, and other protection technologies such as a self-sealing fuel tank. *Active countermeasures* are further comprised of soft-kill countermeasures and hard-kill countermeasures. *Soft-kill countermeasures* change the electromagnetic, acoustic, or other forms of signatures of the platform to be protected, thereby adversely affecting the tracking or sensing capability of the incoming threat. In the context of electro-optic countermeasures, the incoming threat could be a laser-guided bomb or projectile or an infrared-guided missile. A common example of the use of electro-optic countermeasures to protect the platform under attack from a laser-guided threat is through deployment of a smoke or aerosol screen to block the radiation from the laser target designator irradiating the platform. This deprives the laser seeker in the guided munitions the all-important guidance information required to home on to the target. In the case of active countermeasures, means are adopted to counterattack the incoming threat. This may be achieved by temporarily or permanently neutralizing one of the elements such as the laser or infrared seeker responsible for the functioning of the guided threat or by physically destroying the incoming threat by launching a counterprojectile in that direction. The latter involves use of radar for detection of threat and therefore is not considered as electro-optic countermeasures. In the case of laser-based active countermeasures for protection against laser-guided munitions, direction of arrival of laser radiation from the laser target designator is detected by a laser warning sensor suite and then a high-power/-energy laser beam is launched in the same direction to neutralize the designator. Another technique in serious contention is to use the high-energy laser radiation to illuminate a dummy target 100 to 200m away from the platform to be protected, thereby forcing the incoming laser-guided threat to move away from the intended target and land on the dummy target. This is a very attractive proposition for protection of critical and high-value military assets such as aircraft shelters and ammunition depots. While laser radiation scattered off the dummy target is used to misguide the laser-guided munitions, directional infrared countermeasures employing in-band mid-infrared lasers are used to confuse the infrared seekers in infrared-guided missiles such as MANPADS to move away from the targeted aircraft.

4.2.3 Neutralization of Optoelectronic Sensors

There are two broad categories of laser-based active countermeasures used for neutralization of optoelectronic sensors constituting the front end of many electro-optic devices and systems. These include both handheld and portable devices as well as those mounted on weapon platforms. The first type employs lasers with pulse energies in the range of hundreds of millijoules to a few joules. These systems in most cases saturate and in some cases permanently damage the front-end photosensors such as PIN and avalanche photodiodes in laser warning sensors and laser range finders, quadrant photodiodes in laser seekers, CCD and CMOS sensors in imaging sensors, and photocathodes in image intensifier tube-based night vision devices. In the second type of laser-based active countermeasures, kilojoule or kilowatt classes of lasers are used. The lasers are used to damage the front-end optics of the sensor system. Damage threshold power or energy is relatively higher than the corresponding saturation values. As well, saturation and damage thresholds are different for different sensor types. For example, damage thresholds for CMOS and CCD sensors are much lower than those for silicon photodiodes, and the damage threshold for CCD sensors is lower compared to that of CMOS sensors. In silicon photodiodes, the onset of damage typically occurs for an energy density of greater than 1.0 J/cm^2. The same for CCD and CMOS sensors, respectively, are 0.03 J/cm^2 and 0.05 to 0.1 J/cm^2. These are typical values for nanosecond pulses of laser radiation at 1,064 nm.

There are several deployment scenarios where laser countermeasures are employed in an antisensor role. These include neutralizing front-end sensors of handheld and portable electro-optic devices such as night vision devices, surveillance cameras, and optical target locators used for observation and surveillance; front-end sensors of the receiving channels of laser target designators and range finders; platform-mounted laser warning sensors used as a part of a defensive aids suite; and front-end sensors of laser seekers used in laser-guided munitions deployed in semiactive laser guidance mode. Some of the important deployment scenarios are briefly described next.

Most deployment scenarios use either of the following two countermeasure mechanisms. In the first approach, the countermeasure system employs a relatively low-power probing laser beam collinear with the main high-power/-energy laser beam. The probing laser beam is scanned across the targeted electro-optic system. When the probing laser beam comes within the field-of-view of the front-end optics of the targeted system, there is a retroreflection from the focal plane of the front-end optics. The retroreflected laser beam travels back to the countermeasure system, thereby providing information on the location of the targeted system. The main high-power/-energy beam is then transmitted in that direction to achieve the intended effect. This mechanism is employed in the case of passive electro-optical devices and systems such as sighting devices, night vision devices, thermal imagers, and surveillance cameras. Figure 4.1 illustrates the concept.

In the second concept, the countermeasure system targets active electro-optical systems that use laser emission for their operation. Representative examples include laser range finders, laser target designators, and optical target

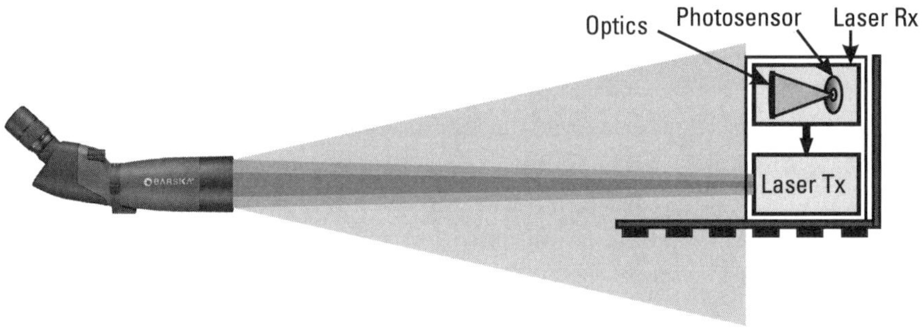

Figure 4.1 Countermeasure approach for passive electro-optical systems.

locators. In this case, the countermeasure system is comprised of a laser sensor or an array of laser sensors and a high-power/-energy laser source. A laser sensor detects the presence and direction-of-arrival of the laser threat emanating from the targeted electro-optical system. The information on the location of targeted system is used to align the high-power/-energy laser beam with the targeted system. The high-power/-energy laser beam is transmitted in the direction of the targeted system to neutralize either the front-end optoelectronic sensor, the front-end optics, or both depending on the power/energy level. Countermeasure systems used to provide protection from laser-guided munitions attack employ this concept. Figures 4.2 to 4.5 illustrate the concept for three different laser-guided munitions (LGM) delivery approaches. Figure 4.2 shows the concept of a ground-based laser target designator (LTD) and laser-guided munitions delivery platform. In Figure 4.3, the target designator is ground-based and the guided munitions is delivered from an aircraft. Figure 4.4 shows a delivery mechanism where both the target designator and guided munitions are on different airborne platforms. In Figure 4.5, the target designator and guided munitions are on the same airborne platform.

Neutralization of the laser target designator used in the delivery operation also deprives the laser seeker head of the incoming threat the guidance signal it relies on to precisely hit the target. In this case, usually due to the narrow receiver field of view of the laser target designator, the laser warning sensor needs to have angle-of-arrival measurement accuracy that is better than 1°. As well, state-of-the-art laser seeker heads are extremely difficult to defeat in the close-in range and continue to follow the trajectory at the time of the last lock.

There are other countermeasure techniques that do not have any stringent requirement of determining the angle-of-arrival of laser threat for executing countermeasure action. These techniques, instead of using directed laser energy to incapacitate the optoelectronic sensors of the targeted electro-optical system, deceive them to follow an unintended trajectory away from the target. One such well-established technique involves the use of a laser warning sensor

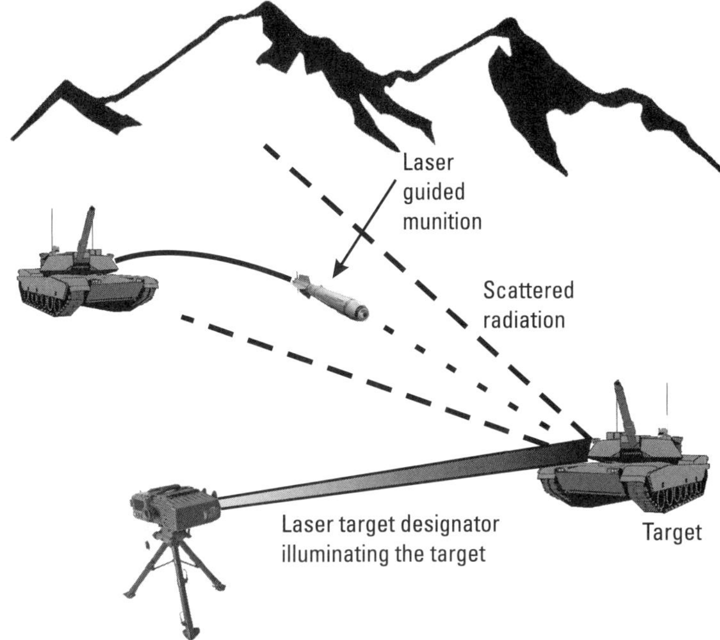

Figure 4.2 Countermeasures: ground-based LTD and LGM launch platform.

system usually with 360° azimuth and desired elevation coverage interfaced with an aerosol-/smoke-screening system. The laser warning sensor detects the laser threat and determines its direction of arrival. It then sends a command signal to the relevant aerosol or smoke grenade/s to deploy in the path of the laser radiation, thereby blocking the radiation from reaching the target to be protected. As a result, the laser-guided munitions guidance system goes out of lock, drastically bringing down its probability of hitting the target. Figure 4.6 illustrates the concept.

Yet another technique uses a kind of laser decoy that can mimic the laser signatures in terms of the pulse repetition frequency code of the actual target to be protected. Figure 4.7 illustrates the concept with a schematic arrangement of the different components of the system. This concept is particularly suitable for protection of critical assets from a laser-guided munitions attack from an airborne platform. The arrangement shown in Figure 4.8 illustrates the case of protection of an asset—an aircraft shelter in this case—against a laser-guided munitions attack when the illumination is from a ground-based laser designator. The concept is equally valid for designation from an airborne platform. The laser warning sensor suite comprised of multiple laser detection sensors mounted atop the shelter has a spectral response band of 400 to 1,100 nm to cover the desired operating wavelength of 1,064 nm of the laser target

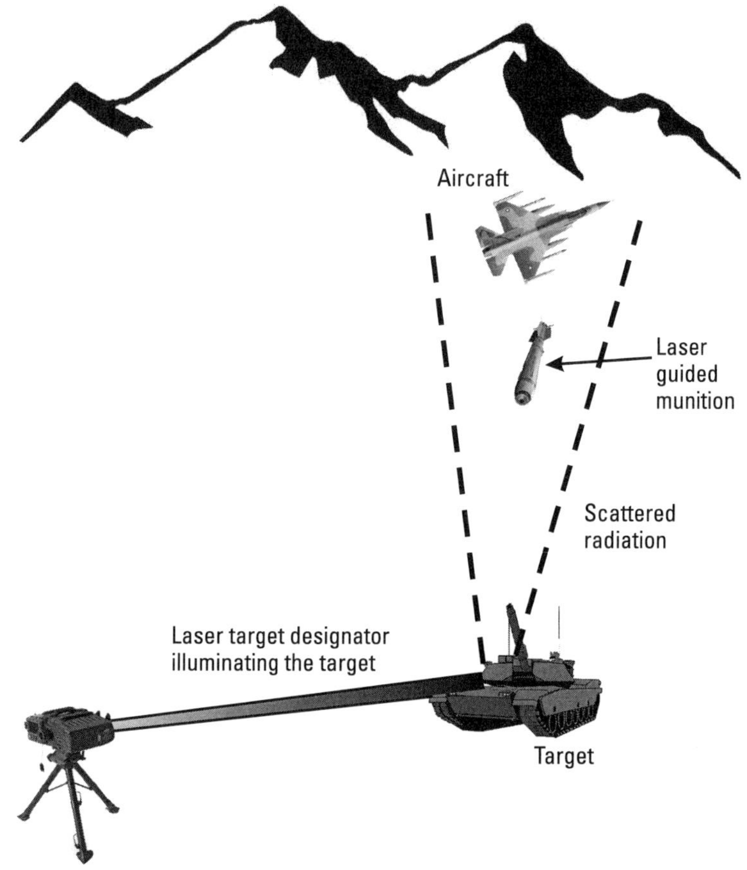

Figure 4.3 Countermeasures: ground-based LTD and airborne LGM launch platform.

designators used in laser-guided munitions delivery, angle-of-arrival accuracy of better than 1°, and a built-in PRF decoding module that generates a train of TTL pulses at the operating PRF code with desired accuracy, typically better than ±1 μs in time interval between two consecutive pulses. The PRF decoding module also performs the task of doing edge matching of decoded pulse train and the replicated pulse train. As well, PRF is decoded with an accuracy that is typically around ±1 μs in the time interval between two consecutive pulses. Depending on the size and location of the asset to be protected, it may be desirable to use multiple laser sensor modules. The PRF code information is fed to a high-energy laser of the same type as the laser target designator operating at same wavelength and decoded pulse repetition frequency code. The laser is used to irradiate the dummy target that is usually 200 to 300m away from the asset to be protected. The dummy target is designed to scatter the laser radiation in a wide cone toward the incoming threat. Simultaneous to the laser illumination

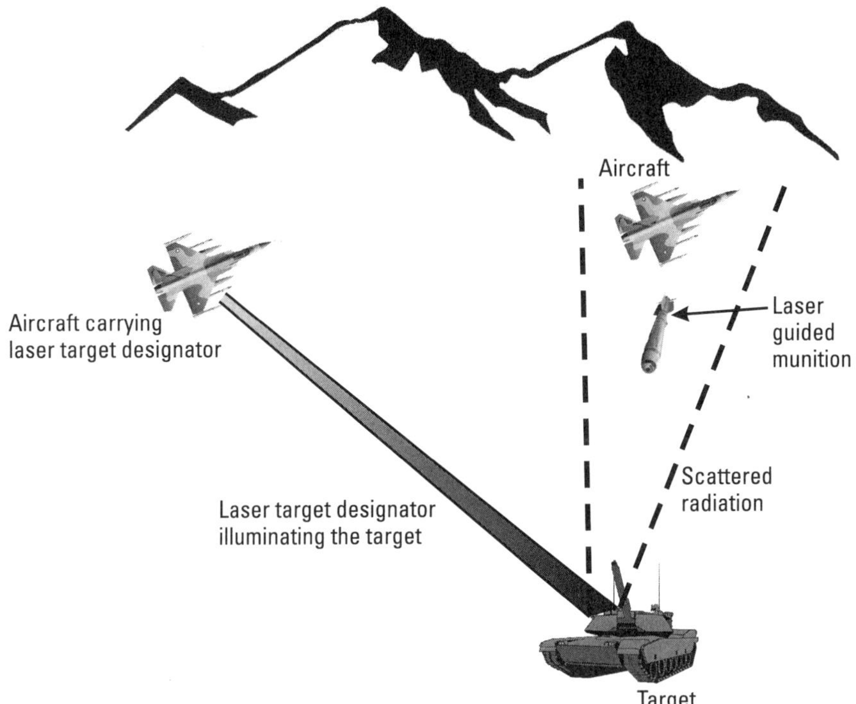

Figure 4.4 Countermeasures: LTD and LGM on separate airborne platforms.

of the dummy target, a smoke grenade is fired in the direction of the incoming laser radiation from the adversary's laser target designator. This blocks the laser radiation from reaching the seeker head of the incoming laser-guided munitions. The seeker head now receives radiation only from the dummy target. The laser seeker locks on to the radiation from the dummy target and guides the weapon to the dummy target. Most state-of-the-art seeker heads will lock on to the first genuine pulse repetition code. It is therefore important to break lock from the genuine radiation only momentarily and force the seeker head to switch lock-on to the mimicked radiation. Once lock-on to the mimicked radiation is established, recurrence of genuine radiation does not restore the original lock condition. This phenomenon is used to the advantage in the laser decoy concept.

4.2.4 Laser-Blinding Shells and Grenades

An emerging concept in the field of laser countermeasures to neutralize opto-electronic sensors is the one-time *laser-blinding shell* or laser grenade that generates extremely intense flashes of light on activation. The shell contains explosive material together with inert gases like argon, neon, and xenon. When the ex-

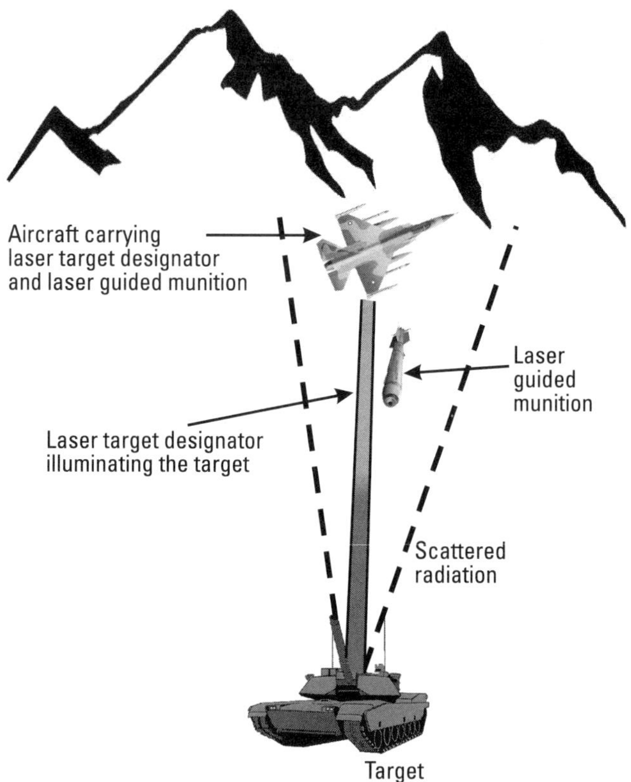

Figure 4.5   Countermeasures: LTD and LGM on the same airborne platforms.

plosive material is ignited, high-pressure, high-temperature gases are generated, leading to formation of plasma at a high temperature of about 1,000°C. The plasma is accompanied by emission of an extremely intense flash. The flash generates omnidirectional and directional radiation in the wavelength band from ultraviolet to infrared. These light-blinding shells can be fired through conventional howitzers, mortars, or gravity bombs. These can also be thrown as grenades. Two types of laser grenades—a scattering-type and a directional type—have been developed. These grenades can generate high efficiency laser beams to render optoelectronic sensors dysfunctional. Laser-blinding shells have the operational advantages of longer operational ranges that could be as much as 20 km. As well, unlike other laser countermeasure systems, they do not require precise aiming and do not expose launch position.

4.2.5 Representative Countermeasures Systems

There is a wide range of electro-optic countermeasure systems including platform-specific and application-specific systems. An overview of some representa-

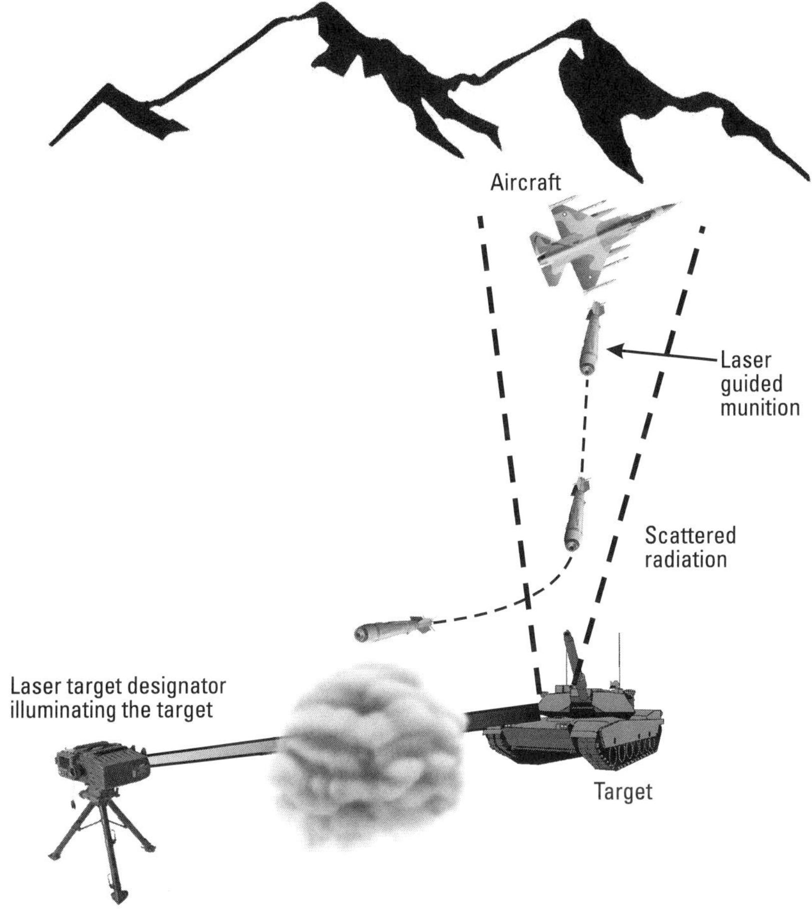

Figure 4.6   Laser warning and countermeasures system.

tive countermeasures systems is presented here. These EOCM systems are designed and built to neutralize an adversary's optoelectronic systems and protect the host platforms or high-value assets from laser-guided munitions attack.

1. The *Rapid Obscurant System (ROSY)* from Rheinmetall Defence has two variants: *Rapid Obscurant System–Land* (ROSY-L) for providing protection on land, and Rapid Obscurant System–Navy (ROSY-N) for protection on sea. ROSY-L (Figure 4.9) is the smoke/obscurant protection system designed to provide protection to the crew and passengers of light military and civilian vehicles against surprise attacks, such as those experienced during reconnaissance patrols or while moving in convoy, from a wide range of guided-weapon threats, includ-

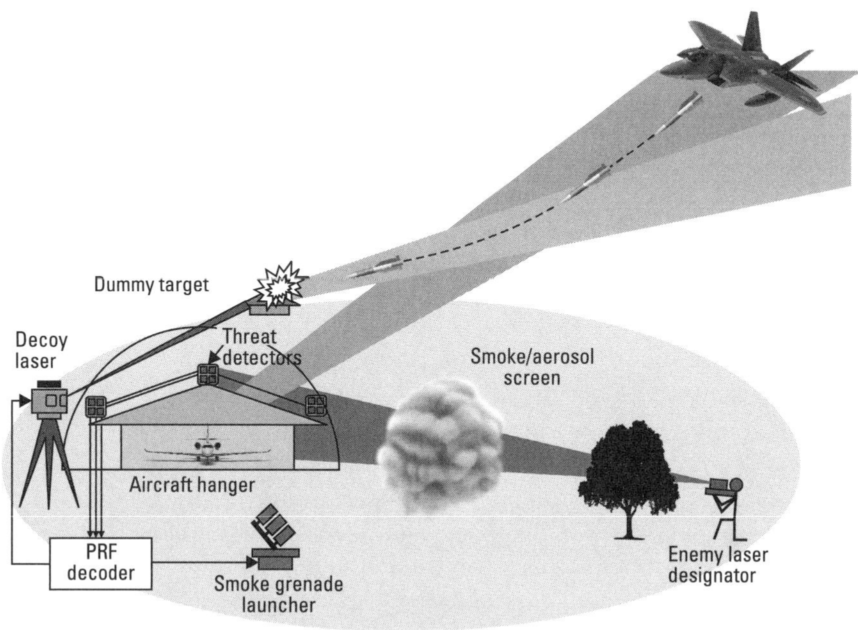

Figure 4.7   Laser threat detection and decoy concept.

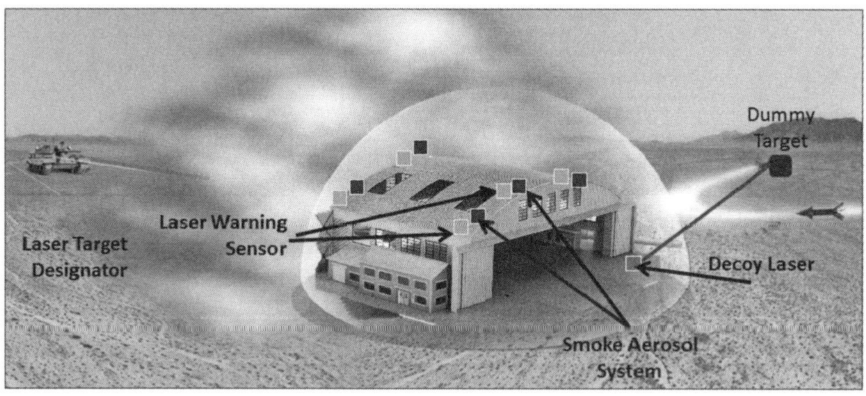

Figure 4.8   Laser threat detection and decoy deployment.

ing TV, electro-optic (EO), infrared (IR), imaging infrared (IIR) and semiactive command to line-of-sight (SACLOS) guided weapons. A variant of the ROSY-L system called ROSY MOD is designed for use on small weapon stations and light vehicles used by special operations

Figure 4.9 ROSY-L countermeasure system.

forces. Its integration with the platform without the need for a surface-mounted launcher makes it undetectable. ROSY-N is designed to protect naval and coast guard units from missiles and asymmetric attack in littoral zones and inland waters. It is particularly suited to installation on patrol vessels, speedboats, fast-attack craft, and landing craft.

2. The *Multi Ammunition Soft Kill System (MASS)*, a soft-kill countermeasure system from Rheinmetall Defence, is a compact and fully automatic shipborne soft-kill launcher system originally designed to provide protection against antiship missiles and asymmetric threats employing RF, IR, and/or EO seekers through deployment of programmable multispectral *omni-trap decoy rounds* with radar, IR, laser, EO, and UV payloads. MASS can be used in stand-alone mode and as an integrated setup with a command and control system.

3. The *Laser Countermeasure (LaCM) system,* jointly developed by the Office of Naval Research (ONR), Naval Research Laboratory (NRL), and Northrop Grumman under the Electronic Warfare Integrated System for Small Platforms (EWISSP) program, is designed to protect small surface platforms from attack by laser-guided munitions. Major subsystems of the LaCM system include a countermeasure laser to confuse the threat weapon, suitable optics to direct the countermeasure laser energy toward the approaching laser-guided threat, and an erectable mast to provide an improved defensive geometry.

4. The *DHY-322 Laser Decoy System* from CILAS, France, is mainly comprised of a laser detection system and a high-energy laser. The high-energy laser is mounted on a controlled turret. The laser detection system detects the origin of hostile laser radiation, deciphers its PRF code, and then synchronizes the high-energy laser to operate at the decoded PRF code. The high-energy laser radiation is then used to illuminate a dummy target. Laser radiation scattered off the dummy target diverts approaching laser-guided munitions to a false laser spot on the chosen field area. The DHY-322 is specifically designed to protect ships, strategic sites, and other high-value assets.

5. The *Stingray Laser Detection and Countermeasures System* from Lockheed Martin is another EOCM system designed to protect the frontline forces by accurately locating and neutralizing optical and electro-optical fire control systems of an adversary. The Stingray system captures optical and electro-optical systems with a probing laser using a cat's-eye effect. The system with its kilowatt-class CO_2 laser and Nd-YAG and frequency-doubled Nd: YAG lasers are capable of blinding photoelectric sensors up to a distance 8 km and antipersonnel operation over even greater distances.

6. The *AN/VLQ-6 HARDHAT* is a vehicle-mounted missile countermeasures device (MCD) developed by Loral, Inc. The system is used as a part of a comprehensive warning and threat response countermeasures system that detects and intercepts laser signals to provide warning of an imminent attack. The system then emits infrared energy to disrupt the threat's missile/command unit tracking loop. The system is typically mounted high above the turret, which allows it to scan the frontal arc to detect and decoy away most of the widely used anti-tank guided missiles (ATGMs).

7. *LARC* is an airborne laser ranging and countermeasures system developed jointly by the United States and Britain. The system is comprised of four sensors that cover the lower hemispheric region beneath the aircraft to compute the angle of arrival of the laser threat. It employs high-energy countermeasure beams to neutralize the threat.

8. The *Corolla Prince,* developed by the United States, is an airborne laser weapon based on Stingray technology. It features higher output power and larger operational ranges than Stingray and is designed to primarily blind ground-based optical and optoelectronic tracking systems. Laser grenades are another significant development in the field of laser countermeasures.

4.3 Directed-Energy Lasers for Tactical Applications

The operational scenario of directed-energy laser weapons is broadly categorized as short- and medium-range *tactical missions* and long-range *strategic missions*. These are outlined in the following section.

4.3.1 Application Scenarios

Some of the important application areas of tactical-class laser weapons include ground-based defense against RAMs, ground-based capability to destroy UAVs of an adversary, airborne defense of aircraft against MANPADS, such as shoulder-fired surface-to-air missiles, and ship defense against maneuvering cruise missiles and tactical ballistic missiles.

Long-range strategic applications of laser-based DEW systems mainly include ballistic missile defense and space control, such as space-based lasers and antisatellite applications. In all these applications, operational ranges are generally in the hundreds to thousands of kilometers and required power levels are of the order of 1 to 20 MW depending on the actual mission. Space control applications such as antisatellite applications require relatively much higher power than the power level needed for ballistic missile defense. Table 4.1 outlines the mission capabilities of laser weapons for relevant operational ranges and laser power levels for the tactical class of directed-energy laser weapons.

4.3.2 Representative Systems

Several directed-energy laser weapon technology demonstrators have been developed and field-tested for tactical roles in the last couple of decades. A well-known laser-based DEW system in the tactical class of laser weapons is the Northrop Grumman's THEL. This laser is built in two configurations: baseline static HEL and the relocatable mobile version *Mobile THEL* (M-THEL). THEL systems are point defense weapon systems designed to engage and destroy artillery rockets, artillery shells, mortar rounds, and low-flying aircraft. The system uses a DF laser operating at 3.8 μm. The THEL demonstrator was successfully tested repeatedly between 2000 and 2004, destroying a number of 122- and 160-mm Katyusha rockets, multiple artillery shells, and mortar rounds, including a salvo attack by mortar.

The *Advanced Tactical Laser* (ATL) uses an 80-kW COIL and is mounted on a modified Boeing C-130H Hercules aircraft (Figure 4.10) with the most obvious visual difference being a rotating turret protruding from the aircraft's underside through a hole. This chemical laser is like the one developed for the Airborne Laser (ABL) program with much lower output power. The ATL is envisioned to offer the mobility of a small aircraft, high-resolution imagery for

Table 4.1
Laser Mission Capabilities and Required Power Levels

Series Number	Mission	Typical Ranges (km)	Laser Power (kW)
1	Counter UAV/RPV/drone	1–3	10
2	Ship surface threat defense	1–3	10
3	Short-range tactical applications; counter-RAM, missiles	1–3	50
4	Airborne precision strike of ground targets	5–10	100
5	Ground-based air and missile defense;counter-RAM and MANPADS	5–10	100
6	Anti-tank missiles, rocket propelled grenades (RPGs), and cruise missiles	5–10	100
7	Area denial to aircraft, helicopters, and UAVs	5–10	100

Figure 4.10   C-130H aircraft with ATL onboard. (Source: Wikimedia Commons.)

target identification, and the ability to localize damage to a small area of less than a foot in diameter from a range of 8 to10 km.

Several directed-energy laser systems based on solid-state and fiber lasers are being developed and tested for tactical mission needs ranging from ordnance neutralization to antimissile and anti-RAM applications. Raytheon has developed and successfully tested a directed-energy laser system called *Laser Phalanx* employing a 20-kW industrial fiber laser. The system has been successfully demonstrated against a static mortar from 0.5 km. Raytheon has also successfully tested a ship-mounted solid-state laser weapon called *Laser Weapon System* (LaWS) to shoot four drones. Figure 4.11 shows LaWS on a U.S. naval platform.

Figure 4.11   LaWS aboard naval vessel USS Ponce. (Source: Wikimedia Commons.)

4.4   Directed-Energy Lasers as Antisatellite Countermeasures

It has been observed that a directed-energy pulsed laser with a moderate average power level is good enough to inflict partial or total damage to optics such as filters and sensors onboard surveillance and earth imaging satellites in the low-earth orbit provided certain conditions are met. Optoelectronic sensors on some U.S. surveillance satellites were reportedly hit by ground-based directed-energy lasers inflicting partial damage to filters in the optical front-end and photosensors. Reportedly, these incidents were caused unintentionally by the Chinese while carrying out satellite laser ranging experiments. The following sections briefly describe satellite laser ranging and the satellite orbital parameters that can possibly lead to such an incident. Even though the incidents might have been unintentional and without the objective of testing an antisatellite weapon, it does establish the vulnerability of sensors onboard low-earth orbit earth imaging satellites to ground-based directed-energy lasers of relatively much lower average power level as compared to the CW power used in the case of high-power directed-energy laser weapons.

4.4.1   Satellite Laser Ranging

Satellite laser ranging (SLR) is a technique of measuring the range of an earth-orbiting satellite using a laser with the objective of determining the orbital parameters of satellites and their variation from predicted values and through this accurately determining temporal variation of the earth's center of mass. In satellite laser ranging, a global network of ground stations measures the instantaneous time of flight of ultrashort laser pulses from a laser receiver to a satellite equipped with special reflectors and back to the laser receiver to compute the instantaneous range information with the precision of few millimeters, which can be used to generate data on the satellite's orbital parameters. A mode-locked Nd-YAG laser with a Q-switched envelope generating laser pulses having a pulse width of the order of a few tens of picoseconds is the preferred laser source. In most cases, the laser pulse with highest peak power in the train of pulses in the Q-switched envelope is used. Although the average power level may be of the order of a few watts, peak power is generally of the order of a gigawatt assuming pulse energy of a few tens of millijoules to a few joules, a pulse width of 10 to 20 ns, and a pulse repetition frequency of 10 to –20 Hz. As well, frequency-doubled output at 532 nm is generally used as compared to infrared output at 1064 nm due to relatively much higher speed and higher quantum efficiency of detectors in the visible spectrum as compared to infrared.

4.4.2 Vulnerability of Optics and Sensors

Satellite laser ranging experiments are carried out with cooperative satellites in low-earth orbits. These satellites have a retroreflector mounted on the satellite. The retroreflector performs the function of reflecting the transmitted laser energy back toward the SLR station to allow computation of time of flight and hence the range on a periodic basis. SLR stations are also sometimes equipped with lasers used to direct laser energy at uncooperative space debris in low-earth orbits to determine their orbital parameters. Optoelectronic sensors such as sensitive imaging sensors could become prone to being adversely affected by such an exercise. Let us examine the conditions under which satellite sensors could become vulnerable. If the SLR lasers were directed at low-earth orbit satellites to determine their range at a time when the satellite was not overhead, the probability of any sensor damage would be nearly zero. This is because when the satellite is not overhead or not close to zenith, the sensors cannot view the laser. If the satellite were overhead and the sensor viewed that region of earth housing the laser, then there would be a finite probability that the filters covering the sensor or even a part of sensor would be affected if the satellite does not have any shutters or other mechanisms to protect the sensors from high-intensity laser light. As well, the moderate average power level of SLR lasers is too low to interfere with satellites through heating effects. It is also too low to cause any physical damage to satellite components other than imaging sensors. Although lasers used on SLR stations cannot be considered as antisatellite weapons, these ground-based lasers used for satellite laser ranging could be potentially dangerous to sensitive imaging sensors onboard low-earth orbiting surveillance and spy satellites. Note that a sensor with a field of view of 1° located onboard a low-earth orbit satellite would be looking at the ground-based laser for a period of typically 1 to 2 seconds as it passes overhead. During this time, the sensor would receive 10 to 20 laser pulses if the laser were operating at 10 Hz. This number could be large enough to inflict damage to the filters and/or the sensor.

4.5 Directed-Energy Lasers for Debris Removal

The ever-increasing density of debris in low-earth orbits (LEOs) by runaway collisional cascading is a big threat to use of this space by satellites and other spaceborne platforms. Collisional cascading, also known as ablation cascade or Kessler syndrome, refers to a scenario where collisions between objects produce a cascade effect due to each collision generating space debris that further increases the probability of more collisions. This problem if not addressed could render a large part of low-earth orbit space unusable by satellites and other space probes. As well, the probability of occurrence of collisional cascading increases with more and more space vehicle launches. A closing velocity of about 10 to 12 km/s at which a piece of debris is expected to collide with the spacecraft has 10 times the energy density of dynamite. A 100 gram bolt is certainly likely to cause lethal damage to the International Space Station if it struck the crew chamber.

4.5.1 Space Debris Particles: Size and Effect

Space debris particle size is spread over a large range from a fraction of a centimeter to larger than 10 cm. As outlined in Chapter 1, the debris sized between 1 and 10 cm is the most hazardous to space vehicles. This is because debris particles smaller than 1.0 cm can be handled by making space vehicles more rugged, for example, by use of Whipple shields. A Whipple shield, named after Fred Whipple, is a thin shield that protects a spacecraft from damage due to collision with hypervelocity micrometeoroids and other fast-moving minor debris. Also known as a meteor deflection screen, it is based on the principle that small meteoroids or minor debris fragments explode when they strike a solid surface. Therefore, when the spacecraft is protected by a Whipple shield outer skin that is about a tenth of the thickness of its main skin, an impinging body will be destroyed before it causes any real damage. As well, debris particles larger than 10 cm are less likely to cause damage to LEO spacecraft as these particles are large enough to be tracked to avoid collision using required maneuvers. There are already several hundreds of thousands of debris particles larger than 1.0 cm in LEO space and this number is increasing because on an average, one satellite is destroyed every year, contributing to its share of orbital debris. The debris density peaks in 800- to 1,000-km orbit, which is the most widely exploited orbit range for low-earth orbit satellites and other categories of spacecraft. Figure 4.12 illustrates the crowding of LEO space with space debris.

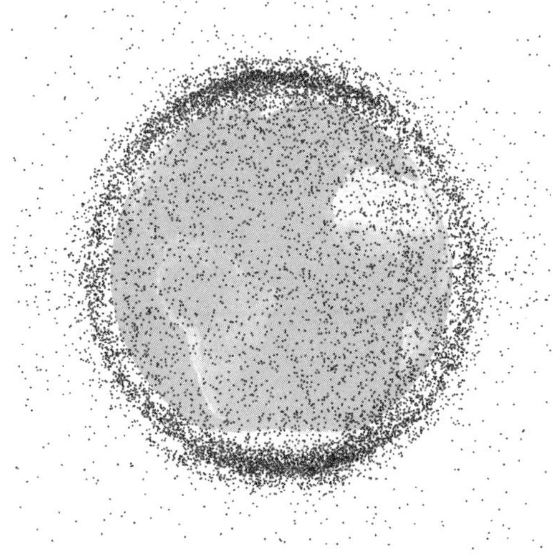

Figure 4.12   Photograph of space debris. (Courtesy: Wikimedia Commons.)

4.5.2 Representative Incidents of Debris Collision

There are numerous incidents to cite that have either endangered the operation of LEO spacecraft or significantly contributed to an increase in debris density, thereby endangering the safety of many spacecraft. There have been incidents where LEO spacecraft have had to take shelter in a capsule to avoid catastrophic collision. There have been two incidents in the recent past, first in March 2009 and then in June 2011, when it became imperative for Space Station astronauts to take cover in a Soyuz capsule that was incidentally docked with the space station to minimize the likelihood of the chance of penetration by an object with unacceptable track uncertainty. In another incident in February 2009, an American Iridium satellite collided with a Russian Cosmos satellite. The resultant debris combined with that from the Chinese Feng Yun 1C ASAT test in January 2007 to significantly increase debris density around the Earth, thereby raising concerns about the safety of the final Hubble servicing mission. Going by the Kessler syndrome, debris–spacecraft collisions are on track to become the most dominant debris-generating mechanism. On August 23, 2016, the European Space Agency's earth imaging satellite called Sentinel-1A suffered an impact from a millimeter-sized object that slammed into one of its solar panels and left a visible dent nearly a half meter across. The dent caused a drop in electrical energy generated by a few percent. It also altered the orientation of the spacecraft a bit and changed the orbit a slight amount as well. Images taken by onboard cameras quickly confirmed the collision. On January 24, 2017, there was a debris collision alert and a likely need for a maneuver to avoid collision.

The flight control team of the Swarm mission was alerted that the Swarm-B satellite would have a close call from a 15-cm chunk of the former Cosmos 375.

Although the debris object was being tracked, the uncertainty in the track combined with the uncertainty in the satellite's orbit meant that a collision could not be completely ruled out. The only solution was for mission controllers to boost the satellite out of the trajectory of the debris object. And even though the collision avoidance maneuver was planned, it ultimately proved to be unnecessary. However, the close encounter did highlight the growing risk of space debris. In another incident, a piece of space debris left over from a 2007 Chinese missile test collided with a Russian satellite on January 22, 2013, rendering the satellite unusable.

While use of Whipple shields can counter debris particles smaller than 1 cm and improved debris tracking and orbit prediction can reduce the likelihood of collision and improve threat avoidance via maneuvering, effective debris removal or clearing strategies will eventually be necessary given the pace with which the LEO space is being populated by a variety of spacecraft, including surveillance and spy satellites that are a vital component of homeland security.

4.5.3 Laser Orbit Debris Removal

The concept of laser-based orbit debris removal is based on the use of directed-laser energy to alter the trajectory of debris in low-earth orbit. The directed-laser energy pulse on hitting the object ablates and vaporizes a thin layer of the material surface, thereby creating plasma and an exhaust plume. The exhaust plume leaves the object surface at such high velocities that it generates enough force to push the object either into a new orbit or to cause it to reenter the atmosphere. In the case of a laser orbit debris removal operation, the targeted debris particle or object is first detected and tracked. A pulsed laser either from a ground facility or a spaceborne platform is transmitted in the direction of the debris particle. The laser beam is in fact focused on the particle generating a plasma jet. Most of the laser energy goes into the jet. The engagement is so designed as to point the jet in the right direction to slow the target by about 100 to 150 m/s on average. The decrease in velocity is just right to drop a perigee of the debris particle to 200 km, which is adequate for rapid reentry. Figures 4.13 and 4.14 illustrate laser orbit debris removal from a ground-based laser facility and spaceborne laser, respectively. The fluence on a target required for debris clearing by a laser is typically 5 to 10 J/cm^2. The lasers used for this purpose are high-energy pulsed lasers generating pulse energy in the 50- to 150-kJ range, with a pulse width of 5 to 10 ns at a repetition rate of 5 to 10 Hz. The required pulse energy and average laser power primarily depends on the size of the targeted debris particles.

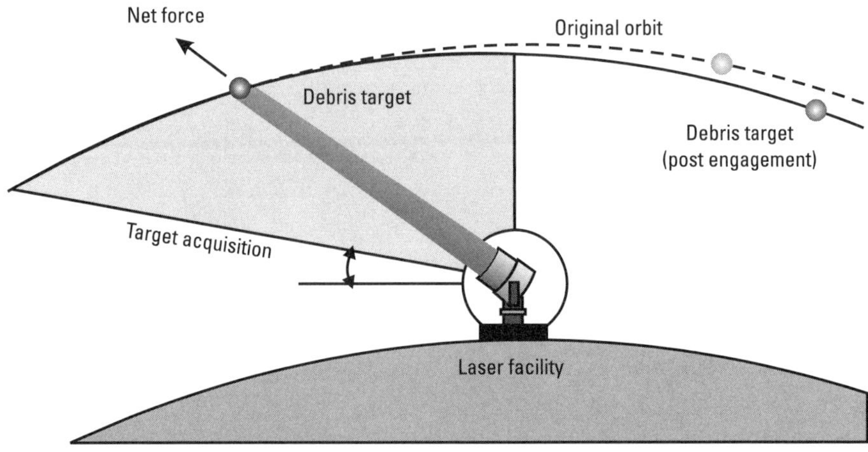

Figure 4.13   Laser orbit debris removal with a ground-based laser.

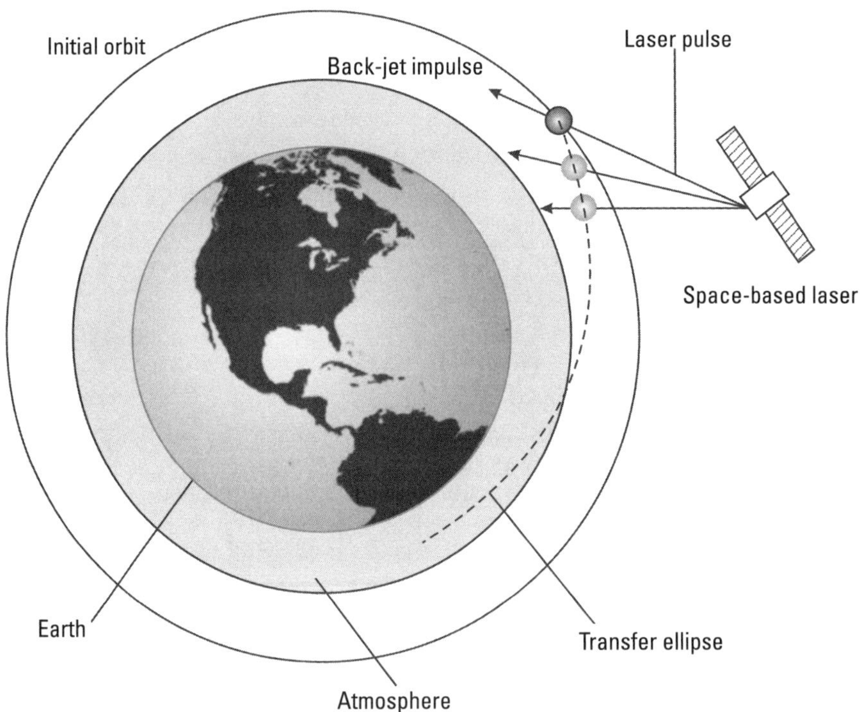

Figure 4.14   Laser orbit debris removal from a spaceborne laser facility.

Laser orbit debris removal uses two possible modes of operation: *direct ablation mode* and *ablation back-jet mode*. Direct ablation mode is used in the case of smaller debris particles. The laser energy is used to burn down debris particles. In the case of ablation back-jet mode of clearing larger debris of the centimeter class, the laser beam is aimed at the debris and the laser energy is used to transfer the orbit of debris to facilitate its reentry into the atmosphere to be subsequently burned down. Laser-based debris removal has the advantage that high-power laser technology is mature. Both ground- and space-based laser equipment have already been used for the purpose. The ground-based facility has the disadvantage of attenuation in pulse energy due to atmospheric propagation. This limitation is overcome in the case of a space-based laser. In addition, a space-based laser can track and target the debris with a much larger field of view. However, a space-based laser would be far more expensive to launch and operate.

4.6 Directed Energy Lasers for Ordnance disposal

Disposal of unexploded ordnances, including surface-laid mines, IEDs, grenade shells, artillery/mortar rounds, and cluster bombs, from safe stand-off ranges using a high-power laser beam is an emerging application of directed-energy laser systems in homeland security related applications. Ordnance is disposed of by focusing a high-power laser beam on the ordnance casing, thereby heating it until the temperature of the backplane of the casing exceeds the ignition temperature of the explosive filler. The explosive filler ignites and begins to burn, and the process is independent of the type of fusing used by the target explosive. This leads to a low-level detonation or deflagration rather than a full-power detonation. The advantages of using laser energy for ordnance disposal include a large magazine, high precision, controllable effects with reduced collateral damage, and assured and fast disposal from safe standoff ranges.

4.6.1 Application Scenarios

A laser ordnance neutralization system is being considered for a range of application scenarios. One possible application relevant to homeland security is disposal of IEDs, also known as roadside bombs. A laser-based disposal system offers a quick, safe, and reliable method of neutralizing these roadside bombs from a safe standoff distance more than 150m. Another possible application is in disposal of friendly explosives that have outlived their shelf life. Since full-scale conventional wars do not occur very often, there is a large inventory of ammunition stored in ammunition depots that have exceeded their shelf lives and are waiting to be neutralized. Neutralization of such large quantities of ammunition is not only a cumbersome exercise, but it also poses a safety hazard. Laser-driven disposal of unexploded ordnances is safe and also allows rapid neutralization of large quantities of explosive materials.

4.6.2 Deflagration and Detonation

The behavior of chemical explosives can be different from what is expected from their explosive nature depending on the conditions to which they are exposed. The decomposition processes of an explosive compound include combustion, deflagration, and detonation. While combustion is a chemical process in which a fuel reacts with an oxidizer and deflagration and detonation are the two ways by which energy may be released

Deflagration is an exothermic process in which the transmission of decomposition reaction is based on thermal conductivity. The reaction proceeds radially outward in all directions through the available fuel away from the ignition source. More and more surface area meets the fuel as the volume of the reaction zone expands. The reaction gathers energy with time. This process occurs at speeds depending largely on the chemistry of the fuel and may vary from 1 to 10 m/s in gasoline vapors mixed with air to hundreds of m/s in black powder or nitrocellulose propellants. Deflagrations do not exceed 1,000 m/s and remain subsonic, and therefore may be termed as thermally initiated reactions propagating at subsonic speeds. The pressure in the case of the deflagration process depends on the nature and geometry of fuels involved and the failure pressure of confining structure. It may vary from 0.1 psi to a few thousands of psi. Times of development are of the order of a thousandth of a second to a half-second or more. Maximum temperatures are of the order of 1,000°C to 2,000°C. The effect of a deflagration process is to push, shove, and heave, often with very limited shattering and little production of fragmentation. In this case, as the reaction proceeds, the constituents are likely to have sufficient time to move in response to the pressure buildup and get discharged or expelled.

Detonation is a reaction involving chemically unstable molecules that when energized instantaneously split into many small pieces, subsequently recombining into different chemical products releasing very large amounts of thermal energy. The speed of reaction in this case is supersonic with 1,000 m/s considered as the minimum speed that distinguishes deflagration from detonation. Detonation speeds, times of development, involved temperatures, and pressures, respectively, are of the order of 1,000 to 10,000 m/s, a millionth of a second and 3,000°C to 5,000°C, and 10,000 psi to 100,000 psi. High explosives such as TNT, nitroglycerine, picric acid, and dynamite are materials that are intended to function by detonation. Detonation tends to shatter, pulverize, and splinter nearby materials with fragments propelled away at a very high speed. However, these high explosives can be made to deflagrate rather than detonate by intentionally slowing down the process, thereby making their neutralization or disposal safe.

4.6.3 Representative Systems

There can be various possible variants of the ordnance neutralization system. One configuration is the stand-alone high-power laser system mounted on a vehicular platform. ZEUS-HLONS and Laser Avenger are examples of this category. In another possible variant, the system may be mounted on a remotely controlled vehicle with a robotic arm. While the laser system along with its power and thermal management systems could be mounted on the vehicular platform, the laser delivery system could be mounted on the robotic arm. In addition to these configurations, a high-power laser system may be integrated with other equipment such as ground penetration radar for detection of buried explosives. The integrated system allows for detecting and neutralizing both surface-laid as well as buried explosives. In some cases, a laser ordnance disposal system is mounted on a remotely controlled weapon station to get both defensive as well as offensive capability. Rafael's high-energy laser weapon system called THOR, developed in Israel to defeat IEDs, roadside bombs, unexploded ordnance (UXO), and other potentially hazardous explosives, is an example of this type of system.

The *ZEUS-HLONS* (HUMMWV Laser Ordnance Neutralization System) of the United States, is a vehicle-mounted laser ordnance neutralization system. The concept of neutralization of live ordnances using laser energy was first demonstrated in the field in 1994 with the development and field testing of MODS that employed a 1.1-kW arc lamp driven solid-state laser mounted on an M113 A2 armored personnel carrier. The ZEUS system initially employed a 500W laser. The latest version of the ZEUS system, the ZEUS-II, consists of a multikilowatt-class laser, beam director, operator's station, and all support subsystems on a single vehicle. The system can be integrated with M1114 HMMWV, M-ATV, or other mine resistant ambush protected (MRAP) vehicles. The ZEUS has been extensively field evaluated against more than 40 different types of ordnance, which includes landmines, improved conventional munitions, mortar rounds, rifle grenades, rockets, and artillery projectiles, ranging from small plastic landmines to large, thick-walled 155-mm projectiles and 500-pound general-purpose bombs. The system has demonstrated its effectiveness in a variety of missions including counter-IED operations, clearing landmines, clearing UXO from battlefields or during peacekeeping missions, clearing active and formerly used defense sites, and clearing active, test, and training ranges of exposed UXO.

Another laser system designed for disposal of unexploded ordnances is the *Laser Avenger* system from Boeing combat systems (Figure 4.15). Boeing successfully tested a 1-kW solid-state laser weapon mounted on a converted antiaircraft vehicle in Redstone Arsenal in Huntsville, Alabama, in September 2009 by neutralizing multiple types of IEDS, including large-caliber artillery munitions, smaller bomblets, and mortar rounds. The system was operated at

Figure 4.15   The Laser Avenger ordnance neutralization system.

safe distances from the targets under a variety of conditions including different angles and ranges. The test follows earlier field tests in which Laser Avenger was used to neutralize five targets representing unexploded ordnance and IEDs and after that used to shoot down UAVs. The ordnance neutralization test was carried out in 2007 with an earlier relatively lower-power version of the system. The UAV shoot-down tests were carried out in 2008 and 2009 at White Sands Missile Range, New Mexico, in which *Laser Avenger* shot down a small unmanned aerial vehicle in each event. *Laser Avenger* used a 1-kW solid-state laser system mounted on a military Humvee that is usually equipped with Stinger antiaircraft missiles. During the tests, the Laser Avenger's advanced targeting system acquired and tracked three small UAVs flying against a complex background of mountains and desert, shooting down one of the UAVs.

Rafael's *Thor* (Figure 4.16) is a laser ordnance neutralization system with an integrated remotely controlled weapon station. The system is comprised of

Figure 4.16   Thor ordnance neutralization system.

a high-energy laser along with its beam director and a coaxial 12.7-mm M2 machine gun. Thor uses an air-cooled 700W solid-state laser that offers continuous laser engagement with no cooldown time requirement. The upgraded version of Thor employs a 2-kW water-cooled laser with a continuous duty cycle. The M2 machine gun is used to perform twin functions: as a standoff disrupter, destroying fusing, thick-cased munitions, and booby traps, and to provide accurate, direct fire on enemy forces and targets in either an offensive or defensive role. This dual capability enables Thor to be used for offensive and defensive purposes, as well as for safe standoff removal of explosive obstacles by laser-directed energy or projectile kinetic energy. While use of a directed high-energy laser gives the operator ability to neutralize the IED's content by means of burning, deflagrating, or detonating the explosive, the machine gun may be used to neutralize the IEDs by targeting the operating device by cutting a wire or detonating cord. The IED can be subsequently retrieved by a robot for further neutralization and investigation. The system is modular and can be installed on a variety of vehicles and weapon stations as an add-on system.

Explosive ordnance disposal (EOD) robots with a variety of payloads such as cameras, GPS, motion detectors, and X-ray scanners have been in use around the world. These systems are used to assist the explosive ordnance disposal units to detect and dispose IEDs and bombs. Some representative EOD robots include the iRobot 510 PackBot from iRobot Corporation, the TALON family of robots from QinetiQ North America, the Andros range of military robots produced by Remotec, a subsidiary of Northrop Grumman Laser, the Cobham tEODor (Telerob's explosive ordnance disposal and observation robot), and Dragon Runner, originally designed by the U.S. Marine Corps (USMC).

The *iRobot 510 PackBot* robotic platform (Figure 4.17) executes nondestructive inspection and detection of explosives or bombs. More than 4,500 PackBots are in service with military and security forces around the world with

Figure 4.17 iRobot 510 PackBot EOD.

a large number in operation in Iraq and Afghanistan. The TALON family of robots are designed for explosive ordnance and improvised explosive device disposal, reconnaissance, communications, CBRNE, heavy lift, defense, and rescue operations. More than 4,000 TALON robots are in use. *Andros robots* are also in operation in large numbers around the world to counter evolving threats such as IEDs. The *tEODor system* (Figure 4.18) is another well-established and widely used EOD robot that integrates a programmable six-axis manipulator, firing systems, a built-in diagnostic system, and storage space for additional EOD systems. The *Dragon Runner* EOD robot available in small and micro versions is an all-terrain robot capable of assisting bomb disposal units to detect and dispose of IEDs and bombs.

Explosive ordnance disposal systems of the future will be configured on remotely controlled robotic platforms employing lasers to dispose of explosive devices. RE2 Robotics is developing such a robotic system to inspect airfields for unexploded IEDs under the service's rapid airfield damage repair program. Reportedly, it is proposed to position the lasers on top of a MRAP vehicle.

4.7   High-Power Lasers for Neutralization of Hideouts

Another possible application of kilowatt-class high-power lasers commonly used for ordnance disposal can be in flushing out of terrorists and militants taking refuge in hideouts. Quite often, these miscreants are engaged with armed forces and security agencies in a close-quarter battle. Such engagements often continue unabated for hours. Being inside the hideout, the terrorists seem to be in an advantageous position in this engagement, which increases the probability of more casualties on the part of armed forces.

Figure 4.18   tEODor EOD system.

Figure 4.19 Use of a laser to set a hideout on fire

A kilowatt-class laser, typically 1 to 5 kW, if focused on the fire-sensitive parts of the hideout, such as the exit points and windows, could set it on fire. This would force the terrorists to either perish or surrender. Figure 4.19 illustrates the operation. Although high-power solid-state or fiber lasers can be used for the purpose, carbon dioxide lasers emitting at 10.6 μm are preferred for such an operation.

Selected Bibliography

Accetta, J. S., and D. L. Shumaker, *The Infrared and Electro-Optic Systems Handbook,* Volume 7, Revised Edition, Bellingham, WA: SPIE International Society for Optical Engineering, 1998.

Hecht, J., *Understanding Lasers: An Entry Level Guide*, 3rd Edition, Hoboken, NJ: Wiley-IEEE Press, 2008.

Maini, A. K., *Handbook of Defense Electronics and Optronics*, Hoboken, NJ: John Wiley & Sons, 2018.

McAulay, A. D., *Military Laser Technology for Defense*, Hoboken, NJ: Wiley-Interscience, 2011.

Perram, G. P., *An Introduction to Laser Weapon Systems*, Directed Energy Laser Society, 2009.

Titterton, D. H., *Military Laser Technology and Systems*, Norwood, MA: Artech House, 2015.

Wilson, C., *Improvised Explosive Devices (IEDs) in Iraq and Afghanistan: Effects and Countermeasures*, Congressional Research Service Report for Congress, 2008.

5

Sighting Observation and Surveillance Devices

In an overview of sighting, observation, and surveillance devices presented in Chaper 1, emphasis was placed on the importance of different types of weapon sights, laser range finders and laser fencing, different types of security cameras, covert listeners, sniper locators, and devices for detection of concealed weapons. This chapter describes at length devices under different categories of sighting, observation, and surveillance devices. Beginning with different types of weapon sights for use on small arms and armored fighting vehicles, the subsequent sections discuss at length observation, monitoring, and surveillance devices such as laser range finders, laser fencing, security cameras, sniper locators, and detectors of concealed weapons.

5.1 Sighting Devices

Sighting devices are integral to all weapon platforms, which includes small arms, antiaircraft guns, mortars, armored carriers, and main battle tanks where they are used for alignment and aiming of weapons. Common types of weapon sights include iron sights, telescopic sights, reflex sights, holographic sights, red dot sights, panoramic sights, periscopic sights, thermal sights, and laser sights.

5.1.1 Iron Sights

An *iron sight* typically is comprised of two components: the front sight fastened toward the front of the barrel and a rear sight fastened near the back of the barrel (Figure 5.1). The front and rear sights may be fixed as was the case in the past or adjustable to accommodate for range and windage. The name iron sight has its origins in the early days when the two parts of the sight were made of iron. Steel or polymer plastic are the materials of choice in present-day iron sights. In most iron sights, the two pieces are in the shape of a bead, a post, or a ring, or a combination of these. For aiming, the operator needs to align the front sight, the rear sight, and the target. The two main categories of iron sights include *open sights* and *aperture sights.*

In an *open-type sight,* the rear sight is merely a piece with a notch cut in the middle. The notch is typically V-shaped, square-shaped, or U-shaped. The front sight is usually some kind of ramp or post or a bead on post. To use the sight, the post or bead is positioned both vertically and horizontally in the center of the rear sight notch. Figure 5.2(a–e) show different types of open sights. In an *aperture sight,* the rear sight employs an aperture usually in the form of an extremely thin ring as in the case of a ghost ring sight or a large disk with a pin-sized aperture. While the former type of aperture sight is faster, the latter is more precise. In general, the thicker the ring, the more precise the sight, and the thinner the ring, the faster the sight. Figure 5.2(f) shows a ghost ring sight. Iron sights are simple with no associated optics or electronics, and are weather-resistant, durable, lightweight, and low cost. The limitations include lower precision as compared to other types of sights, lower range of operation due to absence of any magnification, and reduced field of view.

5.1.2 Telescopic Sight

A *telescopic sight* is an optical sighting device based on the principle of operation of a refracting telescope and is mounted on firearms for target viewing and aiming. A graphical image pattern called a reticle placed at an appropriate position in the optical system is used as an aiming point. Important components of a telescopic sight include the ocular lens or eyepiece, eye relief, reticle, magnification assembly with first and second focal planes, objective lens, and a tube. The

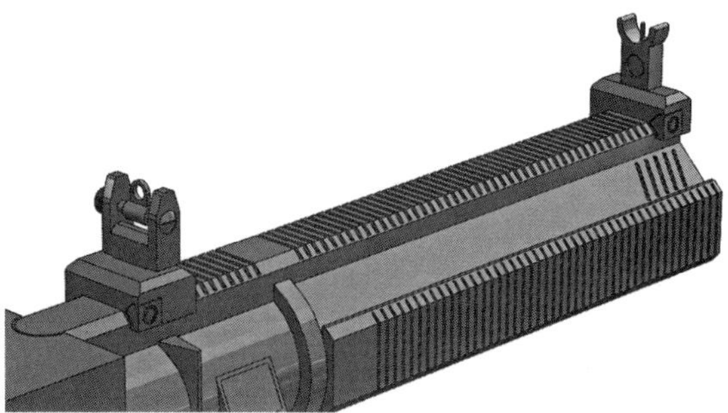

Figure 5.1 Iron sight.

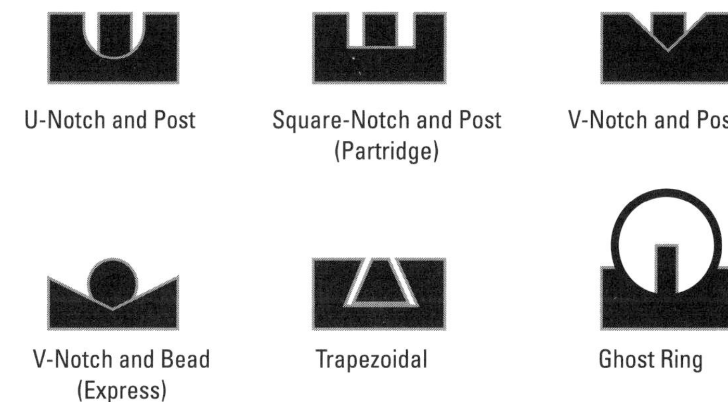

Figure 5.2 Iron sights: (a) U-notch and post, (b) square notch and post, (c) V-notch and post, (d) V-notch and bead, (e) trapezoidal notch and ramp, and (f) ghost ring.

ocular lens generally has a screw for fine adjustment to enable users to focus the scope to their eyesight. The eye relief allows some distance, typically 25 to 100 mm, between the eye and the ocular lens to get a good and unblurred sight picture. Reticle, which could be a simple crosshair, mil dot, half mil dot, or MTC-SCB reticle, is used to aim the firearm. The magnification system is comprising a set of lenses, which generally has first and second focal planes. If it only has a first focal plane, then the size of the reticle also increases with magnification. If it has first and second focal planes, then the size of the reticle remains the same irrespective of magnification. The primary function of the objective lens is to gather as much light as possible. However, a smaller objective gives a larger depth of field. The tube holds together the objective, the ocular lens, and other components, such as the reticle and various adjustment controls. The adjustment controls mainly include *Dioptre control, elevation control* used for vertical adjustment of reticle, *windage or horizontal adjustment control* of the reticle, *magnification control, illumination adjustment control* of the reticle, and *parallax compensation control.* The scope is filled with dry nitrogen to allow operation at lower temperatures.

There are two types of telescopic sights: *Keplerian telescopes* and *Galilean telescopes.* In both types, the objective is converging as a whole even though it is made up of both converging and diverging elements. The two telescopes differ in the type of lens used as the eyepiece, which is converging in the Keplerian telescope and diverging in the Galilean telescope. Telescopic sights used on firearms are mostly Keplerian telescopes. Figure 5.3(a) illustrates the principle of operation of a Keplerian telescope. Angle (α) here is the angle the target would subtend at the eye. If the target were at infinity, angle (α) would be zero. The objective forms a diminished inverted image. The eyepiece with its short focal length magnifies the inverted image. The distance between the objective and

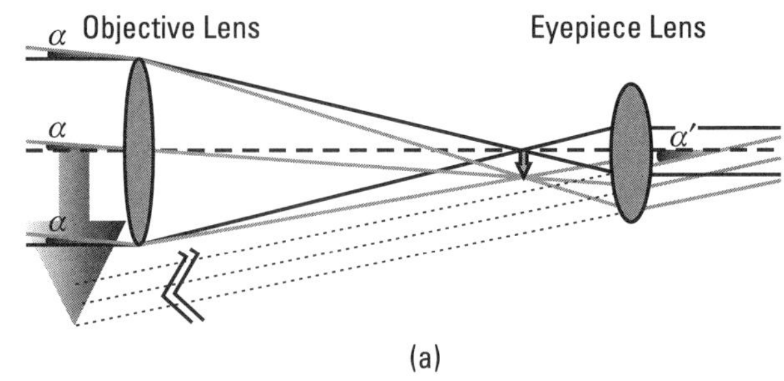

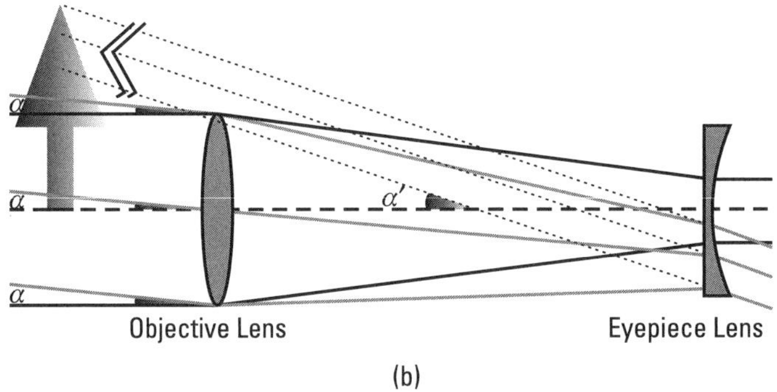

Figure 5.3   Operating principle: (a) Keplerian telescope, and (b) Galilean telescope.

eyepiece is the algebraic sum of the two focal lengths. Keplerian telescopes usually have an optical system between the eyepiece and the objective (not shown in the figure) to invert the inverted image again to produce an erect image. The Galilean telescope has a converging lens as the objective and a diverging lens as the eyepiece. The distance between the objective and eyepiece is again equal to the algebraic sum of the two focal lengths. Note that the two focal lengths have the opposite sense. The distance between the objective and eyepiece in this case is therefore equal to the difference of the two focal lengths. Figure 5.3(b) illustrates the principle of operation of a Galilean telescope. As is evident from the ray diagram, the telescope produces a magnified erect virtual image. A Galilean telescope offers an extremely narrow field of view, thereby limiting its usability to a magnification of about 30.

Figure 5.4 shows a packaged Keplerian telescope. This is how a typical telescopic weapon sight would look like from inside. Different components of the telescopic sight such as the objective and ocular lenses, the image reversal assembly and the associated first and second focal planes, the elevation adjustment control, and the tube can be seen in the diagram. The windage and other controls are not visible in this schematic. Figure 5.5 shows a typical telescopic sight.

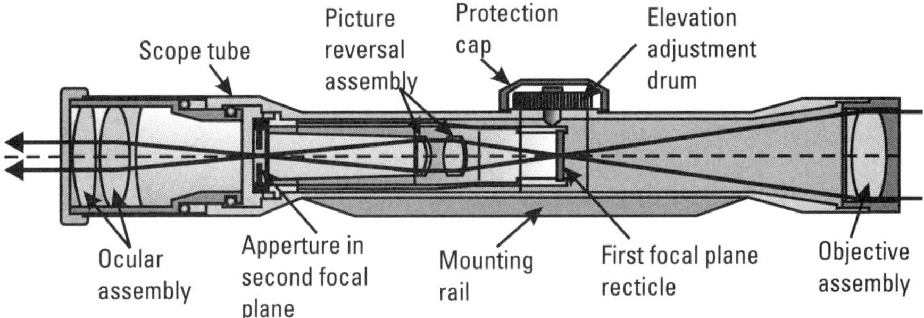

Figure 5.4 Inside a telescopic sight.

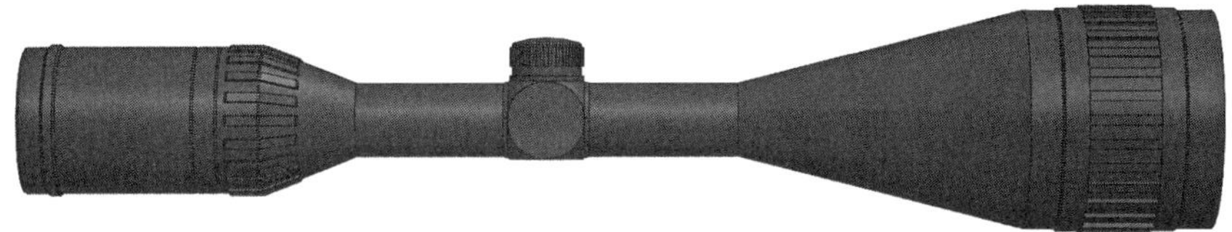

Figure 5.5 Packaged telescopic sight.

5.1.3 Reflex Sight

A *reflex sight,* also called a *reflector sight,* allows the user to see a projection of an aiming point or some other image superimposed on its field of view at infinity. The reticle or image pattern to be projected is placed at the focus of a lens or curved mirror. A type of reflector is used to allow the user to see a virtual image in front of him or her at infinity and the field of view at the same time either by bouncing the image created by the lens off a slanted glass plate (Figure 5.6(a)), or by employing a curved glass reflector, as shown in the two commonly employed arrangements of Figure 5.6(b, c), which is mainly transmitting and only partially reflecting. The reflex sight is immune to parallax and other sighting errors mainly because the reticle at infinity stays in alignment with the firearm or any other device the sight is attached to regardless of the viewer's eye position. Figure 5.7 shows a typical reflex sight.

5.1.4 Red Dot Sight

A *red dot sight* is a type of reflex sight in which the aim point is in the form of an illuminated red dot type reticle produced by a red emitting diode placed at the focus of collimating optics (Figure 5.8(a)). Figure 5.8(b) shows a cutaway view of a typical red dot sight using concave reflective optics. The concave optics here reflects red light but transmits all other wavelengths. As the user looks through the tube, he or she sees a projection of a red dot on the target. Like

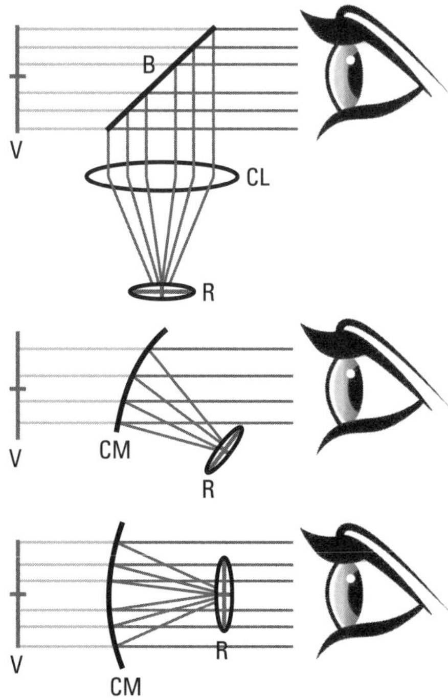

Figure 5.6 Optical arrangement in reflex sight (a) collimating lens and (b) partially reflecting slanted glass, and (c) partially reflecting curved mirror.

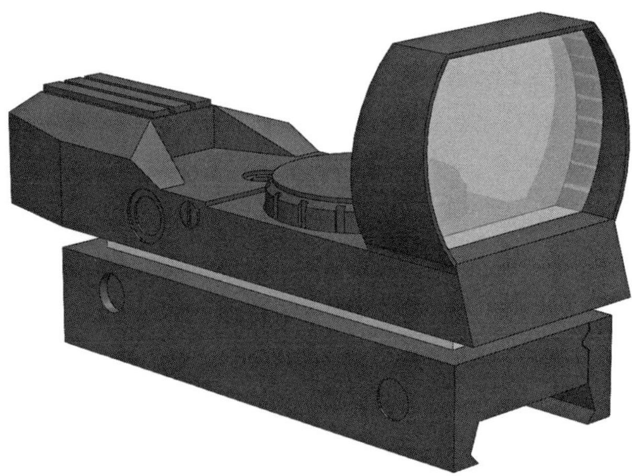

Figure 5.7 Reflex sight.

reflex sight, in this case the reticle also stays aligned with the weapon or any other device the sight is attached to irrespective of eye position. To make sight or point-of-impact adjustments, the inner tube housing with the light-emitting

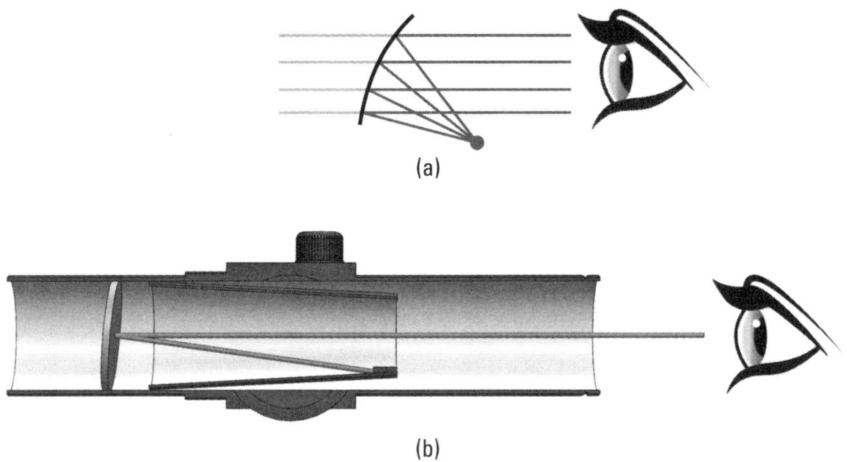

Figure 5.8	Red dot sight (a) operating principle, and (b) cutaway view of a typical red dot sight.

diode is moved either horizontally or vertically to adjust elevation and windage. While dot sights come in both tube and open designs, the former design has the advantage of being more durable and precise with superior zero holding and fewer stray reflections and offering better mounting options to adjust elevation and windage.

5.1.5 Holographic Sight

The functioning of a *holographic weapon sight* is based on the use of laser-driven holographic technology. A laser transmission hologram of a 2-D or 3-D image of a reticle is recorded and embedded into an optical glass window that forms a part of the holographic sight. The operator while looking through the glass window sees an image of the reticle when illuminated by a laser at a distance on the target plane within the field of view. The sight uses a holographic grating in the optical path to compensate for any changes in laser wavelength due to temperature. The grating disperses the laser light by an equal amount in a direction opposite to the one dispersed by the hologram forming the aiming reticle. The sight can be adjusted for range and windage by simply tilting or pivoting the holographic grating. The laser diode and the optics are also part of the holographic sight. Figure 5.9(a) shows the optical arrangement of a typical holographic sight, which is self-explanatory. Figure 5.9(b) shows a typical packaged holographic sight. In comparison with the telescopic and red dot sights, a holographic weapon sight is the preferred choice for shooters. Holographic sights give to the shooter ability to quickly lock in on a target even if the shooter's head position is not aligned with the sight. With a telescopic sight, a shooter needs to align his or her eye with the sight to place the reticle over the target; otherwise,

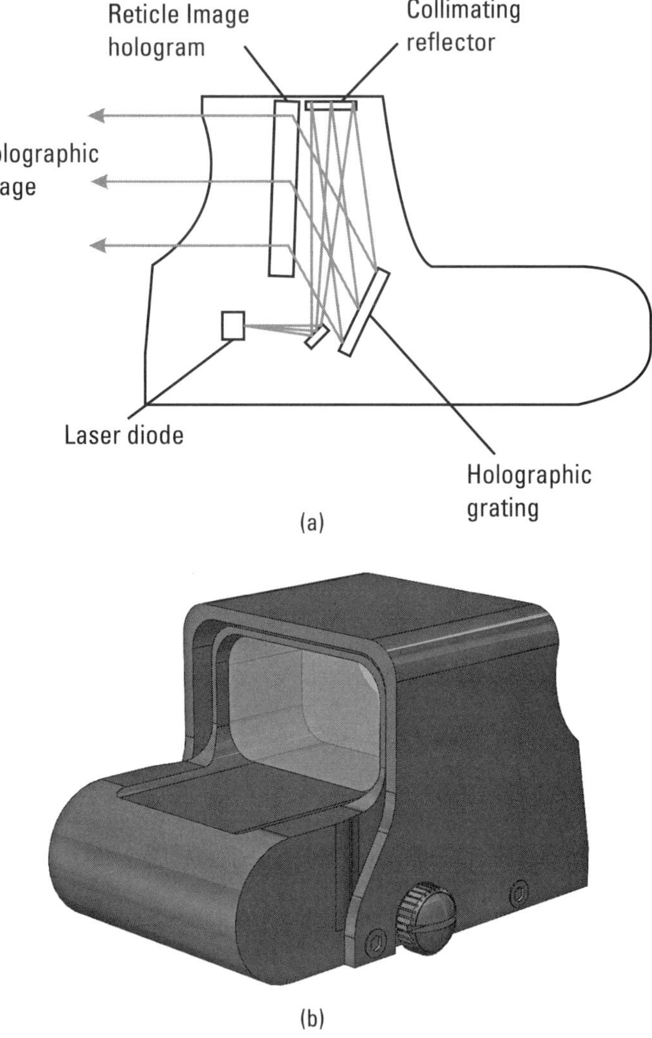

Figure 5.9 Holographic sight (a) optical arrangement, and (b) typical package.

the sight's aiming reticle and the weapon would be pointing at two different spots on the target. In a holographic sight, the reticle always stays aligned with the weapon as it is moved left or right.

5.1.6 Panoramic Sight

Panoramic sight is a type of artillery sight with a large field of view. It permits the gunner to sight in all directions without moving his or her head while laying an artillery piece for direction. The sight can be rotated full circle through 360°(or 6,400 mils), enabling the operator to view the scene all around him or her. The artillery panoramic sight is fitted with derotation optics to prevent the image from being inverted after 180°(or 3,200 mils) of rotation. Panoramic sights with their gyro systems and additional derotation optics are more complicated and expensive than fixed sights. As well, these additional optical components reduce transmission. Nevertheless, panoramic sights are essential equipment in a modern battle tank if it were to acquire and engage moving targets while itself being in motion.

5.1.7 Periscopic Sight

A *periscopic sight* enables an operator to observe intended surroundings in the absence of direct line of sight. It also allows the operator to remain under cover or behind armour or in a submerged location while viewing the scene around him or her. In its simplest form, a periscopic sight is comprised of two mirrors placed parallel to each other and at 45° to the axis of the tube, or two reflecting prisms to bend the direction of light coming from the scene to be observed twice by 90°. Figure 5.10(a, b) shows the basic periscope using mirrors and right-angle prisms, respectively. The simple periscope shown in Figure 5.10 produces no magnification and gives no crossline or reticle pattern. Most periscopic sights include a telescopic optical system to give magnification and wide field of view and a crossline or reticle pattern to establish the line of sight to the object under observation. In some cases, the sight also has devices for estimating the range and course of the target and for taking photographs through the periscope. Periscopic sights are extensively used in armoured fighting vehicles

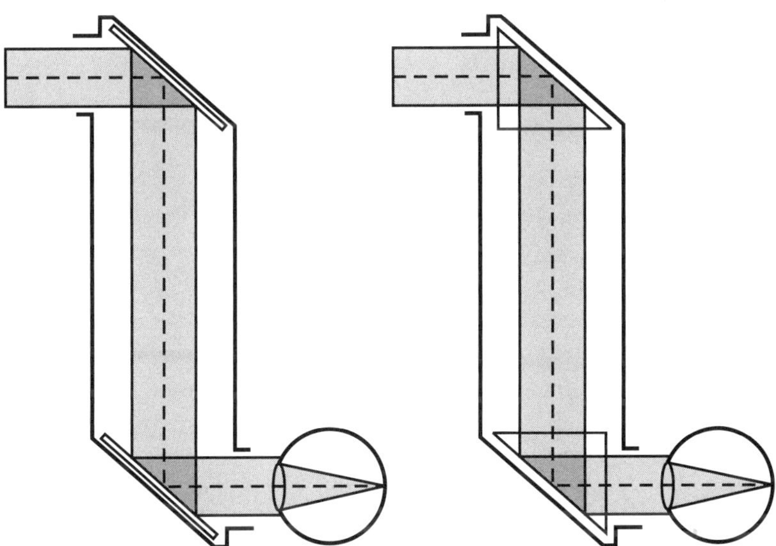

Figure 5.10 Basic periscope (a) using mirrors, and (b) using prisms.

and battle tanks enabling a vehicle crew to observe their surroundings through the rooftop. These sights, when fitted with an auxiliary telescopic gun sight, can also be used for pointing and firing of guns. Periscopic sights are also used in submarines submerged at a relatively shallow depth to observe their surroundings on the surface of the water and in the air for any threats.

5.1.8   Thermal Sight

Thermal sight operates on the principle of infrared thermography that converts thermal information and the associated infrared wavelengths emitted from the object into an image. All objects that have temperature above absolute zero emit heat. A thermal weapon sight combines a compact thermal camera and an aiming reticle. A thermal sight or thermal camera makes use of the contrast between the thermal signatures of the intended object and its surroundings to construct an image of the object. Thermal riflescopes and weapon sights read off thermal signatures of all objects and present the user with an image showing a gradient map of the heat signatures. These are extensively used on a variety of small arms as well as some heavier weapons. Thermal sights not only offer the flexibility of daytime and nighttime use, they can also be used in adverse weather conditions such as haze, fog, and smoke when other optical devices simply fail. Most modern thermal riflescopes feature several color palates as well as the standard black and white. Another useful feature of thermal riflescopes or sight is their ability to output video signal. Thermal sights are extensively used by security and law enforcement agencies. With an integrated portable recorder, the thermal sight has the capability to record the action, which is of relevance in a homeland security scenario. Thermal imaging technology is discussed further in Chapter 6.

5.1.9   Laser Sight

A *laser sight,* also called a *laser aiming device,* mounted on a firearm is a very common and well-established application of a laser device. Laser aiming modules are usually configured around low-power CW semiconductor diode lasers emitting in visible (usually red or green wavelength) or infrared. The CW power level is in the range of 5 to 10 mW. Most laser sights use red laser diodes emitting at either 635 or 650 nm. Green laser-based, diode-pumped solid-state lasers emitting at 532 nm are also presently in use. One of the limitations of using visible lasers for aiming and targeting is that they are visible to the naked eye and thus inhibit a covert operation. Laser aiming modules configured around infrared diodes to produce an aim point on the target invisible to the naked human eye are now available. These aim points are detectable with night vision devices usually fitted to the firearm.

The aiming module is aligned to emit a laser beam parallel to the barrel. Due to an extremely low value of divergence of the emitted laser beam, it makes very a small spot even at long distances up to hundreds of meters. As an illustration, full-angle divergence of 0.5 mrad would produce a spot diameter of 50 mm at a 100m distance to the target. The user places the spot on the desired target and the barrel of the gun is aligned, not necessarily allowing for bullet drop, windage, distance between the direction of the beam and barrel axis, and the target mobility during the travel time of the bullet. Dual-wavelength laser aiming devices emitting at a visible wavelength, usually red, and a near-infrared wavelength are also commercially available. These modules are equipped with a mode select switch that allows the user to select either one wavelength at a time or both wavelengths simultaneously. Some devices are also equipped with a mechanism to provide adjustment for windage and elevation. LAM-10D, LAM-3G, LAM-10M, IRIL-1000M, and NCFL-9 from Newcon Optik and AN/PEQ-6 from Insight Technology are some representative examples of laser sights. Note that most devices intended for a similar role have comparable technical specifications. LAM-10D is an infrared laser aiming module emitting at 850 ± 10 nm with a range of 1,000m. LAM-3G is a three-channel laser aimer/illuminator with a visible laser aimer emitting at 532 ± 10 nm, a beam divergence of 0.5 mrad and a maximum range of 500m, an infrared laser aimer emitting at 830 ± 15 nm, a beam divergence of 0.5 mrad with a maximum range of 2000m, and an infrared illuminator emitting at 830 ± 15 nm with a maximum range of 2000m and a beam divergence of 1 to 105 mrad. LAM-10M is a dual-wavelength laser aiming module that offers a visibility range of greater than 1000m, both infrared (830–850 nm) and visible (650 nm) wavelengths of operation, a beam divergence of 0.5 mrad, and a windage/elevation adjustment of ± 20 mrad. The IRIL 1000M is designed to reach out to extreme distances to aid in target identification and engagement from ground- and air-based platforms. The module emits at 810 ± 10 nm and has an adjustable beam divergence between 1.5 and 60 mrad, allowing for use as either an infrared laser pointer or illuminator. NCFL-9 (Figure 5.11) is a tactical LED flashlight with a built-in infrared laser aiming module emitting at 830 ± 10 nm to assist with

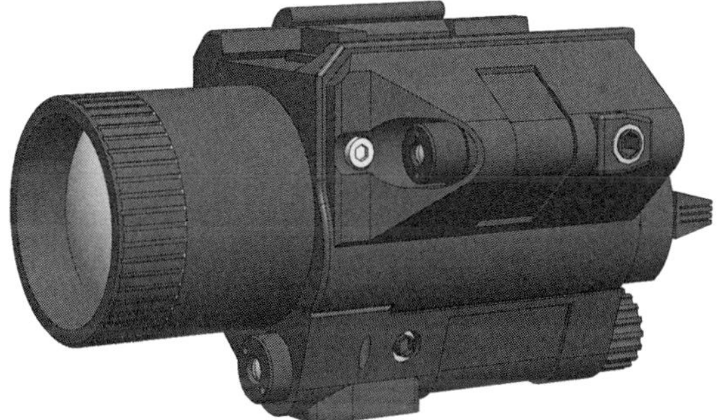

Figure 5.11 NCFL-9 flashlight with built-in laser aiming module.

precision aiming. AN/PEQ-6 from Insight Technology is a dual-wavelength laser aiming module with settable visible and infrared outputs. It is also equipped with an infrared illuminator and a white light illuminator/flashlight. It is designed for use with the Mark-23 pistol.

5.2 Surveillance Cameras

Surveillance or *security cameras* with their wide range of package configurations and features are integral to any security setup, be it in homes, offices, industries, or critical infrastructures such as ammunition depots, oil refineries, nuclear installations, highways, airports, and railway stations. Applications range from being a deterrent to theft and vandalism to traffic monitoring and crowd control and crime prevention in an event management scenario; from remote monitoring of suspicious activities to gathering evidence; and from offering perimeter security solutions in critical infrastructures to monitoring seacoasts and borders for any suspicious activities including infiltration or intrusion. The following sections briefly describe common applications, types, and selection criteria of surveillance cameras.

5.2.1 Applications

Surveillance cameras are extensively used for a variety of application areas including infrastructure protection, monitoring industrial operations, public safety, patrolling borders and seacoasts, monitoring traffic, video surveillance of public places such as railway stations, airports, highways, hotels, and restaurants, managing parking lots, vandalism deterrence, capturing and recording evidence in cases of unlawful activities, video surveillance of events, and perimeter security. As well, with video analytics usage on the rise, security surveillance has become a proactive solution rather than just a reactive elucidation. Body-worn cameras are also being used to monitor the behavior of security personnel on duty and for isolating security personnel from being dragged into unnecessary controversies. Some of the common and well-established application areas of surveillance cameras include

1. Infrastructure protection;
2. Vandalism deterrence;
3. Patrolling borders and coastal areas;
4. Event surveillance;
5. Monitoring traffic and parking lots;
6. Public safety.

5.2.1.1 Infrastructure Protection

Round-the-clock video surveillance of various locations inside the premises as well as the perimeters of businesses, industrial buildings, and critical infrastructure against intrusions and any suspicious activity is a well-established and important application of surveillance cameras. Real-time monitoring of perimeters and operations inside the premises allows security setup to quickly and appropriately respond to any suspicious and unlawful activity. Use of IP-enabled cameras allow monitoring from remote locations, which is particularly relevant to the safety and protection of critical infrastructure.

5.2.1.2 Vandalism Deterrence

Surveillance cameras have proven to be an effective deterrent to vandals and rogue elements because of the possibility of being identified on video. Use of rugged and tamperproof camera housing and high-definition cameras for facial recognition are a great asset to law enforcement agencies in keeping a vigil on suspicious activities. Video surveillance also provides law enforcement agencies with much-needed electronically recorded evidence for interrogations and for use in courts of law. In these ways, security cameras help law enforcement agencies in reducing vulnerability to vandalism, catch criminals in the act, and prevent trespassers from loitering.

5.2.1.3 Patrolling Border and Coastal Areas

Video surveillance can be a very effective tool for keeping a round-the-clock vigil on vast borders and seacoast usually spread across thousands of kilometers. A network of thermal cameras can monitor suspicious activity and keep track of intrusions on a day and night basis. State-of-the-art thermal cameras can detect an intrusion or infiltration up to few tens of kilometers away and can offer video evidence. For example, a FLIR HDC thermal camera can detect man-sized targets beyond 18 km and vehicles beyond 22 km. Real-time detection of unlawful activity such as drug and narcotics trafficking and intrusion can be promptly and appropriately responded to by security forces to keep the area safe.

5.2.1.4 Event Surveillance

Surveillance cameras play a vital role in ensuring safety and security at public events by keeping a vigil on the people gathered for the event. The use of surveillance cameras provides effective crowd control and faster response times, thereby preventing violence, harassment, and theft at events. In addition, they can keep track of the number of attendants, zoom in on suspicious acts or behavior, enhancing the evidentiary value of video in a court of law. IP cameras on PTZ mounts with their wide-swath coverage and optical/digital zoom are usually the preferred choice for such applications.

5.2.1.5   Monitoring Traffic and Parking Lots

Video surveillance cameras are being used extensively for traffic monitoring and monitoring of parking lots. If the areas are sensitive, live monitoring of security camera footage is being taken up by law enforcement agencies. Installation of a reliable video surveillance system within the confines of a parking lot enhances security and safety by keeping a tab on vehicles coming in and going out of the lot, especially if it is poorly lit. The key benefits of video surveillance in parking lots include prevention of vehicle theft, identification of unpaid and prohibited vehicles, and reduction of liability and avoidance of lawsuits in the event of collisions such as a car crash or an errant shopping cart denting a car.

5.2.1.6   Public Safety

Video surveillance can be a very effective monitoring system for ensuring public safety by monitoring a crowd at large gatherings, processions, demonstrations, and other such events. Safety is an important component to consider in public parks, communities, and neighborhoods. Video surveillance helps enhance public safety in multiple ways, including deterring harassment, vandalism, and other violent acts, thereby providing an unbiased picture of the public's behavior and making the public feel safe and secure. It also allows security personnel to respond faster to any suspicious activity or behavior.

5.2.2   Types of Surveillance Cameras

Surveillance cameras are classified in many ways, such as type of camera mount, camera housing, and technological features. Camera mount types include fixed cameras, PTZ cameras, and virtual PTZ or 360-degree cameras. With regard to camera housing, surveillance cameras are classified as outdoor, dome, bullet, and discreet (or unobtrusive) cameras. Many technological features distinguish different cameras, including resolution, day or day/night operation, visible or infrared, optical/digital zoom, and wired or wireless. Based on the above-mentioned criteria, common types of surveillance cameras include

1. Fixed cameras
2. PTZ cameras;
3. Virtual PTZ or 360-degree cameras;
4. Box-style camera;
5. Dome cameras;
6. Bullet cameras;
7. Discreet cameras;
8. Thermal cameras;

    9. IP cameras;

  10. Webcams.

5.2.2.1 Fixed Cameras

A *fixed camera* is always pointing in the intended direction. It is positioned to capture the image of a certain area. Different camera housings are used depending on application. Compared to PTZ cameras, fixed cameras are less prone to failure given the number of moving parts in a PTZ camera. As well, a fixed camera is five to ten times less expensive than a PTZ camera. With improvements in the image resolution of IP CCTV cameras, multiple fixed cameras can provide better coverage at a lower cost than a single PTZ camera. Multiple high-resolution fixed IP cameras can also be configured and deployed to cover the same area as a single PTZ camera with better results. However, a comprehensive video surveillance setup should preferably be layered with the first and second layers often comprising fixed IP cameras and fixed IP zoom cameras, and the third layer using PTZ IP cameras.

5.2.2.2 PTZ Cameras

PTZ cameras have pan, tilt, and zoom features that allow a single camera to monitor large areas as well as the fine detail of different objects in the scene. While pan and tilt features control directional movement of the camera, the optical zoom enables focusing on the fine details like faces and license plates. Use of IP PTZ cameras allows remote control of the camera from any part of the world. Modern PTZ cameras are equipped with a variety of intelligent features that enable movement between preset positions and zooming in automatically in response to detected events. Key features and advantages of state-of-the-art PTZ cameras include large field of view, built-in motion tracking, autofocus, powerful zoom, weather resistance, night vision, and simple installation. Figure 5.12 shows an assortment of PTZ cameras in some common housing and mounting arrangements.

5.2.2.3 Virtual PTZ or 360-Degree Cameras

Virtual PTZ cameras, also called *360-degree cameras* or sometimes electronic PTZ cameras *(ePTZ cameras),* are comprised of several high-resolution fixed cameras in a single, usually dome, housing. The images captured by individual cameras are stitched together to provide a full 360-degree view. A virtual PTZ camera uses a software feature to provide a large field of view and can zoom in on intended objects without any physical movement unlike a conventional PTZ camera, which uses a combination of motors for navigating through the viewing area and providing optical zoom.

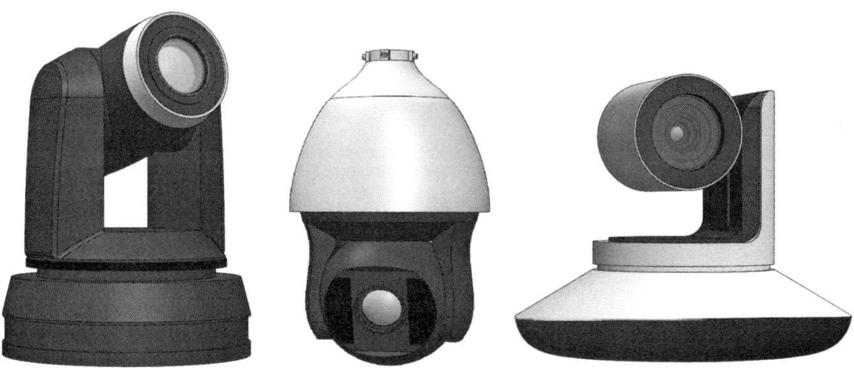

Figure 5.12   PTZ cameras.

5.2.2.4   Box-Style Camera

Box-style cameras have a boxlike shape, are stand-alone cameras, and they function like any other security camera. They are available in both environmentally friendly housings as well as indoor mounting arrangements. They are not preferred for indoor use if aesthetics is a requirement. The key benefits of using a box-style camera include lens customization, with many lens options including fixed and varifocal lens, and the availability of both the camera and the housing to match the requirements of the installation.

5.2.2.5   Dome Cameras

Dome cameras derive their name from their domelike shape. Dome cameras are commonly used in surveillance systems inside of homes, businesses, department stores, hotels, restaurants, and other places where discreet surveillance is an important requirement. Dome cameras suitable for outdoor surveillance are also common. Both fixed focal length and varifocal dome cameras are available. Dome casings not only fit unobtrusively into any setting, they also effectively disguise a camera's viewing direction. A wide variety of surveillance cameras, such as PTZ cameras, IP cameras, night vision cameras, and high-definition cameras, are available in dome housings. Figure 5.13 illustrates some common dome camera housings.

5.2.2.6   Bullet Cameras

Bullet cameras are shaped like a rifle bullet shell. Smaller bullet cameras, also known as lipstick cameras due to their resemblance to a lipstick case, are typically the diameter of a cigar with a slightly shorter length. Most bullet cameras use a triaxis mount and adapt well to mounting on ceilings or walls. Infrared bullet cameras tend to be larger in diameter to accommodate the infrared LEDs. Both fixed focal length and varifocal bullet cameras are available. Bullet

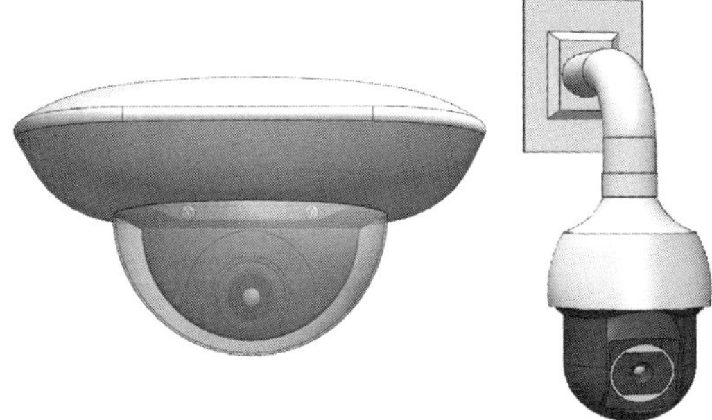

Figure 5.13 Dome cameras.

cameras are also available as IP cameras and night vision cameras. Figure 5.14 shows typical bullet camera housings.

5.2.2.7 Discreet Cameras

Discreet cameras use discreet or covert housings. These cameras are generally used either by architects seeking to achieve a certain aesthetic within a space or in situations where surveillance is done without the knowledge of targeted individuals, such as in sting operations. Another common application of these cameras is their use in offices to monitor activities of employees. These cameras are packaged as common everyday objects like lightbulbs, phone chargers, smoke detectors, and clocks.

5.2.2.8 Thermal Cameras

Thermal cameras operate on the heat radiated by an object rather than the visible light reflected off the object. All natural or manmade objects generate infrared thermal energy with the amount of emitted thermal energy depending on what the object is. Some objects, such as warm-blooded animals and certain machines, create their own heat signatures, while objects such as land, rocks, and water absorb heat from the sun during the day and radiate that heat through the night. Thus, different objects and different parts of a given object emit infrared energy with minute differences in the thermal energy. Thermal

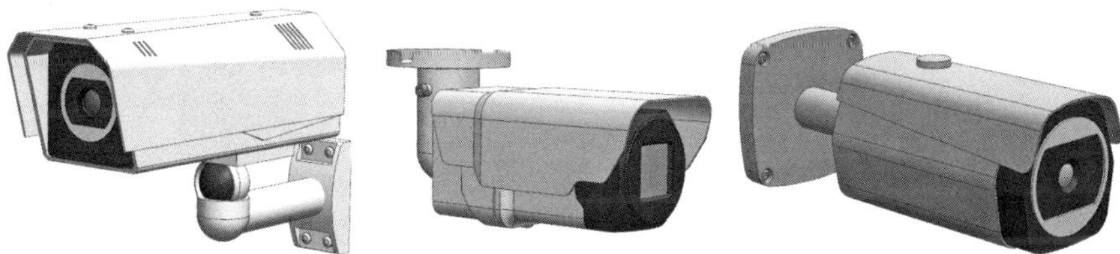

Figure 5.14 Bullet cameras

cameras construct images from these minute differences in thermal energy. The performance of a visible camera is adversely affected by atmospheric conditions such as smoke, heavy fog, or rain. Thermal imagers do not rely on visible light for their operation; these obstructions and even modest foliage do not affect their ability to create high-contrast images. With their operation depending on heat signatures, thermal security cameras are capable of day/night operation, thereby providing an undeniable tactical advantage around the clock. Thermal energy generally travels more effectively through the atmosphere as opposed to visible light, enabling it to detect potential intruders from much greater distances, giving people much more time to react and respond. Thermal cameras make it difficult for targets to remain undetected by hiding in and around trees, bushes, or foliage. As the motion detection is triggered only by moving sources of heat and not by things like debris or moving branches, the probability of motion detection false alarms is significantly reduced. Although first developed for the military, thermal imaging has now become commonplace for use with police and firefighters and in security systems. Figure 5.15 shows an assortment of thermal cameras in different package styles.

5.2.2.9   IP Cameras

IP cameras are networked digital video cameras capable of transmitting data over a fast Ethernet link. IP cameras, also called network cameras or netcams, are commonly used for IP surveillance, which is a digitized and networked version of analog closed-circuit television. Although a webcam is also an IP camera, the term IP camera is usually applied only to those used for surveillance and are directly accessible over a network connection. In a centralized mode

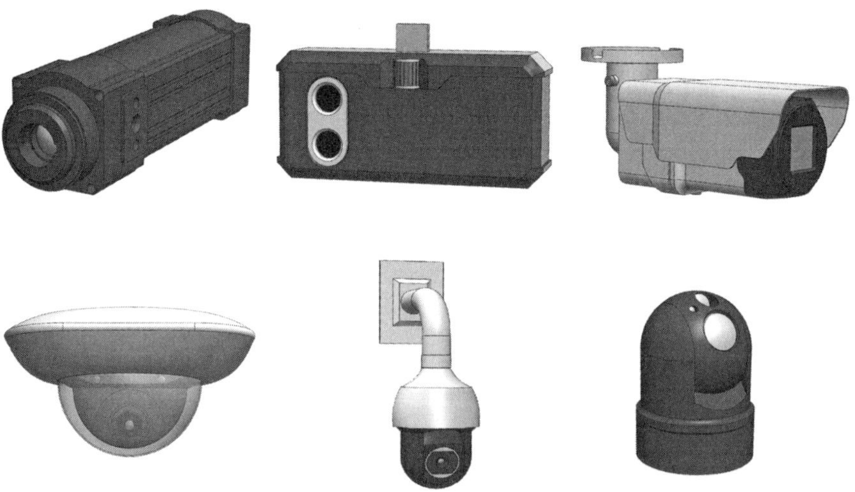

Figure 5.15   Thermal cameras.

of operation, an IP camera requires central network video recorder (NVR) to handle the recording. In a decentralized operational mode, no NVR is needed and the camera records to any local or remote storage media. Most IP cameras have built-in power over Ethernet (PoE) connectivity, making them easy to use with any NVR with a built-in PoE switch. There are a variety of IP cameras. Dome IP cameras are made for outdoor use. PTZ IP cameras provide the best viewing angles. IP cameras equipped with infrared light are meant for night-time surveillance. There are wireless IP cameras and wired IP cameras as well as high-definition and ultrahigh-definition IP cameras. Wireless IP cameras offer flexible installation options and eliminate the need for video cables. They connect to your home network for viewing and remote control using a compatible smartphone or tablet. Wired HD IP cameras with 1080p resolution deliver high-resolution security footage with smooth on-screen movement day and night. Ultrahigh-definition IP cameras with 2K (a horizontal resolution of 2,048 pixels) or 4K (a horizontal resolution of 4,096 pixels) resolution give greater clarity and enable capturing of finer details. A higher number of pixels also provide users with a better digital zoom, thereby enabling them to see further into the distance without drastically degrading image quality.

5.2.2.10 Webcams

Webcams, short form for web cameras, are digital video cameras directly or indirectly connected to a computer or computer network via internet. A webcam captures still pictures as well as motion video of the user or another object to send from its location to another location via internet. Webcams come with software that needs to be installed on the computer to help users record video on or stream from the Web. Many desktop computer screens and laptops come with a built-in webcam and microphone. An external webcam can also be connected to a computer through a USB or FireWire port on the computer. Wireless (Wi-Fi) webcams are also available. Figure 5.16 shows an assortment of common webcam packages and styles.

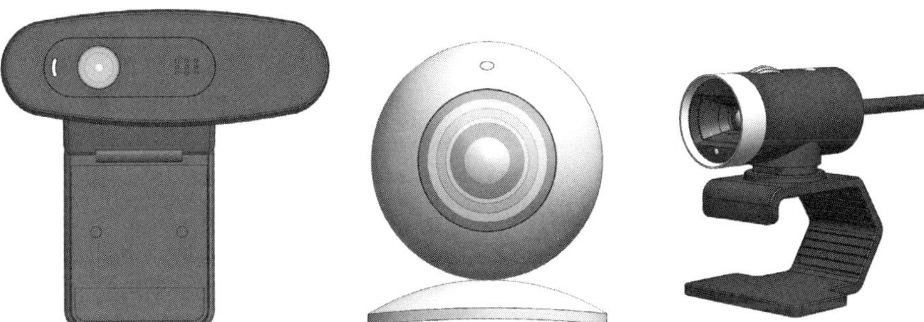

Figure 5.16 Webcams.

5.2.3 Choosing the Right Surveillance Camera

When choosing the right surveillance camera for a given application, the user needs to consider several factors. He or she must make a choice between fixed or PTZ, dome or box-style, wired or wireless, and centralized or decentralized and consider technology-related features such as resolution, capability to operate in low light conditions, and lens features. These aspects of selection criteria are briefly described in the following sections.

5.2.3.1 Fixed versus PTZ Cameras

Fixed cameras are placed in fixed positions and therefore have fixed viewing directions and angles. Because of these fixed viewing angles, the camera surveys the same area without any panning capability. Although it is fixed, this camera may also be equipped with different lenses to enhance its function. A PTZ camera on the other hand can pan, tilt, and zoom using either manual or automatic controls, which allows a single camera to carry out surveillance from different directions and multiple angles. This allows continuous tracking of an object of interest and zooming in on a specific point when required.

5.2.3.2 Dome versus Box Cameras

Both dome and box cameras are available as fixed as well as PTZ cameras. A box camera, with its exposed pointing direction, is not concealed, is mainly suited to indoor installation, requires a special enclosure for outdoor use, has a limited viewing angle, and is usually not equipped with an infrared light source for operation in low light conditions. However, lens switching is easier. On the other hand, a dome camera is suited equally well to indoor and outdoor installation, offers better viewing angles for dome PTZ cameras, and has a hidden direction of focus, integrated infrared light for operation in low light conditions, and built-in weather protection. Lens switching is harder in dome cameras.

5.2.3.3 Wired versus Wireless Cameras

The wired or wireless feature is mostly relevant to IP cameras. While wired cameras are linked to local computer networks or the internet via Ethernet cables, wireless cameras are connected to the network by Wi-Fi or Bluetooth. The former allows the cameras to use maximal network bandwidth and provide a reliable connection, while the latter can be used to survey a wider area and are relatively easier to install.

5.2.3.4 Centralized versus Decentralized Cameras

The centralized or decentralized feature is relevant to IP cameras. Centralized IP cameras are a network of linked security cameras where video recording is carried out by a central NVR that all networked cameras link to. Central storage ensures that the security footage remains safe even if one or more cameras are damaged or stolen. On the other hand, decentralized IP cameras are stand-alone recording devices with each camera recording its footage in a local storage medium such as a hard drive or thumb drive. A decentralized storage mode has very little probability of losing all surveillance data footage.

5.2.3.5 Technology-Related Features

Technology-related features to be considered when choosing a surveillance camera mainly include camera resolution, low light performance, lens features, and other features such as the need to have explosion-proof housing or thermal imaging technology. Resolution is measured in millions of pixels, or megapixels. A higher-resolution camera enables greater clarity and capability to see finer details and provide better digital zoom. However, higher resolution demands more storage space to store the images and higher processing power to manipulate them. As well, higher-resolution cameras do not respond as well to low light situations as lower resolution cameras. Therefore, higher-resolution cameras should be used only when necessitated by application. Cameras with 720p, 1080p, and 2K and 4K resolution specifications are common. The capability to operate in low light conditions is more than often a requirement. Cameras with infrared illuminators provide their own light source, allowing better images in dark areas as well. Thermal imaging technology allows operation in both daylight as well as in complete darkness. In terms of lens features, while varifocal lenses allow manual adjustment of the image magnification during camera installation, many fixed cameras now come with remote zoom and focus capability. This allows the user to adjust the camera without physically going to the camera site.

5.3 Laser Fencing

The technologies of laser fencing for protection of critical infrastructure against intrusion by mischievous and rogue elements and for checking infiltration through borders are well established. Three key technologies around which laser fencing is implemented include the use of multiple pairs of laser transmitters and receivers along the perimeter of the infrastructure or the border to be protected, the use of laser scanners or lidar sensors that create multiple customized infrared barriers throughout the area to be protected at key locations or even follow the contours of the perimeter, and the use of an over-ground or buried fiber-optic cable along with a sensing unit. The three types of intrusion detection technologies are briefly described in the following sections.

5.3.1 Laser Transmitter/Receiver Pairs

A laser fence is created by using multiple pairs of laser transmitters and receivers between two fixed points usually 100 to 200m apart. Multiple laser transmitter and receiver pairs, usually three or four, are used to create multiple lines of sight between the two fixed points as shown in the schematic arrangement of Figure 5.17. Multiple lines of sight ensure that likely attempts of intrusion, including normal walking and crawling through the fence, get detected. Near-infrared wavelengths are generally used for the laser transmitter to make it covert and also facilitate easy detection by silicon PIN photodiodes. As well, the laser beam is coded in pulse repetition frequency to make it immune to jamming. Pulse coding coupled with phase-locked detection with a carefully chosen detection bandwidth can be used to minimize false alarms due to slow-moving animals or rapidly flying birds through the fence. The section of multiple transmitter/receiver pairs between two fixed points may be repeated to cover the whole perimeter. In Figure 5.17, four such sections have been used to cover four sides of a perimeter. The use of different pulse repetition frequency codes in different sections allows detection of not only an intrusion but also of the zone that has been breached. Any breach of line of sight generates an electrical pulse. The information is transmitted to a control room to sound an alarm for the security agencies to respond to.

5.3.2 Laser Scanners

The conventional methods of providing security against unauthorized intrusions and trespassing such as the use of structural barriers, deployment of security guards, and the use of surveillance cameras have their limitations when

Figure 5.17 Laser fence using laser transmitter/receiver pairs.

it comes to protecting critical infrastructure and guarding borders. Laser scanning technology overcomes many of these limitations to provide a reliable and easy-to-install and easy-to-use alternative solution. Laser scanning technology uses a lidar sensor emitting a series of laser pulses to create a diverging cone of laser radiation. When reflected from a target, these laser pulses, if present, are processed to quickly and accurately detect intrusions into a protected area. The time-of-flight method is used to measure the distance to the target. Laser scanning technology is usually integrated with other security equipment and software tools such as auto-tracking cameras and GPS mapping to effectively respond to any security threat. While auto-tracking security cameras could be used to track the movements of an intruder after an intrusion has been detected, thereby allowing for more accurate identification of the intruder; GPS coordinates may be used to track and display the location of the threat on a map in the case of border security or on the blueprint of a building in the case of hidden intruder.

Laser scanning technology is particularly suited to both perimeter protection of a large area as well as individual sensitive locations within that large area. This technology has several key advantages, including being safe and covert due to the use of class-1 eye-safe infrared lasers; having the flexibility to make it suitable for indoor and outdoor applications; immunity to inclement weather conditions such as rain, snow, sunlight glare, and low light conditions by using multiecho technology for minimizing false alarms due to reflections from raindrops and snow and special filters to prevent dazzling of laser scanners by bright sunlight; and portability allowing configuration of temporary and portable perimeter protection.

5.3.3 Fiber-Optic Intrusion Detection System

A fiber-optic intrusion detection system operates on the principle of transmitting pulses of laser light into a single-mode optical fiber and accurately measuring the minute light reflections that occur along its length. The light backscattered from different points along the length of the fiber-optic cable is affected by disturbances along the perimeter fence or wall in the case of over-ground-laid fiber-optic cable and by ground vibrations in the case of buried cable. The backscattered light is detected and analyzed by a suitable sensing unit to detect potential intrusions and their locations. The total system is mainly comprised of fiber-optic cable and a centrally located sensor unit. The key advantages of the fiber-optic system include the capability to pinpoint intrusions with an accuracy of a few meters and to detect multiple simultaneous intrusions, immunity to lightning and electromagnetic interference, no requirement for outdoor power or other electronics infrastructure, software-configurable detection zones, and uninterrupted operation up to the point of cut if a cable is cut. In addition, buried fiber-optic cable provides covert detection of intruders, vehicles, and tunneling.

5.4 Laser Range Finders

The most common of all the tactical military applications of lasers has been as laser range finders for observation, surveillance, situational awareness and fire control applications, and laser target designators for munitions guidance. The fire control system of modern armored fighting vehicles and battlefield tanks utilize the services of a laser range finder for generating data on the target coordinates, which in turn is used for gun control. Short-range semiconductor diode laser-based rangefinders are finding use on squad weapons like assault rifles and light machine guns used by security agencies for observation and situational awareness. There are also other applications that exploit the principles of laser range finding. Laser-based proximity sensors, gap measuring devices, and obstacle avoidance systems are some examples employing the laser range finding principle. Laser tracking is another example where a laser target designator and range finder mounted on a two-axis gimbal platform can be used to determine the 3-D coordinates of a remote target and used to track the target. In the following sections, we briefly describe different range finding methodologies, typical applications for homeland security, and salient features of some of the representative range finder systems.

5.4.1 Laser Range Finding Methodologies

Laser range finders exploit one of the following four techniques for determining the distance to a target. These include time-of-flight, triangulation, phase-shift, and FM CW techniques. Each of the four techniques is briefly described below.

In the *time-of-flight technique*, a narrow pulse-width laser beam is transmitted toward the intended target. The target range is measured from the time taken by the laser pulse to travel to the target and back, as shown in Figure 5.18. Target range or distance to target (d) is given by (5.1):

$$d = (c \times \Delta t)/2 \tag{5.1}$$

where

$d =$ target range (m)

$c =$ speed of light $= 3 \times 10^8$ m/s

$\Delta t =$ time interval between transmitted and received laser pulses

Range accuracy in this case depends on the receiver processing speed and rise and fall time of the laser pulses. Range here is measured as the time interval

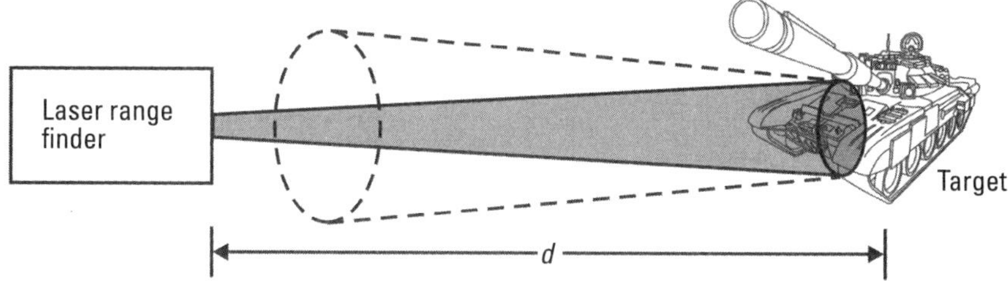

Figure 5.18   Time-of-flight laser range finder.

between the rising or falling edge of the transmitted pulse and the corresponding edge of the received pulse. Uncertainty due to finite values of rise or fall time causes range inaccuracy.

In the *laser triangulation* technique, laser range finders use simple laws of trigonometry to compute distance to the target. A triangulation sensor is typically comprised of a solid-state laser source, which is generally a semiconductor diode laser, and a position sensing detector (PSD) or CMOS/CCD detector. A laser beam is projected on the target whose distance to the sensor is to be measured. A fraction of laser beam power is reflected toward the sensor. The reflected radiation, whose intensity depends on the reflectivity of the target surface, is focused onto the detector through focusing optics. As the target moves, the laser beam proportionally moves on the detector as shown in Figure 5.19. The detector signal is suitably processed to determine the relative distance to the target. CMOS- and CCD-type sensors detect the peak distribution of light quantity on a sensor pixel array to identify target position, whereas PSD-type sensors calculate the beam centroid based on the entire reflected spot on an array.

In the *phase-shift technique* of range finding, the laser beam with sinusoidal power modulation is transmitted toward the target and the diffused or specular reflection from the target is received. The phase of the received laser beam is measured and compared with that of the transmitted laser beam. The phase shift is 2π times the product of the time of flight and modulation frequency. This allows us to compute the time of flight and hence the distance to the target from the known values of phase-shift and modulation frequency. The range distance (d) can be computed from (5.2) for the known values of phase shift ($\Delta\phi$) and modulation frequency (f):

$$d = c.\Delta\phi/4\pi f \tag{5.2}$$

The *FM-CW laser range finding technique* is like the one followed in the case of its radar counterpart (i.e., FM-CW radar). In this technique, the

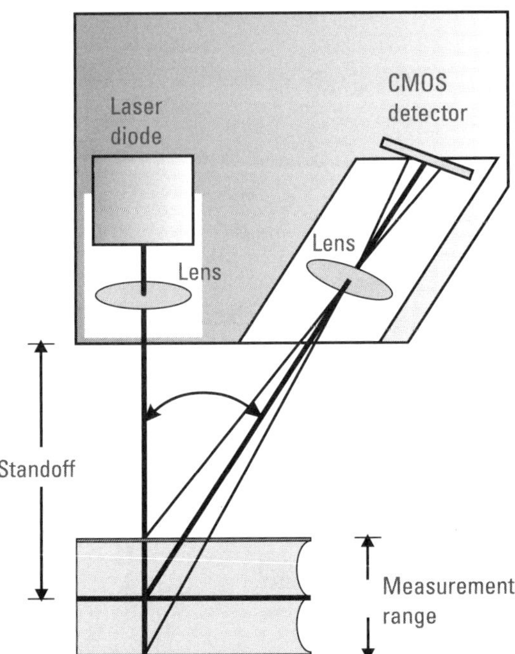

Figure 5.19 Triangulation laser range finder.

frequency of a narrow linewidth laser is modulated with a ramp or sinusoidal signal, collimated, and then transmitted toward the target. The received signal corresponding to the reflected laser beam, specular or diffused, is mixed with the reference signal representing the transmitted laser beam to produce a beat frequency. The received signal is time-delayed from the transmitted signal. The beat frequency corresponding to the time delay along with the ramp time period and modulation frequency bandwidth represent the distance to the target that can be computed from (5.3):

$$d = \frac{\left(f_B \times c \times T_R\right)}{\left(4 \times \Delta f\right)}$$

(5.3)

where

f_B is the beat frequency

T_R is the time period of the ramp waveform

Δf is the modulation frequency bandwidth

5.4.2 Applications for Internal Security

Conventional warfare applications of laser range finders mainly include their use for observation and situational awareness of an adversary's movement of personnel and military assets, in the integrated fire control system of armored fighting platforms for gun control and in laser trackers and 3-D scanners. Obstacle avoidance systems, gap measuring devices, and proximity sensors are other systems that use the laser range finding principle.

Most applications of laser range finders with law enforcement and security agencies also involve their use for observation, surveillance, and situational awareness. They are used as an integral part of most sensor systems for aircraft and drones, remote-controlled defense systems, fire-control-systems, military ground vehicles, and ships. They are also used by manufacturers of sighting systems to supplement or upgrade telescopic sights and weapon-mounted systems and for integration into handheld observation systems such as night vision devices and thermal imagers. Yet another application of laser range finders is in detection of contrabands. Persons trying to transport illegal drugs, weapons, or other contraband use a variety of ways to conceal these items. The appropriate law enforcement and security agencies use laser range finders to quickly determine if a vehicle has been modified to create false walls and hidden compartments for concealing contrabands. Laser range finders take measurements on trucks, truck beds, trailers, cargo containers, and other transportation vessels to find false walls by comparing internal and external dimensions, which helps in identifying hidden compartments.

5.4.3 Some Representative Laser Range Finder Systems

The salient features of some representative laser range finder systems, including handheld Nd-YAG laser range finders, eye-safe (Er-glass, OPO-shifted) laser range finders, and semiconductor diode laser range finders are presented here. While most solid-state laser range finders employing Q-switched lasers and the time-of-flight principle have maximum range measuring capability equal to or greater than 20 km and a range accuracy of 1 to 5m, semiconductor diode laser range finders have limited ranging capability extending to a few hundreds of meters.

The *Model LH-30* handheld Nd-YAG laser range finder manufactured by Bharat Electronics, India, has a maximum operational range of 20 km with a range accuracy specification of ± 5m. Other features include a beam divergence of 1.0 mrad, pulse repetition rate options of 10 pulses per minute and 30 PPM, pulse energy of 6 to 12 mJ, built-in magnification of 6X, an RS-422A serial interface, and remote triggering and bite readout.

The *Model LRF-3M* is a high repetition rate Nd-YAG laser range finder capable of operating at pulse repetition frequencies up to 20 Hz. The system has a maximum operational range of 30 km. Range accuracy and range resolution specifications are 1.0m and 15m, respectively.

The *LRB-21K* (Figure 5.20) manufactured by Newcon Optik is a hand-held Er-glass laser range finder operating at a 1,540-nm wavelength with a laser beam divergence of 1 mrad. It has an operational range of 50 to 21,000m with a range accuracy of ± 5m and a maximum pulse repetition frequency of 2 Hz. It has a built-in three-axis compass and an embedded GPS receiver to facilitate instant calculation of target coordinates, enabling a more precise target triangulation.

The *MELT* from Thales Australia is an eye-safe laser range finder designed for a military environment. It is configured around an Er-glass solid-state laser emitting at 1,535 nm with a transmitted beam divergence of 0.4 to 0.6 mrad. This range finder based on a single shot time-of-flight principle is designed to be used as a handheld electro-optic surveillance device and is also easily inte-grable to OEM systems. The MELT is designed to be integrated within turrets, payloads, handheld sighting systems, light armored vehicles, remote weapon stations, fixed wing aircraft, UAVs, and other state- of-the-art multisensory suites. The salient features of the range finder include a maximum ranging dis-tance of 20 km with a range accuracy of ± 5m, a range resolution of 5m, and a maximum pulse repetition frequency of 1 Hz.

The *LIORA* from Elbit systems Electro-optics (ELOP) Ltd. is a multifunc-tional handheld device that incorporates a monocular day optical telescope, an eye-safe laser rangefinder, a digital compass, and a GPS receiver. Designed and built as a compact unit for infantry units and special operations forces, it can also be integrated as part of a modular target acquisition or observation system.

The *G-Force DX* and *Scout DX* configured around semiconductor diode lasers belong to a wide range of compact laser range finders from Bushnell. Both have similar performance specifications with a maximum ranging dis-tance in the range of 900 to 1,200m and a range accuracy of better than 50 cm. Equipped with selectable bow and rifle modes, each with their own specialized

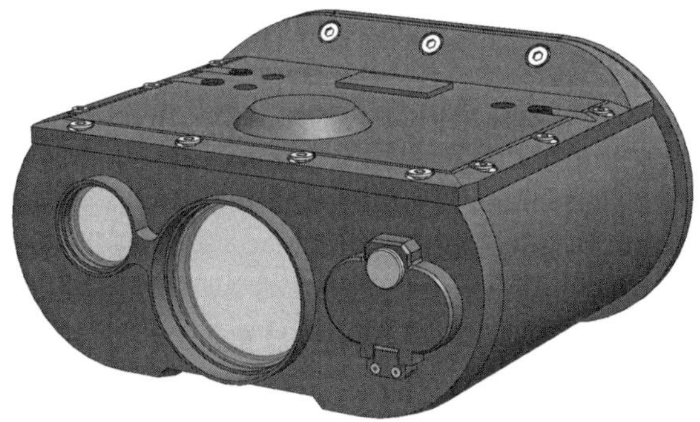

Figure 5.20   LRB-21K laser range finder.

capabilities, these range finders perform like two units in one. They are small enough to fit in a pocket and tough enough to withstand any weather or terrain.

Laser range finder modules from Newcon Optik including LRF MOD 3/3CI employ the same optics and electronics as used by the monoculars and binoculars offered by the company. These modules are typically designed to be integrated with thermal imagers, day/night surveillance systems, and a variety of aircraft optical systems to add range finding capabilities to bigger systems. All modules offer distance and speed measurement and support the RS-232 interface.

Leica laser range finders designed for contraband detection have a measuring range of 0.05 to 60m with a range accuracy of 1.5 mm. The ELEM-series (ELEM 10K and ELEM-DP 10K) are 1,540-nm Er-glass laser range finders with a maximum range of 40 km and a range accuracy better than 5m. The DLEM-series (DLEM-20, DLEM-30, DLEM-45, and DLEM-SR) encompass semiconductor diode laser range finders ideally suited for integration into handheld and weapon-mounted optronic systems such as portable observation devices and weapon-mounted sights. Maximum measurement range is 5 km (DLEM-20 and DLEM-SR), 14 km (DLEM-30), and 20 km (DLEM-45) (Figure 5.21). The range accuracy is in the range of 0.5 to 1m.

5.5 Detection of Electro-Optic Targets

Detection and identification of battlefield optical and optoelectronic sighting and observation systems including weapon sights, night vision devices, thermal imagers, laser range finders, and target designators is an emerging application of lasers. The device involved is commonly known as an *optical target locator* (OTL) and is also called a sniper locator. In the following sections, we briefly describe the operational principle of an optical target locator followed by the salient features of some representative devices.

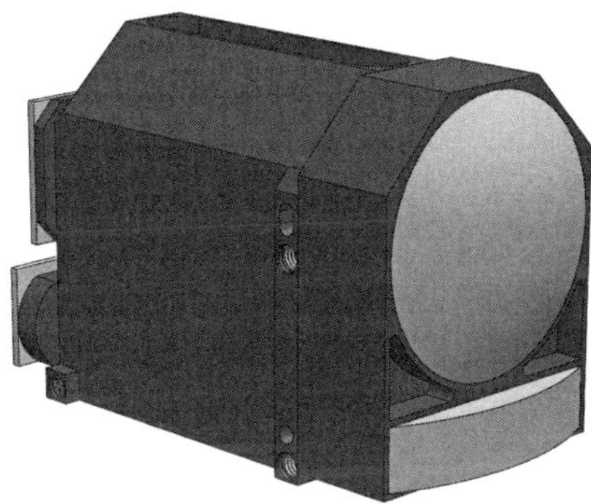

Figure 5.21 DLEM-45 laser range finder.

5.5.1 Operational Principle

An optical target locator operates on the principle of the cat-eye effect, as shown in Figure 5.22. It is illuminated by a laser beam. The optical system returns a fraction of it as backscattered energy, which is then received by a sensitive receiver. The backscattered signal is processed to indicate the location of the optical device. This device could be very useful to homeland security for detection of optical scopes employed by snipers. Another homeland security related application could be surveillance of sensitive areas, particularly in urban territories. Figure 5.23 shows a typical deployment scenario of such a device in an urban environment when used for sanitization of the area. A security personnel is using an optical target locator to scan a building for a possible hidden sniper or a suspect observer. The scanning diverging laser beam gets retroreflected from the front-end optics of the optical device to be received by the receiver of optical target locator to indicate the location of the targeted optical device.

5.5.2 Representative Systems

Many counter-sniper systems capable of detecting optical scopes used by snipers and other observation devices with front-end optics are commercially available for use by security agencies and armed forces.

The *SLD-400* from CILAS is designed to detect and locate optical scopes used by snipers and other optronic sighting systems used on the battlefield. Based on the cat-eye effect, it has a detection range of 1,000m during the daytime and 3,000m during nighttime. It can be fitted on a tripod or on a static vehicle. Its highly accurate threat detection and localization capability is compatible with the performance of the associated weapon or fire control system. The *SLD-500* is an improved version of the SLD-400.

The *Mirage-1200* from Torrey Pines Logic Inc. is an active system designed to detect the retroreflected signal caused by the front-end optics of the target system. It uses a combination of a high-sensitivity detector and an optical filter to offer a maximum operational range of 1,200m. It also makes use of the time-of-flight calculation of the transmitted signal and the retroreflected signal

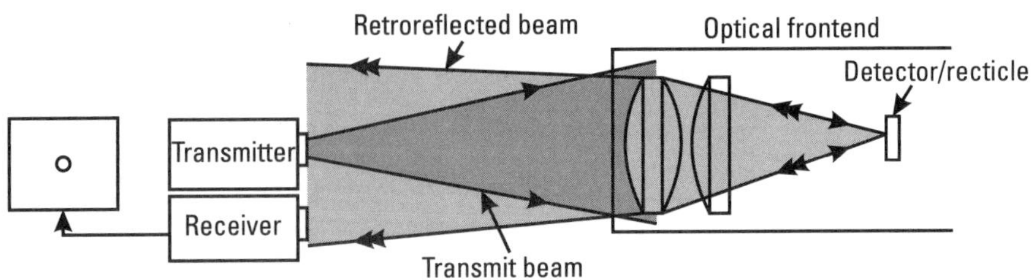

Figure 5.22 Cat-eye effect.

Figure 5.23 Typical deployment scenario for area sanitization in an urban environment.

to compute the target range. The chosen detector offers day and night operational capability. The system allows penetration through bad weather and vegetation by range gating. The *Sentinel S45* from the same company (Figure 5.24) is designed to be used for ensuring security of VIPs, borders, and important and strategic locations. It has a maximum operational range of 2,500m with an achievable range depending on the operating environment and a distance measuring accuracy of 5m. The *Myth-350* also from Torrey Pines Logic Inc. has a detection range of 350m. The *Beam^{TM}85*, *Beam^{TM}210*, and *Beam^{TM}220* are optical target detection systems with detection ranges of 50 to 2,000m, 50 to 2,000m, and 50 to 2,500m, respectively. They are particularly suited to sniper detection, VIP security, border protection, and security of critical assets. The Beam^{TM}210 and Beam^{TM}220 also feature optional SWIR, thermal or CCD payloads, geolocation of targets with GPS data, and multispectral imaging for target verification.

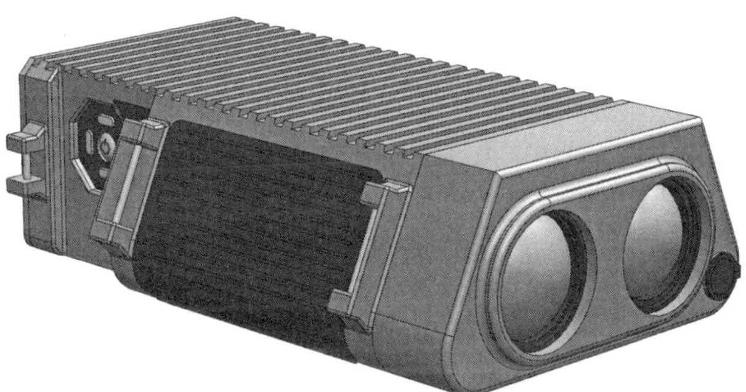

Figure 5.24 Sentinel S45 sniper detection system.

The *OCELOT-3* sniper detection system from FidusCrypt GmbH combines laser range gating and thermal and optical technologies for maximum capabilities in target detection. It integrates a basic module comprised of a laser range gating camera, a multisensor module including a laser range finder, HDTV CCD camera, navigation module, and thermal module. It can be used for a variety of applications including perimeter intrusion detection, border control and protection, VIP protection, protection of critical infrastructures, sniper detection, and by special forces for combat operations. The system can detect optical devices up to a maximum of 1,200m (for 24-mm lenses) and 2,000m (for 100-mm lenses).

The *LAS-1000* from Newcon Optik detects snipers and other forward observers in a variety of tactical scenarios. It pinpoints the location of a threat before it has a chance to act. This system is ideally suited to border and perimeter security as well as VIP protection. The LAS-1000 can detect optical devices in its line of sight even when these objects are covered behind bushes, windows, or windshields. Detection range and distance measuring accuracy are 70 to 1,000m and 10 m, respectively. It is also equipped with an audio alarm that can be set to activate on detection of threat to provide added situational awareness.

The *GCU-OSD10* from JSC Sekotech is designed to detect and locate portable observation systems such as surveillance cameras camouflaged in various forms at distances up to a maximum of 500 to 1000m with an effective range depending on prevailing weather conditions. The transmitter and receiver are configured around an 808-nm semiconductor diode laser and a half-inch 752×582 pixel CCD, respectively. The *GCU-OSD15* is designed for detection of small-sized video surveillance cameras up to a range of 25m. The *GCU-OCD20* (Figure 5.25) allows detection of electro-optical surveillance systems through windows as well as tinted glass, organic glass, and semitransparent mirrors.

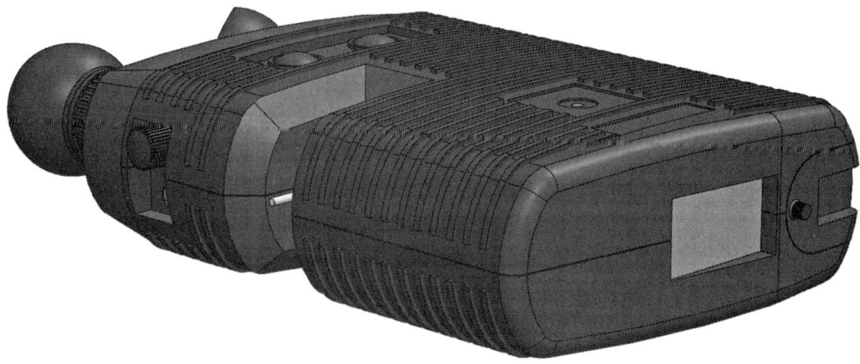

Figure 5.25   GCU-OCD20 optics detector.

5.6 Covert Laser Listening Devices

Laser listening devices, also called *laser microphones,* are covert surveillance devices that utilize an invisible infrared laser beam to eavesdrop on a conversation taking place in a remote room from a distance without entering the room. They are used by law enforcement and security agencies to carry out covert surveillance on suspect and rogue elements when it is not feasible to plant a listening device in the targeted area. These devices are particularly useful in tactical situations when speed of acquiring information is of concern. Laser covert listening devices with operational ranges of 100 to 1,000m are commercially available. In the following sections, we describe the operational principle of laser microphones followed by the salient features of some representative devices.

5.6.1 Operational Principle

A laser listening device is comprised of an *infrared laser transmitter,* an *optical receiver,* and an *electronics processing unit.* Infrared laser beam from the laser transmitter is beamed onto the surface of a windowpane of the room to be monitored. The oscillating frequencies of the sound waves originating from the speech conversations and other sounds inside the targeted room cause microvibrations of the surface of the windowpane. The infrared laser beam on striking the windowpane encounters the microvibrations on the windowpane's surface. The laser beam reflects off the surface of the windowpane back to an optical receiver. The minute differences in the distance traveled by the light as it reflects from the vibrating object are detected interferometrically in the optical receiver. The receiver converts the path difference variations in the reflected laser beam into intensity variations. The intensity variations are then converted into an electrical signal, which is filtered and amplified and then converted back to sound. The audio signal may be simultaneously available for real-time listening or recorded on an external recorder.

There can be several possible optical arrangements that can be used to build a laser microphone. One such arrangement is shown in Figure 5.26(a). In this case, the laser beam is incident on the windowpane at an angle and the optical receiver with the photodetector is placed closer to the windowpane on the other side at an angle equal to the angle of incidence. In this case, the microvibrations in the windowpane are good enough to deflect the laser beam across the photodetector to induce intensity variations. This arrangement does not make use of the interference phenomenon. In another optical arrangement shown in Figure 5.26(b), which is a slight variation of the arrangement of Figure 5.26(a), the incident and reflected laser beams are bore-sighted and the transmitter and receiver are collocated. The third optical arrangement shown in Figure 5.26(c) utilizes the principle of interferometry. It is essentially a Michelson

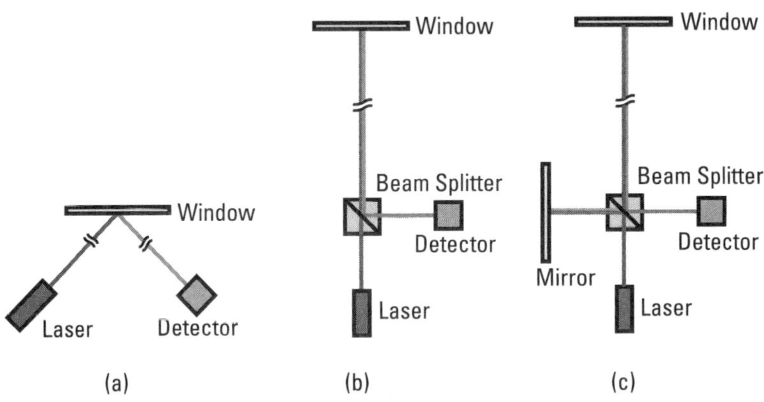

Figure 5.26   Optical arrangements of a laser microphone.

interferometer. The problem with this setup is that the difference between the two legs of the interferometer is too large, leading to loss of phase coherence between the two beams, thereby limiting the sensitivity of the device.

This shortcoming is overcome in the optical arrangement of Figure 5.27. This arrangement has nearly equal path lengths of the two legs of the interferometer, thereby preserving temporal coherence. The interferometer responds to differential movements across a small section of the windowpane and significantly rejects common-mode path disturbances.

5.6.2   Representative Systems

Laser-based listening devices capable of covert surveillance on suspects by listening to their conversations taking place in a remote room from a distance from a few hundreds of meters have been developed and are commercially available.

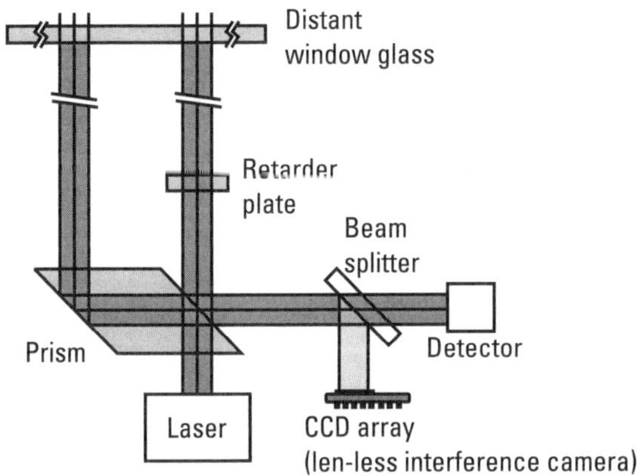

Figure 5.27   Interferometric set-up of a laser microphone with nearly equal path lengths.

The *Opto Acoustical Laser Model SKU-1688* from Discovery Telecom has a maximum operational range of 300m and can be used with any angle of direction. It does not necessarily need a glass windowpane for remote audio monitoring; it can pick up a signal from other materials such as paper, metal, plastic, and textiles. The device is immune to interferences from its surroundings.

The *PKI-3100 Laser Monitoring System* from PKI Electronic Intelligence GmbH, like other laser microphones, is used for covert observations. It is specially designed for acoustic monitoring through glass and windowpanes and has a maximum operational range of 300m. The *LAS-MIC Audio Monitoring System* from Endoacustica is similar to PKI-3100 and has same specifications. *Spectra M Laser Microphone* from Detective Store Ltd. is a portable device with a maximum operational range of 400m and has a built-in sound recorder. *EMAX-3100* from Electromax International, Inc. is the same as the PKI-3100. The *EMAX-2510,* also from Electromax International, Inc., has an operational range of 150m. The salient features of the EMAX-2510 include detection angle independence, usability on a variety of target materials like paper, plastic, metal, and glass, good speech intelligibility, and insensitivity to background noise and interference that is facilitated by the use of an interferometric detection process.

The *Long Range Laser Audio Surveillance Device* from Eyetek Surveillance Ltd. is a covert surveillance system capable of monitoring conversations in a room from over 450m away. It, like other laser microphones, is designed for use in situations when a building cannot be accessed to place a surveillance device in the targeted room. The device is packaged into standard SLR cameras, which makes it look like a photographic setup, thereby allowing for further concealment.

5.7 Detection of Concealed Weapons

While on one hand there has been continuous development in the field of security technologies in terms of screening, detection, and surveillance equipment to counter a variety of threats in different deployment scenarios, there has also been an emergence of countermeasure techniques that have enabled rogue and suspect elements to evade detection by conventional equipment. Detection of concealed weapons is one of the greatest challenges facing security and law enforcement agencies. Concealment of guns as purses, camcorders, and mobile phones and of knives as pens, combs, and books are a commonplace. Polymer knives are also used to evade detection by metal detectors. As well, explosives are disguised as everyday items such as books, briefcases, and vests. Screening procedures are in place for detecting weapons such as handguns, knives, and explosives concealed in different items of common usage and weapons concealed underneath a suspect's clothing. This sophisticated screening equipment is generally used for security of important public assets and critical infrastructure. It is also desirable to detect concealed weapons from a standoff distance, particularly in situations where it is nearly impossible to arrange the flow of people through a controlled procedure. Different technologies have been developed for the detection of concealed weapons. The following sections briefly describe common concealed weapon detection technologies including X-ray scanners, millimeter-wave imaging, IR imaging, and terahertz imaging sensors.

5.7.1 Concealed Weapon Detection Technologies

Several technologies have been used to build sensors for the detection of concealed weapons. Commonly used sensors for concealed weapon detection include X-ray scanners (baggage and body scanners), millimeter-wave imaging sensors, terahertz imaging sensors, infrared sensors, magnetic sensors, and acoustic sensors. *X-ray scanners* use either material-dependent transmission (in the case of baggage scanners) or scattering (in the case of body scanners) properties of X-rays to construct an image that highlights objects being carried inside baggage or concealed underneath clothing. *Millimeter-wave imaging sensors* use sensing in the millimeter-wave band of 30 o 300 GHz that allows detection of both metallic and nonmetallic objects concealed underneath clothing. Both passive and active millimeter-wave imaging sensors are available. *Terahertz imaging sensors* work in the terahertz or submillimeter-wave band of 0.3 to –3 THz. *Infrared imaging sensors* detect a fraction of the infrared radiation reemitted from clothing after it has absorbed infrared radiation emitted by the human body. Infrared sensors are effective in producing an image of a concealed weapon only when the clothing is tight, thin, and stationary. *Magnetic sensors* use an array of magnetoresistive gradiometer sensors that sense disturbances in the ambient Earth's magnetic field due to the presence of any ferromagnetic material. They are configured as walk-through weapons detection portal and the detection system accurately pinpoints the number, location, and size of the weapons. *Acoustic imaging sensors* use ultrasound frequencies in the range of 30 to 200 kHz. Ultrasound waves scattering off the objects are used to construct an image. The chosen frequency band can penetrate clothing to detect hidden objects. The following sections briefly describe X-ray, millimeter-wave, infrared, and terahertz sensors.

5.7.2 X-Ray Scanners

X-ray scanners use the intensity of transmission and scattering of X-rays through different materials to construct an image of hidden objects. The image not only describes the shape of the objects but also the material. There are two types of X-ray scanners: baggage scanners and full body scanners. The intensity of transmitted X-rays is used in baggage scanners and the intensity of scattered

X-rays is used in full body scanners. In the case of baggage scanner, some of the energy of the X-rays is absorbed by the various objects in the bag and the X-rays pass through unattenuated through empty spaces. These waves then hit the first platelike detector. The material between the two detectors blocks the low-energy waves with the result that only the high-energy waves hit the second plate. The outputs from these two detectors are compared and used to digitally construct an image. Different materials are represented by different colors with denser materials like metal or glass represented by darker colors and less dense materials such as food and clothes depicted in lighter colors. Several advances have been made in the design of X-ray baggage scanners to improve the quality of inspection. One such scanner is a double-angle X-ray scanner that uses two X-ray sources, one from the side and one from the bottom. This scanner looks at the objects from two directions simultaneously to provide a clearer picture. These are extensively used at airports, railway stations, and ports for security checks. There are also dual-channel high-energy X-ray baggage scanners and large emergence angle scanners.

The *full body X-ray scanner* for humans, also known as a backscatter X-ray scanner, uses the intensity of scattered X-rays. This is less harmful, as the rays do not have to fully pass through the object. Again, the intensity of scattered rays depends on the material encountered. The scanner takes two pictures; one of the front and one of the back (Figure 5.28).

Figure 5.28　Full body scanner.

5.7.3 Millimeter-Wave Imaging

The operational principle of a *millimeter-wave imaging body scanner* is same as that of a backscattered X-ray scanner except that the former uses electromagnetic radiation called millimeter-waves lying in the 30- to 300-GHz band of the spectral region between radio waves and infrared waves and not X-rays. Millimeter waves have the unique property of being able to pass transparently through lightweight materials such as clothing. The millimeter-wave scanner beams low-powered millimeter waves over the surface of the body using two rotating antennas. The radiation bounces off the skin of the person under scan and off any suspicious object concealed underneath the clothing and is received by two receivers. The receivers process the information and generate front and back images to locate any objectionable items for the operator to see on the display. Millimeter-wave scanners are believed to be far less harmful than their backscatter X-ray counterparts as the former emit nonionizing radiation and presumably do not have the potential for cancer-causing DNA damage. The established health-related effects of nonionizing radiation are limited to thermal effects. The emitted power density in the range of 0.00001 to 0.0006 mW/cm^2 is well below that required to produce tissue heating.

5.7.4 Infrared Imaging

Infrared imaging for concealed weapon detection depends for its operation on the infrared radiation emitted by a human body and any objectionable objects concealed therein, since there is a temperature difference between different parts of the body and concealed objects. The emitted infrared radiation is detected typically in the mid- or long-wavelength infrared band. Both active and passive infrared imaging technologies are available. In the case of passive infrared imaging, there is no illumination of the target. The infrared radiation emitted by a human body and a concealed object is absorbed by clothing and then reemitted. A concealed object is colder than a body and appears darker than the surrounding body in the image due to a temperature difference between it and the body. Fusion of two images, one from a visible sensor and the other from an infrared sensor, reveals greater detail. In the case of active infrared imaging, the target is illuminated by a source of radiation in the near- or short-wavelength infrared region. Infrared imaging is further discussed in Chapter 6.

5.7.5 Terahertz Imaging

Terahertz radiation possesses certain unique characteristics that make terahertz imaging attractive for a variety of applications: in quality control for nondestructive testing of the internal structure of objects; in wireless communication for implementing wireless local area networks (WLANs) and wireless personal area networks (WPANs) of new generations as well as for creating entirely secured dedicated channels of wireless communication; in medicine for terahertz tomography; and above all in security for screening people and baggage. Terahertz radiation is both coherent and spectrally broad with the result that terahertz images contain far more information than a conventional image formed with a single-frequency source. Terahertz radiation can penetrate fabrics, ceramics, and plastics, which enables detection of objects concealed underneath clothing. It is also nonionizing and therefore harmless to living tissue or DNA, making it absolutely safe for humans, plants, and animals. These unique properties make terahertz rays much more attractive and informative than X-rays and near-infrared radiation (NIR). Terahertz imaging scanners intended for the screening of letters, envelopes, and small packages and for the screening of humans for detection of concealed objectionable objects are commercially available. One such system is the *Terasense security body scanner* from TeraSense Technology, which operates in reflection mode. It is capable of standoff detection of weapons, including cold steel and firearms, bombs and grenades, explosive belts, and various contraband items hidden under clothes from 3m.

Selected Bibliography

Accetta, J. S., and D. L. Shumaker, *The Infrared and Electro-Optic Systems Handbook,* Volume 7, Revised Edition, Bellingham, WA: SPIE International Society for Optical Engineering, 1998.

Desmarais, L., *Applied Electro Optics*, Upper Saddle River, NJ: Prentice Hall/Pearson, 1998.

Hecht, J., *Understanding Lasers: An Entry Level Guide*, 3rd Edition, Hoboken, NJ: Wiley-IEEE Press, 2008.

Kasap, S. O., *Optoelectronics & Photonics: Principles and Practices*, Pearson, 2012.

Lee, Y.-S., *Principles of Terahertz Science and Technology*, New York: Springer, 2008.

McAulay, A. D., *Military Laser Technology for Defense*, Hoboken, NJ: Wiley-Interscience, 2011.

Rosencher, E., B. Vinter, and P. G. Piva, *Optoelectronics*, Cambridge, UK: Cambridge University Press, 2002.

Sayeedkia, D. (ed.), *Handbook of Terahertz Technology for Imaging, Sensing and Communications*, Oxford, UK: Woodhead Publishing Limited, 2013.

Uiga, E., *Optoelectronics*, Englewood Cliffs, NJ: Pearson, 1995.

Waynant, R., and M. Ediger (eds.), *Electro-Optics Handbook,* New York: McGraw-Hill, 2000

Webb, C. E., and J. D. C. Jones (eds.), *Handbook of Laser Technology and Applications: Volume III*, Boca Raton, FL: CRC Press, 2003.

Wilson, J., and J. Hawkes, *Optoelectronics: An Introduction,* Second Edition, Englewood Cliffs, NJ: Prentice Hall, 1989.

Woolard, D. L., J. O. Jens, R. J. Hwu, and M. S. Shur (eds.), *Terahertz Science and Technology for Military and Security Applications,* Hackensack, NJ: World Scientific, 2007.

6

Night Vision Technologies

Night vision devices in different forms such as night vision scopes, night vision goggles, and night vision cameras are crucial to the efficacy of security and law enforcement agencies and night-fighting capability of the armed forces. These devices are extensively used by the military for the location of enemy targets, surveillance, and navigation. They are also used by law enforcement and security agencies for surveillance. Night-vision-enabled cameras are being used even by private businesses and military establishments to monitor the surroundings of their critical assets. This chapter discusses the two major night vision technologies: image enhancement, and thermal imaging along with their capabilities and limitations. This is followed by a discussion on different types of night vision devices and their typical applications.

6.1   Basic Approaches to Night Vision

Night vision technologies and the associated night vision devices enable users see in low light conditions. Contemporary devices allow viewing even in near total darkness. The ability to see in low light conditions is governed by two basic requirements: sufficient spectral range and sufficient intensity range. Low values of spectral range and intensity range in the case of human eyes therefore become the limiting factors for their ability to see with an acceptable level of contrast in low light conditions. The use of technology to enhance both these parameters makes night vision possible. The two basic and widely different approaches to night vision include *image intensification* (or *enhancement*) and *thermal imaging*. Both techniques are briefly described in the following sections.

6.1.1 Image Intensification

Image intensification or *enhancement* works on the principle of collecting small quanta of light reflected off the target scene to be viewed in visible and near-infrared bands of electromagnetic spectrum in low light conditions. The collected photons are amplified through the processes of photon-electron conversion, electron multiplication, and electron-photon conversion. These processes take shape in an image intensifier tube. The other important constituent parts of an image intensifier tube-based night vision device include the objective lens used for collection of photons, an eyepiece for viewing an intensified image, and a power supply that generates the required DC voltages for electron acceleration.

6.1.2 Active Illumination

Active illumination is often used in conjunction with an image intensifier tube in what is known as *active night vision technology* to enhance image resolution in very-low level-light conditions. Illumination is generally provided by infrared diodes emitting in the 700- to 1,000-nm spectral band. Active night vision technology has the disadvantage that it can be detected by night vision goggles and is therefore prone to giving away the location of the user. This is particularly undesirable in tactical military operations.

6.1.3 Thermal Imaging

Thermal imaging night vision technology works on the principle of detecting the temperature difference between the objects in the foreground and those in the background. All objects above absolute zero emit infrared energy. The magnitude of infrared energy emitted by a hot body is proportional to the fourth power of its absolute temperature (Stefan-Boltzmann law) and the peak emission occurs at a wavelength that is inversely proportional to its absolute temperature (Wien's displacement law). Therefore, the hotter the body, the higher the magnitude of the infrared energy emitted by the object is and the lower the wavelength of the peak emission is. A thermal imaging device is essentially a heat sensor capable of detecting tiny differences in temperature of different points on the surface of the object to be viewed. The information on the temperature difference available in the form of infrared energy is collected by the thermal imaging device and converted into an electronic image. The ability to detect tiny temperature differences that always exist not only between the desired object and the surroundings, but also between different points of the object itself coupled with emission in the infrared region, allows a thermal imaging device to see in near total darkness.

6.1.4 Digital Night Vision

A variation of the conventional night vision device is the *digital night vision* device. While in a conventional night vision device, available light is collected through the objective lens and focused on an intensifier, most digital night vision devices process and convert the optical image into an electrical signal through a highly sensitive CCD image sensor. This electrical signal is then transferred onto a microdisplay, which is a type of LCD flat-panel display screen. The microdisplay usually takes the form of an eyepiece that you look into to view the image rather than on an LCD screen found on most digital cameras. Digital night vision offers several advantages over conventional night vision. Digital night vision devices have relatively lower costs, are free from image distortions of photocathode and blemishes of phosphorescent screen, are immune to damage by bright light exposure, and offer image recording facility.

6.1.5 Image Intensification versus Thermal Imaging

Both image enhancement and thermal imaging night vision devices have similar application areas. While choosing one or the other for a given application, important considerations include the cost, lighting conditions, and type of environment. In terms of cost, image intensifier devices have a decisive edge over thermal imaging devices. A good night vision device, including weapon-mountable variants, would cost a few hundred U.S. dollars. On the other hand, the price of a thermal imager may be anywhere from a few thousands of U.S. dollars, with military-qualified devices costing as much as tens of thousands of U.S. dollars. As well, lighting conditions must be considered before investing in one device or the other. A night vision device needs light to operate even though a small amount of light may be adequate for achieving desired results. Thermal imagers on the other hand can work in total darkness. In addition, the type of environment plays a big role. A thermal imager is the only choice for heavy fog or dense foliage as the long wavelength band (8–14 μm) penetrates much better through smoke, smog, dust, and water vapor. It can also see through sandstorms, a common occurrence in desert operations, which makes them an ideal choice for such operations. Extreme cold makes night vision the better choice. Finally, thermal imaging is great for detection, but not ideal for recognition. Night vision devices allow for better recognition once detection occurs, but if the person detected is wearing camouflage or the animal detected stands stationary at a distance, they can be difficult to find.

6.2   Image Intensifier Devices

Common image intensification devices include the image intensifier tube and the intensified CCD. Both types are briefly described in the following sections.

6.2.1   Image Intensifier Tube

An *image intensifier tube* amplifies a low-light–level image to a level that can be seen with the human eye or detected by digital image sensors. Image intensifier tubes collect the existing ambient light originating from natural sources, such as starlight or moonlight, or from artificial sources, such as streetlights or infrared illuminators, through the objective lens of a night vision device. Low-level light consisting of photons enters the night vision device through an input window and strikes the photocathode. The inside of the input window is generally coated with a thin layer of light-sensitive material. This light-sensitive layer acts as the photocathode. The photocathode is protected from any damage from oxidation by operating the image intensifier tube under vacuum of the order of 10^{-9} to 10^{-10} torr. Photo electrons released by the photocathode are accelerated and focused by a high-magnitude electric field toward the MCP. The MCP has millions of small channels and the electrons entering these channels are both accelerated by another high-magnitude electric field within the MCP and multiplied by a secondary emission resulting from electrons bouncing off the inner walls of these channels. For each electron entering the MCP, approximately 1,000 electrons are generated and subsequently accelerated from the output of a single-stage MCP by a third electrical field toward the phosphor screen. The phosphor screen, which is a thin light-emitting layer deposited on the inside of the output window of the image intensifier tube, converts impinging electrons back to photons. For every photon entering the input window of the intensifier tube, tens of thousands of photons come out of the output window after emission from the phosphor screen. This photon multiplication takes place due to electron acceleration in the region between the input window and the photocathode, electron acceleration and secondary emission within the MCP channels, and electron acceleration in the region between the MCP and the phosphor screen. This multistage process produces an intensified or amplified image of the object that is much brighter than the original image.

6.2.1.1   Construction

The constituent parts of an image intensifier tube include the input window, photocathode, microchannel plate, phosphor screen, output window, and power supply. Figure 6.1 illustrates the composition and features of a typical image intensifier tube.

The *input window material* is selected according to the required sensitivity at shorter wavelengths. Common materials used for an input window are

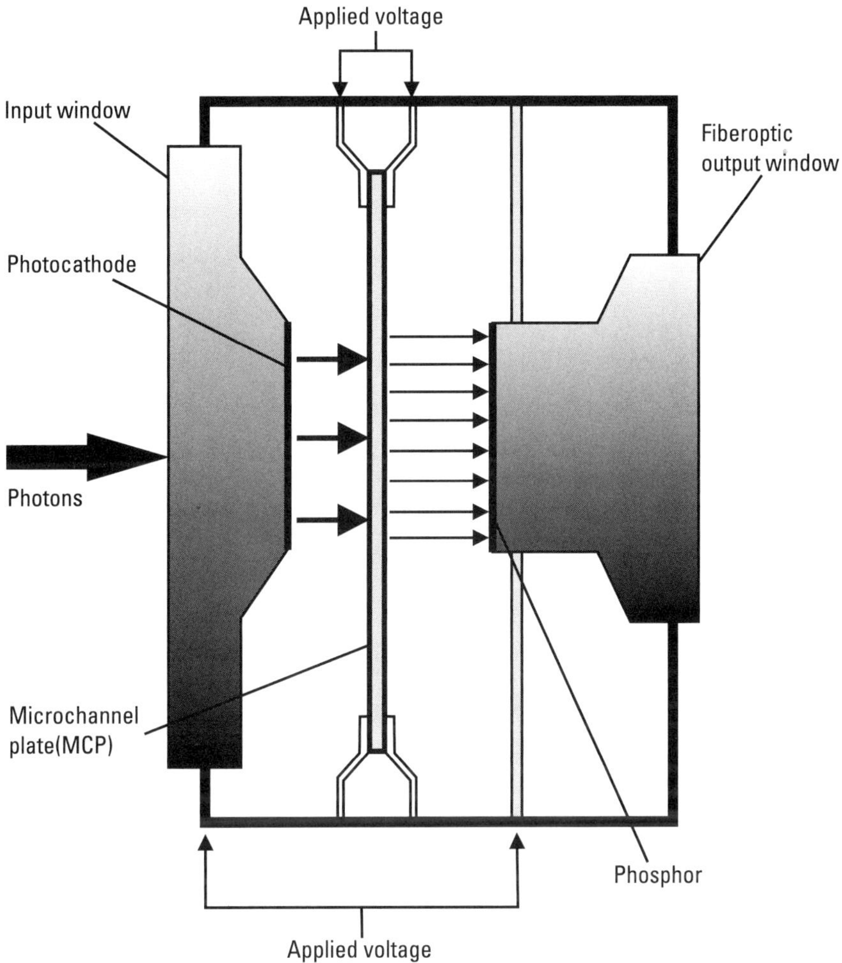

Figure 6.1 Composition of an image intensifier tube.

synthetic silica (transmitting wavelength of 160 nm or longer), fiber-optic plate (with a transmission wavelength of 350 nm or longer), magnesium fluoride (with a transmission wavelength of 115 nm or longer), and borosilicate glass (with s transmission wavelength of 300 nm or longer).

The photocathode that converts photons into photo electrons, the microchannel plate that multiplies the photo-generated electrons, the phosphor screen that reconverts the electrons back into photons, and the power supply that produces the electric field responsible for acceleration of electrons in different regions are all arranged in close proximity in an evacuated ceramic case. The efficiency with which the photocathode converts photons into electrons, also known as photocathode radiant sensitivity or quantum efficiency, depends on wavelength. A number of photocathode materials are in use. Of these, the semiconductor crystals gallium arsenide and gallium arsenide phosphide offer extremely high sensitivity. Photoelectrons are accelerated by an electric field produced by a high voltage applied between the photocathode and MCP input surface.

The MCP is a thin glass disc about 0.5-mm thick consisting of an array of millions of tilted glass channels, each of about 5 to 6 μm in diameter bundled in parallel (Figure 6.2). A single-stage MCP used earlier in second generation intensifier tubes provides electron multiplication of about 10^3. Two-stage and three-stage MCPs produce a gain of 10^5 and greater than 10^6. First generation tubes did not use an MCP. The number of stages to be used in the MCP depends on the required value of the gain. The strip current that flows through the MCP decides the dynamic range or linearity of the image intensifier tube. A low-resistance MCP causing a large strip current to flow through the MCP is desirable for achieving high linearity.

The *phosphor screen* reconverts the impinging electrons back to photons. Commonly used phosphor types include P24, P43, P46, and P47. Phosphor screens are characterized by peak emission wavelength, decay time, power efficiency, and emission color. Phosphor screen decay time is one of the most important parameters to be considered when selecting a suitable phosphor type. The chosen phosphor type is such that its decay time matches the readout method and its spectral emission matches the readout sensitivity. When used with a linear image sensor or high-speed CCD, a short decay time is recommended for the phosphor screen to avoid the appearance of an afterimage in the next frame. On the other hand, a short decay time that minimizes flicker is recommended for nighttime viewing and surveillance applications.

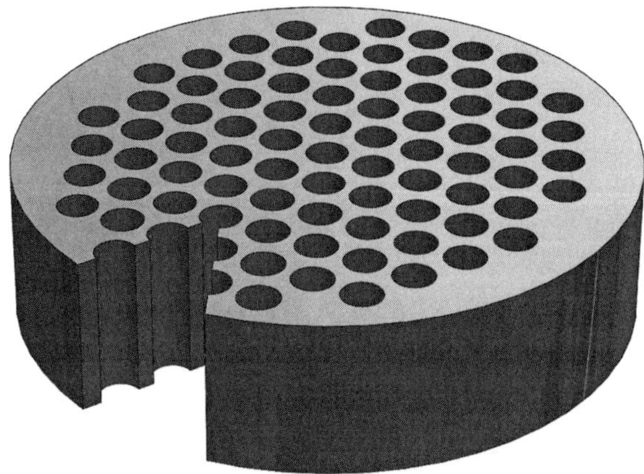

Figure 6.2 Microchannel plate.

The *output window* material is selected to match the readout method. Different output window types include borosilicate glass, fiber-optic plate (FOP), and twisted fiber optics. Borosilicate glass window is used for relay lens readout. In this case, a relay lens is focused on the phosphor screen. A fiber-optic output plate is a standard output window and is ideal for direct coupling to a CCD with a FOP input window. A fiber-optic plate consists of some millions to hundreds of millions of glass fibers bundled in parallel. A FOP can transmit an optical image from one surface to another without causing any distortion. The diameter of the glass fiber matches the channel diameter of the MCP. Twisted fiber optics as an output window is used for nighttime viewing applications. Using twisted fiber optics reduces eyepiece length, thereby making a night vision device more compact.

6.2.1.2 Operational Modes

There are two common modes of operation of an image intensifier tube: the *gated mode* and the *photon counting mode.* In the gated mode of operation, the intensified image can be gated to open or close the optical shutter by varying the potential difference between the photocathode and the inside surface of the MCP, thereby either allowing or disallowing the formation of an intensified image. In the *gate-on mode*, the potential of the photocathode is lower than that of the MCP. As a result, the photoelectrons are attracted toward the MCP to be subsequently multiplied and hit the phosphor screen, producing an intensified image. In the *gate-off mode*, the potential of the inside surface of the MCP is less than that of photocathode with the result that the photoelectrons revert back to the photocathode. Therefore, there is no intensified image seen on the phosphor screen. In practice, the MCP potential is fixed and the intensifier tube is turned on by applying a negative polarity pulse of about 200V to the photocathode. Gated operation is very effective in analyzing the high-speed optical phenomenon.

Image intensifier tubes using three-stage MCPs have much higher sensitivity than that of image intensifier tubes employing a single-stage MCP. This is particularly important when it comes to operating at extremely low light levels. When the light level is as low as 10^{-4} lux, a three-stage MCP helps in producing an image of acceptable quality. However, when the light level falls below 10^{-5} lux, the incident photons are separated in time and space and it is no longer possible to capture an image with a gradation. At extremely low light levels when only a few light spots appear on the phosphor screen per second, a good-quality image can be obtained by detecting each spot and its position and integrating them into an image storage unit. The brightness distribution in this case, called the *photon counting mode,* is given by the difference in the number of photons at each position.

6.2.2 Intensified CCD

Intensified CCD (ICCD) successfully exploits the optical amplification provided by an image intensifier to overcome limitations of the basic CCD sensor. Two important features of an intensified CCD are high optical gain and gated operation. Both attributes are characteristic features of the image intensifier tube. Although image intensifiers were initially developed for the military and law enforcement agencies for a range of surveillance, targeting, and navigation applications, development of ICCD technology and related devices has extended its usage to many scientific application in spectroscopy, scientific and industrial imaging, and medical diagnostics. In fact, the development of image intensifier tubes is increasingly being driven by scientific applications.

6.2.2.1 Composition

An intensified CCD is primarily comprised of an image intensifier tube whose light output is coupled to a CCD sensor. The output of the image intensifier is coupled to the CCD typically by a fiber-optic coupler (Figure 6.3). Fiber-coupled systems are physically compact with low optical distortion levels. The high-efficiency fiber-optic coupling also allows image intensifier tubes to operate at lower gains, which in turn results in better dynamic range performance. A lens-coupled ICCD uses a lens between the output of the image intensifier and the CCD (Figure 6.4). Lens coupling offers the flexibility of using the ICCD

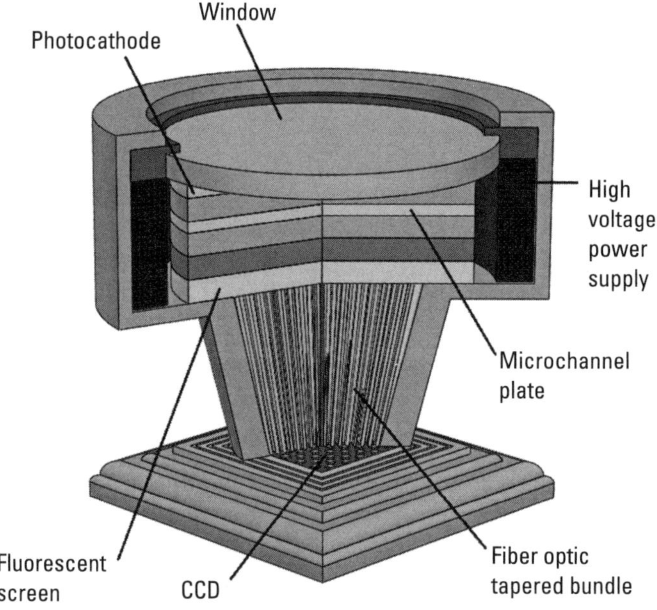

Figure 6.3 Composition of ICCD with fiber-optic output coupling.

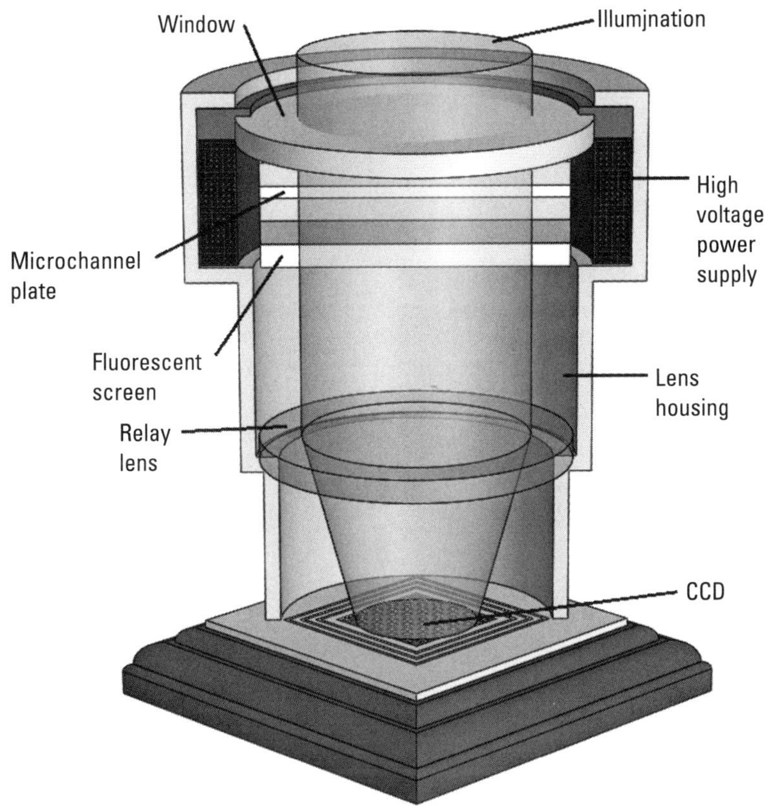

Figure 6.4 Composition of ICCD with lens coupling

sensor in nonintensified mode by allowing the image intensifier to be removed. Disadvantages of lens-coupled ICCDs are larger physical size, lower coupling efficiencies, and increased scatter. Power supply is another important constituent part of the ICCD sensor. The power supply section generates DC voltage (typically 600 to 900V) for MCP to achieve desired gain, DC voltage (typically 4 to 8 kV) for the phosphor screen, and voltage pulses (typically 200V) for gated operation of the photocathode. The gating pulse width and rise/fall time depend on the desired gating parameters. A gating pulse width of less than a nanosecond and rise/fall time of a small fraction of a nanosecond are achievable.

6.2.2.2 Characteristic Features

Important characteristic features of ICCD include its spectral response, spatial resolution, gating time and repetition rate, noise, sensitivity, dynamic range, and frame rate. The *spectral response* of an ICCD camera is primarily determined by the input window and photocathode materials and the photo-cathode size used in the image intensifier tube. Early intensifier tubes (generation 2) used bismuth or multialkali photocathodes, which were the materials used earlier in the photocathodes of photomultiplier tubes. Present-day generation 3 image intensifier tubes employ semiconductor photocathodes made from semiconductor materials such as gallium arsenide and gallium arsenide phosphide. The photocathodes of generation 2 intensifiers were made by coating the material on a quartz window, which extended its spectral response down to 160 nm. The quartz window can be substituted for a magnesium fluoride window to provide response further down to approximately 120 nm. The mix and thickness of the photocathode can be adjusted to optimize the wavelength response in different regions. The spectral response of gallium arsenide, gallium arsenide phosphide, and indium gallium arsenide, respectively, is in the range of 350 to 950 nm, 280 to 820 nm, and 370 to 1,100 nm, respectively.

The *noise* and hence *sensitivity* of the ICCD is also governed by the image intensifier. There exists a noise component called dark current component, also called effective background illumination (EBI), which originates from a thermally generated charge in the photocathode. The dark current is generally not an issue when using short gate times.

A significant advantage of an ICCD over electron multiplying CCD (EMCCD) and CCD is its *optical shuttering properties*. An image intensifier inherently includes shutter functionality enabled by the application of a control voltage between the photocathode and the MCP. The image intensifier can be operated as a very fast optical switch, capturing an optical signal in typically a nanosecond. The minimum gate time depends on a number of factors but principally on the structure of the photocathode and the electronic gating circuitry. Gating operation allows the ICCD to capture instantaneous images of high-speed optical phenomenon while excluding extraneous signals. The intensifier can be repetitively gated at rates of up to 50 kHz for standard operation or up to 500 kHz for specially requested cameras. Although the CCD section of the camera cannot be readout at this rate, there are advantages in operating the optical gating independently. A repetitive signal can be sampled and the output of the intensifier summed on the CCD to integrate a larger signal that otherwise may not be visible.

Spatial resolution is usually defined as a number of line pairs that a camera can resolve per millimeter. Among several methods available for measuring the resolution of an optical system, modulation transfer function (MTF) is in common use. The MTF is a quantitative measure of the ability of an optical system to transfer various levels of detail from object to image. The MCP and the phosphor in the image intensifier tube both contribute toward degrading the MTF compared to the CCD. The modulation transfer function is a measure of the transfer of modulation (or contrast) from the subject to the image. In other words, it measures how faithfully the optical system reproduces (or transfers) detail from the object to the image. When a black-and-white stripe pattern with associated sine wave changes in brightness is focused on the photocathode, the image contrast drops gradually as the stripe pattern density is increased. The relationship between the contrast and the stripe density in line pairs per mm (lp/mm) is the MTF. Recent developments in finer phosphor deposition, reducing gaps, and reducing the bore of the microchannel plates has resulted in much better performance. However, resolution is typically limited to less than 60 lp/mm.

The *dynamic range* of the ICCD is governed by the CCD section and varies inversely with the gain of the ICCD. A higher dynamic-range CCD used in the ICCD will result in a higher dynamic-range ICCD camera. As the gain increases, the smaller signals that can be accommodated are compensated by the lower read noise to keep the dynamic range constant. When the read noise drops below a single photon level, the dynamic range of the ICCD starts dropping as the gain increases further. The frame rates of an ICCD are governed by the CCD specifications, especially the number of pixels and pixel readout rate.

6.3 Thermal Imaging

Image intensifier and CCD-based night vision devices described in the previous sections have a spectral response that is sensitive to the visible region of the electromagnetic spectrum with the trailing part extending slightly to the near-infrared region. Thermal imaging sensors on the other hand use focal plane arrays comprised of infrared sensing elements that respond to mid- (3–5 μm) and long infrared (8–14 μm) regions. The infrared energy radiated by the object to be imaged is incident on the thermal imaging sensor, which uses a series of complex mathematical algorithms to construct the image in a way that is visible to the viewer. A thermal imaging sensor does not need visible light for operation; it is able to see in total darkness. Thermal imaging sensors are far more expensive than their visible spectrum counterparts. In view of their military applications, high-end devices are often export-restricted. In the following sections, we discuss operational basics, types, and applications of thermal imaging sensors.

6.3.1 Operational Basics

A thermal imaging sensor makes use of the thermal radiation emitted by the target or scene of interest to generate its image. Essentially it is comprised of a front-end optical system, a 2-D array of infrared detector elements, and image-processing circuitry to produce output in the desired format. The front-end optical system focuses infrared radiation emitted by all objects in view on a 2-D array of infrared detector elements, which create a detailed temperature pattern of it called a thermogram. The thermogram is generated from several thousand points in the field of view of the detector array. The thermal imager measures very small relative temperature differences and converts otherwise invisible heat patterns into clear, visible images that are seen through either a viewfinder or monitor. Most thermal imaging sensors scan at a rate of 30 times a second. They can sense temperatures in the range of –20°C to + 2,000°C and can sense temperature differences as small as 0.1°C. In the next step, the temperature pattern is translated into electronic impulses. The signal processing unit converts these electronic impulses into data for the display. Figure 6.5 illustrates the concept of thermal imaging.

6.3.2 Types of Thermal Imaging Sensors

There are two distinctive detector technologies: *direct detection* (or photon counting) and *thermal detection*. In the case of direct detection, the detector element translates the photons directly into electrons. The charge accumulated, the current flow, or the change in conductivity is proportional to the radiance of objects in the scene. Detectors in this category include PbSe, mercury cadmium telluride (HgCdTe), indium antimonide (InSb), and platinum silicide (PtSi). All thermal imaging sensors based on direct detection technology, except those working in SWIR, use detectors cooled to cryogenic temperatures close to –200°C. Newer photon-type infrared sensors operating at elevated temperatures are now available. This has allowed solid-state thermal electric coolers and sterling coolers to be used. Cooled thermal imagers are highly susceptible to damage from rugged use, have a long cooling time of typically a few minutes,

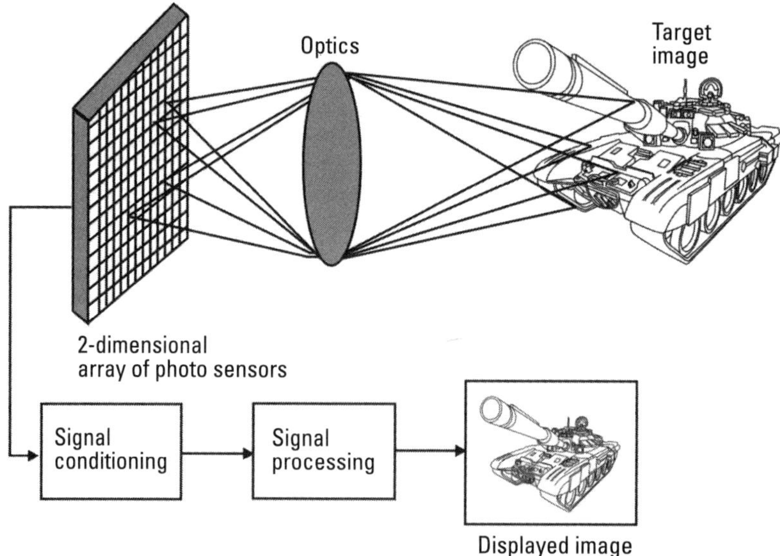

Figure 6.5 Thermal imaging.

a limited MTBF of a few thousand hours, high cost, large size and weight, and high electrical power consumption leading to short battery life. The biggest advantage of these detectors is excellent spatial resolution and sensitivity that results from detector cooling.

Thermal detection on the other hand uses uncooled detectors. They make use of secondary effects such as the relation between conductivity, capacitance, and expansion and detector temperature. Detectors in this category include bolometers, thermocouples, thermopiles, and pyroelectric detectors. These sensors operate at room temperature and are lightweight, which, for example, allows microbolometer thermal imagers to be mounted on helmets.

6.4 Generations of Night Vision Devices

Night vision technology has undergone substantial changes during its existence of more than 40 years. These changes have led to major improvements in the performance standards of night vision devices. Each substantial change in technology is associated with a new generation, and therefore we have seen several generations of night vision devices based on image intensifier tube technology as well as thermal imaging technology.

6.4.1 Image Intensifier Tube Technology

Image intensifier technology-based night vision devices have seen four generations up to now. We are currently in the fourth generation. The generations are briefly described in the following sections.

6.4.1.1 Generation 0 Night Vision Devices

The earliest night vision devices existed during World War II and the1950s. These devices, known as *generation 0 devices*, were based on image conversion rather than image intensification. The night vision device was primarily comprised of a photocathode that converted incident photons into electrons. The electrons were accelerated toward an anode by applying a positive potential to the anode. The device also had an infrared source of radiation called IR illuminator mounted on the device. The IR illuminator irradiated the target scene with infrared radiation. The infrared radiation reflecting off the target back to the night vision device is collected by its objective lens and focused on to the photocathode.

The M1 and M3 infrared night sighting devices called *sniper scope* or *snooper scope* were used by the United States Army during World War II and the Korean War to assist snipers. These night vision devices used an infrared source to illuminate the target and employed an S-1 photocathode primarily made up of silver, caesium, oxygen, and an anode. Electrostatic inversion and electron acceleration were used to achieve gain. Generation 0 night vision devices had several limitations, including the fact that the use of night vision devices was highly prone to being detected by an adversary in possession of infrared viewing devices, and the acceleration of electrons toward the anode produced image distortion and reduced the life of the tube.

6.4.1.2   Generation 1 Night Vision Devices

Genearion 1 night vision devices were an adaptation of generation 0 technology in the sense that both used a photocathode and an anode. A major deviation in generation 1 night vision devices from generation 0 devices was in the absence of an infrared source used in the case of the latter devices. Generation 1 devices depended on the ambient light provided by the moon and stars. Generation 1 tubes were connected in series to enhance sensitivity. These devices suffered from the shortcomings of generation 0 devices; namely, short tube life and image distortion. In addition, they did not perform as well in cloudy and moonless nights. Generation 1 devices were introduced during the Vietnam War in the 1960s. The AN/PVS-2 Starlight Scope and PNV/57 Tanker Goggles are examples of generation 1 night vision devices.

6.4.1.3   Generation 2 Night Vision Devices

Generation 2 night vision devices were the first to use an MCP for electron multiplication, leading to a significant increase in device sensitivity. Generation 2 devices were introduced during the 1970s. Since the MCP increased the number of available photoelectrons for a given number of light photons incident on the photocathode rather than merely accelerating them, the resultant images were relatively brighter and had significantly less distortion. The consequent light amplification was of the order of 20,000X, which resulted in much-improved performance even in very low-light–level ambient conditions of cloudy and moonless nights. The introduction of MCPs into the intensifier tube also obviated the need to connect the tubes in series as was done earlier in the case of generation 1 devices. This significantly reduced the size of the night vision device and made handheld devices and helmet-mounted devices a reality. The AN/PVS-4 and AN/PVS-5 are examples of generation 2 devices. Subsequently, generation 2 devices equipped with better optics, SUPERGEN tubes, led to the arrival of *generation-2+ night vision devices.*

The AN/PVS-4 was the first second-generation passive night sight designed and manufactured by Optic Electronic Corporation of Dallas and was extensively used by the United States Army during the Gulf War and Iraq War. It has now been replaced by third-generation weapon night sights. Equipped with an MX9644 image intensifier tube, the night sight had an image resolution of 32 lp/mm and star light detection and recognition ranges of 600 and 400m, respectively. The AN/PVS-5 is a dual-tube night vision goggle used for aviation and ground support. Manufactured by ITT Industries and Litton Industries and equipped with image intensifier tube MX 9916, the night vision goggles had an image resolution of greater than 20 lp/mm and starlight and moonlight detection ranges of 50 and 150m, respectively.

6.4.1.4 Generation 3 Night Vision Devices

Compared to generation 2 devices, generation 3 night vision devices had two distinctive changes. These included uses of a gallium arsenide photocathode and an ion barrier coating on the MCP. Due to its higher quantum efficiency or radiant sensitivity and its spectral response extending to the near-infrared region, a gallium arsenide photocathode enables target detection at longer ranges and in darker conditions. The light amplification factor increases from 20,000X in the case of generation 2 devices to a maximum of 50,000X in the case of generation 3 devices. The ion barrier film improves the tube life from 2,000 hours in the case of generation 2 devices to 10,000 in the case of generation 3 devices, even though it is at the cost of s slight reduction in radiant sensitivity due to fewer photo electrons being able to reach the MCP.

The AN/PVS-7, ATN NVG7, and AN/PVS-14 are some representative night vision devices employing third-generation image intensifier tubes. Designed by ITT Industries and Litton Industries and manufactured by ITT Industries, Litton Industries, Northrop Grumman, and L3 Communications, AN/PVS-7 is a single-tube passive/active night vision goggle. Active night vision enabled by a built-in LED allows operation in low-light–level situations. The device is waterproof and charged with dry nitrogen, enabling operation in extreme temperature variations. Important technical specifications of the device are as follows: the resolution is greater than 64 lp/mm, field of view is 40°, and starlight detection and recognition ranges are 325 and 225m, respectively. The device was extensively used during the Gulf War, Operation Enduring Freedom in Afghanistan, and armed conflict in Iraq.

ATN Night Vision Goggles NVG7-3 is similar to AN/PVS-7. Available in head or helmet-mounted versions for hands-free use, this device is also equipped with an infrared light source for closeup illumination in total darkness. It is available with a wide range of image intensifier options. Detection and recognition ranges are 180 and 100m, respectively. Resolution and field of view are 40 lp/mm and 40°, respectively.

The AN/PVS-14 is monocular autogated passive night vision device configured around a third-generation image intensifier tube type MX 11769. Designed by ITT Industries and manufactured by Litton Industries (now L3 Warrior Systems) and ITT Industries, it was introduced into service during 2000 and is extensively used by the United States Armed Forces and its NATO allies. It has a resolution greater than 64 lp/mm and a field of view of 40°. Starlight detection and recognition ranges are 350 and 300m, respectively. An autogating feature ensures optimum performance of the image intensifier tube by electronically adjusting the duty cycle of the photocathode gating voltage.

Generation 3+ offers improved performance specifications over generation 3 devices. Two important features associated with generation 3+ night vision devices are automatic gated power supply system and a thinned ion barrier layer. The absence of an ion barrier layer or its thinning improves luminous sensitivity even though it is at the cost of a slight reduction in the life of the tube. Mean time to failure (MTTF) is typically 15,000 hours in generation 3+ devices compared to 20,000 hours in generation-3 devices. Operationally, this is insignificant as an image intensifier tube seldom reaches 15,000 operational hours before it needs replacement.

Another common term encountered while discussing night vision technology is the Omnibus or OMNI. Omnibus or OMNI refers to the multiyear/multiproduct contracts of the United States Army for procurement of night vision devices from Exelis (formerly ITT Night Vision). Under these contracts, the company delivers generation 3 devices with increasingly higher performance. The current contract is for OMNI-VIII.

6.4.1.5   Generation 4 Night Vision Devices

Currently, there is no generation 4 night vision technology. There are only four generations of devices ranging from generation 0 to generation 3. Generation 4 night vision devices were initially conceived to use filmless and gated technology. The proposal was to remove the ion barrier film from the MCP introduced in generation 3 devices. Removal of the film was aimed at reducing background noise and enhancing the signal-to-noise ratio. It also would allow more electrons to reach the MCP so that the images were significantly less distorted and brighter. Introduction of an automatic gated power supply for the photocathode would enable the device to adapt instantaneously to light level fluctuations from a low light level to a high light level or from a high light level to a low light level. Removal of the ion barrier was also intended to reduce the halo effect seen around bright spots or light sources. While device performance improved, the absence of ion barrier film led to increased tube failure rates. Because of this, the idea of film removal was abandoned in favor of using a thinned film, creating what was known as generation 3+ described in the previous section.

6.4.2   Thermal Imaging Technology

There have been different generations of night vision devices based on thermal imaging technology. Each successive generation has incorporated not only a major change in the type of detector but also a major change in the optical systems used to image the target onto the detector.

Four distinct generations of thermal imagers have been designed based on IR detector technologies developed during the last 35 years, and classified according to the number of elements contained in each group. *First generation* thermal imagers contain single-element detectors, or detectors with only a few elements (1×3). A 2-D mechanical scanner was usually used to generate a 2-D image. The sensitivity of thermal imaging sensors of the first generation was limited by background radiation. This problem was overcome in *second generation* thermal imaging sensors by using modified front-end optics that reduced unwanted flux. However, it resulted in a fixed f-number for all fields of view. *Second generation* thermal imagers were vector detectors usually containing 64 or more elements. The 2-D scanner was somewhat simplified in the vertical direction to include only the interlace motion. *Third generation* thermal imagers contained dual-band 2-D arrays with several columns of elements and a dual/variable f-number optical system. These thermal imagers still scanned in one direction and performed time delay integration of the signal in the scanning direction to improve the signal-to-noise ratio. *Fourth generation* thermal imagers contain 2-D array detectors (160×120, 320×240, 680×480), called focal plane arrays, which do not require any scanning mechanism for acquiring the 2-D picture.

6.5 Categories of Night Vision Equipment

Night vision devices are available in different forms and package configurations to suit the requirements of different application environments. These include monoculars, binoculars, goggles, scopes, and cameras. Figure 6.6 shows representative package configurations of night vision equipment. Different forms of night vision equipment are briefly described in the following sections.

6.5.1 Night Vision Monoculars

Monoculars are devices with a single eyepiece and no magnification. They are compact and lightweight, which makes them ideal for head mounting. Many of the state-of-the-art monoculars can be attached to rifle scopes and mounted directly onto a weapon. They can also be mounted onto a weapon in front of a red dot sight to make it night-vision-compatible. They can also be adapted to a camera. Monocular when compared with a night vision goggle has the advantage that it allows the user to switch back and forth between the two eyes when feeling tired and allows the unaided eye to maintain its night adaptation and some of the peripheral vision. However, the night vision goggle feels more natural. The AN/PVS-14 is a representative monocular night vision device de-

Figure 6.6 Common forms of night vision equipment.

signed for use by individual soldiers in a variety of ground-based night operations. While the dark-adapted unaided eye provides situational awareness and vision of close-range objects, the night vision device provides long-range vision of potential threats and targets, thus making it a force multiplier in nighttime warfare.

6.5.2 Night Vision Binoculars

A *night vision binocular* is a device with two eyepieces and built-in magnification. These are too large and heavy to be head-mounted. Binoculars are primarily designed to magnify images at longer distances while standing stationary. They are therefore the preferred choice if the main task requires stationary long-range night viewing. The disadvantage is that most binoculars have fixed magnification, which is a problem while navigating and viewing close-in areas.

6.5.3 Night Vision Goggles

Night vision goggles (NVG) are devices that allow both-eye viewing, have no magnification, and can be head-mounted. There are two types of night vision goggles: single-image tube NVG and dual-image tube NVG. Both types have two eyepieces. Night vision goggles, like monoculars, are a great navigation device. Dual-tube goggles offer stereo vision, which provides still better depth perception allowing for improved navigation ability. Compared to a monocular, the disadvantage is increased size and weight and less versatility. As well, goggles offer limited adaptation to cameras and do not have the option to be mounted on rifles or adapted to rifle scopes. The AN/PVS-7 is a single-tube passive/active night vision pair of goggles. Active night vision enabled by built-in LED allows operation in low-light-level situations. The device is waterproof and charged with dry nitrogen, enabling operation in extreme temperature variations. Important technical specifications of the device are as follows: resolution is greater than 64 lp/mm, field of view is 40°, and starlight detection and recognition ranges are 325 and 225m, respectively. The device was extensively used during the Gulf War, Operation Enduring Freedom in Afghanistan, and armed conflict in Iraq. ATN Night Vision Goggles NVG7-3 are similar to the AN/PVS-7. Available in head- or helmet-mounted versions for hands-free use, this device is also equipped with an infrared light source for closeup illumination in total darkness. It is available with a wide range of image intensifier options. Detection and recognition ranges are 180 and 100m, respectively. Resolution and field of view are 40 lp/mm and 40°, respectively.

6.5.4 Night Vision Scopes

There are two types of night vision scopes. One type is a night vision scope just like a regular rifle scope and is also known as a weapon sight. The other type of scope, also known as a day/night system, attaches to or mounts in front of a regular rifle scope to make it nighttime-compatible. This type may be removed, thereby allowing use of a regular scope during the daytime. Dedicated night vision scopes are generally preferred for better performance. A good alternative to using a dedicated night vision scope or a day/night system in front of regular rifle scope is attaching a monocular like a PVS-14 to the eyepiece of your regular rifle scope with day/night adapter or mounting the monocular directly to the rifle in front of a night-vision–compatible red dot sight. The D-760 and D-790 from Bering Optics are popular third generation night vision rifle scopes. The D-760 is a third generation night vision scope with fixed 6X magnification and is designed for medium- to long-range engagements. The D-790 is an improved update of the D-760 6X scope. Use of state-of-the-art optics and coating technologies improve range, contrast, clarity, and light transference by more than 20% over the D-760.

6.5.5   Night Vision Cameras

The technology of *night vision cameras* and their related aspects were discussed in Section 5.2 of Chapter 5 on surveillance cameras. Applications of night vision cameras are further discussed in Sections 6.6.1 and 6.6.2.

6.6   Applications of Night Vision Equipment

Night vision equipment finds extensive use in a wide range of applications in the military and in law enforcement, surveillance, security, navigation, hunting, and wildlife observation. These devices are used not only by the military, police forces, and law enforcement agencies for navigation, surveillance, and targeting; they are also used by private detectives, firefighters, hunters, and nature enthusiasts. Both image enhancement and thermal imaging night vision equipment have preferred areas of application to which they are better suited.

6.6.1   Image Enhancement Night Vision Devices

Night vision technology is used extensively by the military and police due to the instant tactical advantage they get through the ability to see in the dark and precisely detect and even identify the target at a distance. The AN/PVS-14, a monocular night vision device, is one such versatile night vision scope extensively used by armed forces and special operations units.

Night vision cameras are extensively used for surveillance and security applications, especially for around-the-clock surveillance indoors and in controlled environments. Night vision security cameras provide enhanced security by improving video images in low light conditions. With the help of infrared illumination, they can also be effectively used in near total darkness conditions. Perhaps the greatest feature of night vision cameras is that they are true 24-hour cameras, as they are the only cameras with the ability to record any event in a protected area around-the-clock when connected to a digital video recorder. The resolution typically varies from 400 TV lines to 700 TV lines. Security cameras with long-range night vision find extensive use in monitoring large parking lots, huge and dark warehouses, apartment complexes, and other similar assets on a 24/7 basis. Traffic monitoring and monitoring of suspicious activity in public places, security installations, and critical national assets are the other important application areas of security cameras with night vision capability.

6.6.2   Thermal Imaging Night Vision Devices

Thermal imaging devices are used on land-based armored fighting vehicles and naval vessels and aerial platforms such as aircraft, helicopters, and missiles for surveillance, target acquisition, and tracking applications. Some common nonmilitary applications of these systems include surveillance of living things, search and rescue operations during firefighting, detection of gas leaks, monitoring of volcanoes, and detection of heat in faulty electrical joints.

In military applications, thermal imaging offers several distinct advantages. The first advantage is its immunity to detection by an adversary as it is a passive sensor that does not emit any radiation for generation of image. Second, it is extremely hard to camouflage the target from the sensor as it propagates through atmospheric obscurants better than sensors operating in the visible spectrum. One of the limitations of thermal imaging sensors is that it is hard to discriminate friend from foe. Friendly forces can use heat beacons to overcome this shortcoming.

A thermal imager can be a great asset in firefighting operations. Its ability to see through smoke and debris allows firefighters to find people who have passed out, are fighting for survival, or are too scared to come out. A thermal imager can also tell a firefighter if a door is hot and possibly contains a fierce blaze on the other side. Thermal imagers designed for firefighting missions are designed to operate in very rough and tough environmental conditions.

Thermal imagers are extensively used by law enforcement agencies and armed forces for a variety of applications in navigation, detection, and targeting. Thermal imagers allow them to detect potential threats without exposing their location to the adversary. The state-of-the-art thermal imaging rifle scopes have become rugged enough to withstand the abuse of recoil, making them extremely popular with armed forces. While thermal imagers are incredibly effective when it comes to detecting human beings or animals, discrimination between a friend and a foe may be a challenge. This may be an issue in life and death situations. Thermal imaging cameras are one of the most effective tools for surveillance because they work equally well during the day and night. A regular CCTV camera is limited by its need for light, and night vision does not

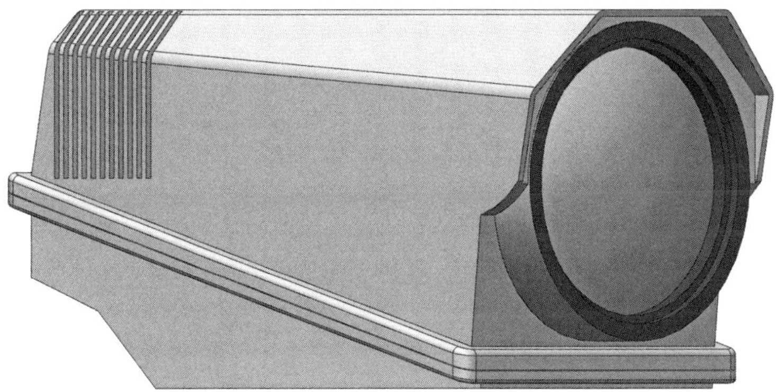

Figure 6.7   HRC-E FLIR.

function during the day. The chance to see through smoke and fog also gives thermal a leg up on other surveillance techniques. Figure 6.7 shows a thermal imaging sensor type HRC-E FLIR designed for border and coastal surveillance. The sensor offers continuous zoom between 25 and 2° and a long-range detection capability. The sensor can detect a man-sized target from greater than 15 km.

State-of-the-art sensors intended for surveillance, target detection, and tracking applications employ multisensor configurations often combining a visible spectrum sensor with an infrared sensor. The final image in this case is the result of the fusion of imaging data from the two sensors.

Selected Bibliography

Biggs, K., M. Burris, and M. Stanley, *The Complete Guide to Night Vision*, CreateSpace Independent Publishing Platform, 2014.

Bantutov, E., *Night Vision Devices? It Is Simple!* Lambert Academic Publishing, 2015.

Holst, G. C., and T.e S. Lowheim, *CMOS/CCD Sensors and Camera Systems*, Bellingham, WA: SPIE Press, 2011.

Hradaynath, R., *Introduction to Night Vision Technology*, Defence Research & Development Organization, India, 2002.

Hradaynath, R., *Selected Papers on Night Vision Technology*, Bellingham, WA: SPIE Press, 2001.

Nakamura, J., *Image Sensors and Signal Processing for Digital Still Cameras*, Boca Raton, FL: CRC Press, 2005.

Vollmer, M., and K.- P. Mollmann, *Infrared Thermal Imaging: Fundamentals, Research and Applications*, Weinheim, Germany: Wiley-VCH, 2017.

Williams, T., *Thermal Imaging Cameras: Characteristics and Performance*, Boca Raton, FL: CRC Press, 2009.

7

Explosive Detection and Identification

Conventional and improvised explosive devices with potential for causing extensive damage to life and property are the most commonly exploited threats in the hands of terrorists and other antinational elements. Reportedly, in the recent past, explosive threats have claimed far greater number of lives than those caused by the combined use of chemical, biological, and radiological substances. Beginning with the explosive threat spectrum and the desirable features that an explosive detector must possess, this chapter goes on to describe various techniques available for detection and identification of explosive agents with a focus on laser-based techniques for standoff detection of explosives. An outline on salient features of some representative explosive detector systems concludes the chapter.

7.1 Introduction

Given the enormous threat that conventional and improvised explosive devices pose to maintaining peace and harmony, it is essential that security and law enforcement agencies are equipped with instrumentation needed for detection, identification, and neutralization of explosive threats. While such equipment is in place in public places involving mass movement of people and transportation of goods, such as airports, railway stations, and ports, other places housing critical infrastructure and strategic assets are also important, as is protection from roadside bombs and monitoring of container transportation. While timely detection and safe neutralization of an explosive saves many a human life and damage to public property, knowing the type of explosive even postdetonation helps in identifying the source of origin of the explosive. Before we discuss the technologies used for the purpose, we briefly describe existing and foreseeable threat scenarios and outline desirable features in explosive detection and identification equipment that would address the challenges arising from these threat scenarios.

7.1.1 Threat Scenarios

An understanding of likely explosive threat scenarios and use of explosive detection equipment is important in the context of deciding on equipment best suited to the intended application. Possible application scenarios vulnerable to explosive threats include

1. *Routine screening of a large number of personnel,* such as at the entrances to airports, railway stations, and other important installations when it is desired to screen all incoming persons for possession of any explosive materials. This would therefore also involve checking hand-carried packages such as parcels, briefcases, and backpacks.

2. *Vehicle screening,* occurring primarily at high-security checkpoints with the activity involving screening both packages and people in the vehicle.

3. *Screening of mailed and shipped items* arriving at a particular location or addressed to specific people.

4. *Screening of small numbers of people, vehicles, or mailed/shipped items,* primarily when investigating a suspicious person or item, such as an abandoned package, in a public place.

5. *Bomb search* involving screening of a building, means of public transport such as buses, trains, aircraft, or other areas, when there is sufficient reason to believe a bomb may have been planted.

6. *Special situations or events,* such as gatherings during festival celebrations, political meetings, or visiting foreign dignitaries, requiring increased security measures with regard to explosives detection.

7. *Protection of critical infrastructure,* including military installations, nuclear power stations, dams, communication facilities, water resources, oil refineries, and space assets.

When it comes to assessing the suitability of an explosive detector for a given application, the involved technology, probability of detection, false alarm rate, capability to operate in real time, portability and field worthiness, adaptability to new and emerging threats, skill level of the operator, radiation hazards, if any, and size and cost are important considerations. In terms of involved technology, there are bulk and trace detection techniques. Both point detection and standoff detection equipment using bulk and trace detection techniques are available. Not all the above-mentioned features are required for all the different threat scenarios listed above. Similarly, not all techniques are suitable for building detection equipment for all applications. For example, some of the bulk detection techniques such as X-ray scatter cannot be used for personnel screening due to health hazards of X-ray radiation, altthough it may be used for baggage screening. Standoff detection equipment is preferred in situations where threat is fast approaching and time is critical, such as finding suicide bombers, roadside bombs, and wide-area surveillance. For screening of human beings at places such as airports and railway stations, detection equipment needs to have high throughput as it involves screening a large routine inflow of people. As well, the detection equipment must necessarily have a high probability of detection without having an unacceptable false alarm rate. Size and cost may not be that important. Adaptability to new explosive threats arising out of homemade explosives is important for almost all scenarios. Field-worthiness and portability are important for equipment to be used for protection against roadside bombs.

7.1.2 Explosive Detectors: Desirable Features

The suitability of explosive detection equipment for a given application scenario is assessed based on certain performance parameters and operational considerations. Key performance parameters include

1. Sensitivity or probability of detection
2. Selectivity;
3. Resolution;
4. Dynamic range;
5. False alarm rate.

Key operational considerations include

1. System throughput;
2. Portability and field-worthiness;
3. Safety issues;
4. Privacy issues;
5. Cost including cost of consumables;
6. Required skill level of operator;
7. Working environment;
8. Types of explosives to be detected.

7.1.2.1 Performance Parameters

The key performance parameters are briefly described next.

Sensitivity is defined as the minimum quantity of an explosive material that the system can detect under a given set of conditions. *Limit of detection* (LOD) is a related parameter. It is the smallest amount of explosive that the explosive detector system will respond to with an alarm. Note that the specified LOD is generally the best-case value and not the value under typical field conditions. Sensitivity or LOD is a measure of probability of detection. *Probability of detection* depends on a large number of parameters, which include explosive packaging, the way the package is scanned, swiped, or vacuumed, skill of the operator, countermeasures used by an adversary to prevent detection, and environment of the explosive.

Selectivity is defined as the ability of the explosive detection system to detect a specific explosive molecule in the presence of interferents. Poor selectivity adversely affects both probability of detection and false alarm performance. Interferents are nonexplosive chemicals that could either be naturally occurring compounds in the environment or added intentionally to the explosive mixture to avoid detection. The interfering molecules could lead to failure to detect an explosive when one is present or indicate its presence when none is present.

Resolution is the smallest concentration variation that can be detected when the concentration is continuously changed.

Dynamic range is the explosive concentration from LOD to maximum concentration that can be reliably detected.

The *false alarm rate* is a critical parameter when evaluating an explosive detector system. There are two types of false alarms: *false negative* and *false positive*. The explosive detection system should desirably have low false alarm rates. A *false negative alarm* occurs when the system fails to detect the presence of an explosive when one is present. A *false positive alarm* is said to occur when the system gives an alarm, indicating the presence of an explosive even in the absence of one. The false positive is the number of positive detections in the absence of explosives divided by the total number of tests. The false negative alarm rate is the number of times the system fails to detect an explosive when one is present divided by total number of tests performed. The false positive alarm rate tends to increase with increase in sensitivity (or decrease in LOD). An increase in LOD threshold may lead to increase in false negative alarm rate.

7.1.2.2 Operational Considerations

Various operational factors to be considered in addition to the performance parameters are briefly described next.

Throughput is the rate at which a sample can be analyzed. In the case of screening of packages or human beings, the throughput of the detection system may be expressed in persons per minute or packages per hour. High throughput rate is important in high-volume applications such as screening at airports where checking of each person and package is mandatory.

Portability and field-worthiness are a requirement for explosive detectors intended for use in the field conditions such as protection against roadside bombs, detection of explosives in trains, buses, and aircraft, and sanitization of specific locations suspected to be rigged. This is not important if the equipment is to be installed in a single fixed location.

Potential health hazards and public safety are highly critical considerations for detection equipment intended for the screening of human beings for explosives. One such example is the health hazards involved in the use of X-ray backscatter equipment for personnel screening to detect concealed explosives and other contraband items. Because of these health hazards, X-ray body scanners have given way to safer technologies such as millimeter-wave imaging.

Privacy is again an important issue in personnel screening. Screening people even with low-dose backscatter X-ray equipment otherwise certified to be safe could provoke strong reservations from people as the X-ray image could show a rather revealing image of the person's body.

Cost of the equipment needs to be considered along with its maintenance costs, which includes cost of consumables. The cost of equipment to an extent depends on the performance parameters and other related features that come along with it. The choice therefore comes down to intended application and its requirements.

Required skill level of the operator tells us how easy it is to use the equipment. It varies from equipment to equipment. There is complex detection equipment used to perform forensic analysis and there are systems including field-portable units used for routine screening of people and baggage. While the former requires handling by experts, the latter may be used by officials after only brief training. Equipment setup and warmup times are also important.

Environmental factors such as temperature, humidity, dust, wind condition, and above all, explosive material contamination must be considered while assessing the suitability of an explosive detection system for a given application. Explosive material contamination could lead to a false positive alarm if the sensitivity of the equipment were too high. In that case, one would need to set the limit-of-detection threshold higher than normal for explosives detection.

The *group of explosives* that the equipment can detect also influences its suitability for a given application. Some explosive types are more commonly used than others for certain specific targets. For example, equipment that can effectively detect black powder would be more suitable for mail screening even if it weren't as effective in detecting plastic explosives. However, it would be desirable that the equipment can detect not only all known explosives but also is adaptable to include new explosive threats due to improvised explosive devices and homemade bombs.

7.2 Explosive Detection Technologies

Explosive detection methodologies are broadly categorized as *bulk detection* and *trace detection*. Both methodologies have their own merits and limitations and are considered as complementary. An overview of bulk and trace detection approaches and different techniques available in each of the two categories is presented in the following sections.

7.2.1 Bulk Detection

Bulk explosive detection involves the detection of a macroscopic mass of explosives material through measurement of material characteristics such as mass, density, and effective atomic number, also called Z-number. Most explosives such as RDX, TNT, and HMX contain hydrogen, carbon, nitrogen, and oxygen. Although these elements are present in all materials, the elemental ratios and concentrations are material-specific and this forms the basis of detecting the possible presence of explosives. Carbon/oxygen and nitrogen/oxygen ratios are used to discriminate between explosive and safe materials. There are further two broad categories of bulk detection technologies: *imaging technologies* and *nuclear-based technologies*. Important imaging technologies include single-energy X-ray, dual-energy X-ray, backscatter X-ray, millimeter-wave imaging, terahertz imaging, computed tomography, fluoroscopy, and dielectrometry. Important nuclear-based technologies include thermal neutron analysis, pulsed fast neutron analysis, nuclear magnetic resonance, and nuclear quadrupole resonance. Each of these technologies is briefly described next.

7.2.1.1 X-Ray Machines

X-ray scanners used for baggage screening produce X-ray images to look for guns and other weapons and the presence of explosive materials. *Single-energy X-ray machines* use high energy X-ray beams of single energy and the image indicates the degree of absorption of the X-rays. These machines look for wiring, fusing, and metal parts and therefore are useful for bomb detection. They do not provide enough information for detection of explosive materials. Typical applications include cargo and vehicle screening. There are two types of *dual energy X-ray machines*: one that uses a broad X-ray beam and two detectors, and one that uses low-energy and high-energy X-ray beams. Two independent images thus produced at two energy levels are processed in a computer to generate a final image that allows identification of materials based on their shape. Multi-axis dual-energy systems also allow discrimination based on material thickness. Automated X-ray detection equipment based on dual-energy and dual-energy/multiaxis technology are commercially available. The operational principle of a *backscatter X-ray machine* was discussed earlier in Section 5.7.2. These machines provide both a standard transmission X-ray image and a backscatter X-ray image of the material in question. While transmission X-ray image is used to identify high-density materials such as metal objects, the backscatter X-ray image highlights organic materials such as plastic explosives. One of the limitations of conventional backscatter X-ray machines is that low Z-number materials could remain hidden behind highly dense materials. Double-beam backscatter machines overcome this problem but at an increased cost of the machine.

7.2.1.2 Millimeter-Wave Imaging

Millimeter wave technology-based imaging allows the operator to see though lightweight materials such as clothing. The operational principle is the same as that of a backscatter X-ray scanner except for frequency or wavelength of operation. Millimeter-wave imaging-based scanners are used for screening of personnel for concealed weapons and explosives. Millimeter-wave scanners were discussed earlier in Section 5.7.3.

7.2.1.3 Terahertz Spectroscopy

Terahertz spectroscopy is a useful tool in the detection and identification of explosives. Terahertz radiation exhibits reasonable penetration in materials such as textiles, plastics, wood, and sand, which allows visualization of substances hidden behind these materials. When these materials are irradiated with terahertz radiation, the absorption and reemission of radiation that passed through the substance is specific to it and can therefore be used for its identification. Terahertz-imaging–based body scanners are used for detection of weapons and explosives concealed underneath clothing. Explosives like RDX, TNT, HMX, and PETN have exhibited characteristic features in 0.3- to 3-THz band.

7.2.1.4 Computed Tomography

Computed tomography (CT) scans use X-ray radiation to produce a 3-D image of the object under screening or examination. The 3-D image in turn is produced by computerized processing of a large number of 2-D images taken at different angles by rotating around the object both the X-ray source and the detector located diametrically opposite to each other. CT scan equipment is also used to take selective CT scans of a suspicious area rather than scanning the object from all possible angles. CT scan images have much improved density resolution although they have coarser spatial resolution. The density profile of the object allows differentiation of explosive-like materials from most innocuous low Z-number materials. Also, a 3-D image allows detection of hidden and obscured objects. CT scan equipment is complex and expensive and the throughput is also low. CT scan instruments used for bulk detection of explosives are commercially available. The CTX-5500 airport baggage scanner from *InVision* Technologies, eXaminer SX airport baggage scanner from *Gate*

Technologies, and ClearScan™ cabin baggage screener from *L3 Communications Security and Detection Systems* are some representative examples.

7.2.1.5   Fluoroscopy

Fluoroscopy is also an X-ray imaging technique that uses low energy X-rays to form a still or dynamic image of the object under screening. The equipment operates on detection of transmitted X-ray radiation through the object. This equipment is portable and can be transported and set up by one person. These machines are generally used for screening of mail and small packages for detection of explosives and contraband material. The MAILGUARD X-ray screening system from Control Screening LLC and SCANMAX-15/20/25 postal X-ray scanners from Scanna MSC Inc. are some representative examples of screening equipment employing fluoroscopy.

7.2.1.6   Dielectrometry

Dielectrometry is an imaging technique that uses a low-energy microwave field to irradiate the object under screening. It is commonly used for personnel screening. It compares the dielectric and loss properties of any concealed objects when placed in a microwave field with the known values of the human body to detect any anomalous areas where the dielectric properties are different. The equipment employs nonionizing radiation and is therefore safe. It also protects the privacy of individuals. The limitation is that it gives information only on anomalies and not explosives.

7.2.1.7   Nuclear Magnetic Resonance

Nuclear magnetic resonance (NMR) is a nuclear-based technology. All nuclear-based technologies interrogate the nuclei of the material for explosive detection. NMR distinguishes between different chemical species based on magnetic resonance signals. NMR works on the nuclei having spin $I \neq 0$ like ^1H, ^{13}C, or ^{14}N with ^1H being the most important NMR probe. Detection is based on measuring changes in orientation of the spins of already orientated spins in the external magnetic field due to application of element-specific radio frequency excitation. Due to the requirement of a large homogeneous magnetic field to make measurements of magnetic resonance signals, it is impractical to have field equipment based on this technique. Use of this technique also runs the risk of certain benign materials such as magnetic recording media getting erased by the external magnetic field.

7.2.1.8   Nuclear Quadrupole Resonance

Nuclear quadrupole resonance (NQR) detects explosives based on nitrogen quadrupole detection when the quadrupole nuclei present in explosive material are exposed to a pulsed radio frequency field. On exposure to the RF field, the quadrupole nuclei move to a higher energy state. They then return to a lower energy state, thereby releasing energy on removal of the RF field. The released energy is characteristic of a type of atom and crystal structure. Advantages include low false alarm rate, probability of detection being independent of explosive shape for a given explosive mass, and absence of any ionizing radiation. Limitations include the requirement of proximity to the to-be-screened object, incapability to detect all types of explosives, and explosives hidden by metal shields. To overcome these limitations, the NQR technique may be used in combination with a complementary technique.

7.2.1.9 Thermal Neutron Analysis

Thermal neutron analysis, also known as *thermal neutron activation,* depends for its operation on the interaction of a low-energy neutron beam with the nitrogen nuclei of the matter under inspection, leading to absorption of the thermal neutron and subsequent emission of a high-energy gamma ray with a characteristic energy of 10.8 MeV. While nitrogen nuclei have a strong interaction with thermal neutrons, interactions with carbon and oxygen nuclei are very weak. Detection of any emitted 10.8-MeV gamma rays indicates the presence of nitrogen, which indicates a high probability of the object containing explosive material.

As both neutrons and gamma rays readily pass through most materials, the technique is used for a variety of applications including baggage screening, vehicle screening, and detection of unexploded ordnances. However, use of a low-energy neutron beam prohibits its use for inspection of large cargo containers. The technique is not suitable for personnel screening due to unacceptable radiation hazards. Other disadvantages include its incapability to detect all kinds of explosives and vulnerability to false alarms due to nitrogen containing nonexplosive materials.

7.2.1.10 Pulsed Fast Neutron Analysis

Pulsed fast neutron analysis, also known as *pulsed fast neutron activation,* uses ultrashort fast neutron pulses of the order of nanoseconds that interact with the nuclei of interest to generate the characteristic gamma ray emission. The pulsed fast neutron analysis technique can produce information on several elements including carbon, hydrogen, nitrogen, and oxygen. Use of ultrashort fast neutron pulses also allows determination of the location of detected explosive material. The capability to detect most explosive-like materials, screen large cargo containers, and determine the 3-D location of the material of interest are the key advantages. Radiation hazards, system complexity, and high cost are major concerns.

7.2.2 Trace Detection

While the bulk detection methods of explosive detection seek the actual explosive material, trace detection methods work on microscopic levels of explosive particles or vapor available in proximity to the explosive material itself, including those present on the skin or clothing of a person who would have handled explosives. Vapor detection methods rely on gas-phase molecules emitted from a solid or liquid explosive with the concentration of explosives in the air depending on the vapor pressure of the explosive material and other factors, such as the packaging of explosive, the time-period the explosive material was present in a given location, and air circulation in that location. Particulate detection methods detect microscopic particles of the solid explosive material that adhere to surfaces either through direct contact with the explosive or indirectly through contact with someone's hands who has been handling explosive material. While particulate sampling requires direct contact to collect a sample from a contaminated surface; vapor sampling requires no contact.

Some of the important trace detection methodologies include ion mobility spectrometry, chemiluminescence, thermo-redox, SAW, chemical reagent, and mass spectrometry, and laser-based spectrometric techniques, such as infrared spectroscopy, Raman spectroscopy and its variant ultraviolet resonant Raman spectroscopy, LIBS, LIF, and laser photoacoustic spectroscopy (LPAS). Except for the last four techniques, which are discussed in the next section under the heading of standoff detection, the other techniques are briefly described below.

7.2.2.1 Ion Mobility Spectrometry Sensors

IMS detects explosives by measuring the drift time of ions moving through the drift region of the instrument. Ions are produced in the ionization region of IMS detector where the explosive molecules are made to interact with electrons. The explosive molecule sample is collected by the operator either by drawing in air near the object or by swiping a surface and then delivered to the ionization region of the IMS detector. Most IMS detectors have both particulate and vapor sample collection features. The drift time is a complex function of the charge, mass, and size of the ion and is of the order of a few milliseconds. IMS systems are portable and easy to use. Throughput is a few samples per minute and warmup time is of the order of 10 minutes for portable systems and higher for portals. IMS systems need periodic recalibration. The resolution of peaks in the IMS spectrum resulting from two different materials forming ions of similar size and mass is another limitation. However, this is overcome by using a GC column. The GC column separates each material in the sample before it is introduced into the IMS. Stand-alone IMS, GC/IMS, and other IMS-based explosive detection systems are commercially available for applications including package, personnel, and vehicle screening.

7.2.2.2 Chemiluminescence Sensors

Chemiluminescence is the emission of light from a chemiluminescent reaction. Two chemicals react to form an excited intermediate whose subsequent transition back to the ground state releases energy in the form of light. One such chemiluminescent reaction leading to emission of infrared light is from excited nitrogen compounds. As most explosive materials contain nitrogen in the form of nitro (NO_2) or nitrate (NO_3) groups and most explosive material taggants used in plastic explosives contain nitro (NO_2) groups, chemiluminescence is used for detection of explosives by measuring the amount of emitted infrared light. The infrared light is a measure of the amount of NO present in the material, which in turn is indicative of the amount of original nitrogen containing explosive material. Chemiluminescence sensors cannot be used to detect explosives that are not nitro-based. As well, chemiluminescence sensors cannot identify explosives. For this, a GC/chemiluminescence (GC/CL) sensor is used. A GC/CL sensor is a chemiluminescence sensor with a GC column used at the front-end of the sensor. CL sensors are portable, easy to use, and require a setup time of less than a minute to a few minutes. Most CL sensors have a vapor/particulate sample collection feature. The E-3500 from Scintrex Trace Corporation and EGIS-II/EGIS-III from Thermo Electron Corporation are some representative explosive detection systems using CL and GC/CL technologies, respectively.

7.2.2.3 Thermal Redox Sensors

Most military-grade explosives are nitro compounds containing an abundance of NO_2 groups. *Thermal redox* sensor draws in the sample, which is passed through a concentrator tube to selectively trap explosive-like materials. Thermo-redox technology of explosive detection is based on thermal decomposition of explosive material to release NO_2 molecules; which are subsequently reduced and detected. Thermal redox explosive sensors are portable and easy to set up and use. Since these sensors detect the presence of NO_2 groups in explosive material, they cannot distinguish explosives from other potential interferents containing NO_2 groups. As well, having detected the presence of a nitro-based explosive material, they cannot identify the explosive. The EVD-2500 and EVD-3000 from Scintrex Trace Corporation are common explosive sensors using thermal redox technology.

7.2.2.4 SAW Sensors

SAW sensors are piezoelectric crystals with chemically selective coatings on the sensor surface. SAW sensor detects explosive materials by measuring the change in the resonant frequency of the sensor caused by absorption of chemical vapors into chemically selective coatings on the sensor surface. Change in resonant frequency is proportional to concentration of chemical vapors. Many SAW sensor coatings have unique physical properties that allow reversible adsorption of chemical vapors. In a combined GC/SAW-based system with a GC as the front-end to the SAW sensor, unique retention times of the GC allows identification of detected explosive molecules. SAW sensors can also be used to detect chemicals other than explosives. The Znose models 4200, 4300, 4600, and 7100 from Electronic Sensor Technology, Inc. are common chemical and explosive sensors using GC/SAW technology.

7.2.2.5   Chemical Reagent-Based Sensors

Reagent-based explosive detection, also called *color-change–based explosives detection,* is based on observation of a color change after a specific chemical reagent in aerosol or liquid form is deposited on the residue sample. Sample collection is through particles only and is taken by wiping the contaminated surface with a special test paper. The operator deposits chemical reagents in a specific order and observes color changes with each reagent addition to detect the presence of explosive material. The order in which a reagent is added is critical. Reagent in aerosol form is used for detection of explosives such as RDX, PETN, Semtex-H, and smokeless powder. Liquid reagents are generally used for detection of improvised chlorate/bromate and peroxide-based explosives. The sensor is portable, easy to use, and low cost. The disadvantage is that a positive detection of explosive is highly dependent on color interpretation of the operator and sample concentration. The EXPRAY kit (Model: M1553), Drop-EX kit (Model: M1584), and PDK kit (Model: M1582) from Mistral Security Inc. are common explosive sensors using color- or reagent-based technology.

7.2.2.6   Mass Spectrometry

Mass spectrometry (MS) is a potent laboratory tool for detection and identification of explosives. A mass spectrometer when used for detection and identification of explosives functions as follows. Trace molecules are sampled and ionized. The ionized molecules are passed through a mass filter and then identified based on their charge-to-mass ratio. Most MS systems used for explosive detection use a gas chromatograph as front-end. Gas chromatography and mass spectrometry are compatible techniques in the sense that both take a sample in the vapor phase and deal with about the same amount of the sample; of the order of less than 1 ng. The GC/MS explosive sensor allows different molecules detected with the mass spectrometer to be specifically identified based on their unique GC retention times. MS has excellent specificity for identification. The disadvantages include relatively long analysis time and some MS systems requiring a gas supply or vacuum pump. The MM2 from Bruker Daltonics and CT-1128 from Constellation Technology Corporation are some representative GC/MS explosive sensors.

7.3 Standoff Detection of Explosives

Both bulk detection methods such as X-ray imaging, terahertz spectroscopy, and millimeter-wave imaging and laser-based trace detection methods have been used for detection of explosives from standoff distances. Bulk detection methods capable of standoff detection of explosives were discussed earlier in Section 7.1. Standoff detection of explosive agents using laser-based trace detection methods is one of the most widely researched technologies internationally. It is a great technological challenge for the technology to mature to an extent where it can be transformed into a product usable in the kind of environment and field conditions usually encountered in homeland security related applications. One of the major problems comes from a decrease in the intensity of a backscattered light signal due to wavelength-dependent absorption and scattering losses and the intensity decreasing inversely with distance squared. The problem is compounded by the fact that the trace levels associated with common explosive agents are extremely low, in the range of a fraction of parts per billion to a few parts per million for common explosives. The second major problem pertains to unique identification of a targeted explosive agent in a background of interferents. Many chemical agents have atomic compositions including sulfur, phosphorus, fluorine, and chlorine in addition to nitrogen, oxygen, hydrogen, and carbon present in organic molecules. Thus, the detection methodology needs to be highly sensitive and selective. Laser-based spectrometric methods have the potential of being fast, sensitive, and selective with the ability to detect and identify a wide range of explosive agents. These are also upgradeable to handle new threats. Atmospheric transmission at the wavelengths concerned is an important factor while assessing suitability of a given standoff detection methodology. Commonly used techniques are LIBS, Raman spectroscopy and its variants, LIF spectroscopy, IR spectroscopy, and laser photoacoustic spectroscopy. These are discussed next.

7.3.1 Infrared Spectroscopy

Infrared spectroscopy based on transmission, reflection, or absorption properties has long been a potent technique for detection and identification of chemical species including atmospheric pollutants, hazardous chemical agents, and trace explosives. Fourier transform infrared (FTIR) spectroscopy is the preferred technique of infrared spectroscopy used for molecular fingerprinting of chemical species. It overcomes the limitations of dispersive instruments like slow scanning process by allowing simultaneous measurement of all infrared wavelengths. In FTIR, infrared radiation from a black body source is first passed through an interferometer where the spectral encoding of the infrared radiation takes place. The resulting interferogram signal has the unique property that every data point that makes up the signal has information about every infrared wavelength coming from the source. This implies that as the interferogram is measured, all wavelengths are being measured simultaneously. The interferometer uses a reference laser such as a helium-neon gas laser for precise wavelength calibration, mirror position control, and data acquisition timing. The interferogram signal then exits the interferometer and irradiates the sample under test. A part of the radiation is either reflected off the surface of sample or transmitted through it. Certain specific wavelengths unique to the sample under test are absorbed. The transmitted or reflected radiation, as the case may be, is detected. The detectors used are specially designed to measure the interferogram signal. The detected analog signal is digitized and then processed in a computer to generate its Fourier transform. The Fourier transform of the detected signal represents the infrared spectrum showing absorption peaks unique to the molecular fingerprint of the sample. Since no two molecular structures produce the same infrared spectrum, the spectrum constitutes the fingerprint of the sample. This forms the basis of detection and identification of specific chemical species. The fingerprint of the material under testing is compared with the known fingerprints of various chemical species stored in the memory of the instrument for positive identification of the unknown species. The size of absorption peaks is used to determine the concentration of identified species in the sample under test. Figure 7.1 shows a simplified block schematic arrangement of an FTIR spectrometric setup.

Resonant infrared photothermal spectroscopy is another promising technique of infrared spectroscopy used for rapid and selective standoff detection of trace explosives. In this technique, light from a tunable mid-infrared laser source such as a QCL is directed at the target surface. The laser wavelength is tunable over the absorption fingerprint of the targeted explosive materials. As the laser wavelength is tuned, there will be resonant absorption at one or more wavelengths unique to the targeted explosive substance. The absorption of laser radiation at wavelength/s unique to the explosive substance under examination produces a thermal contrast to the surrounding surface. The thermal emission signal from the surface is directly proportional to the absorption coefficient.

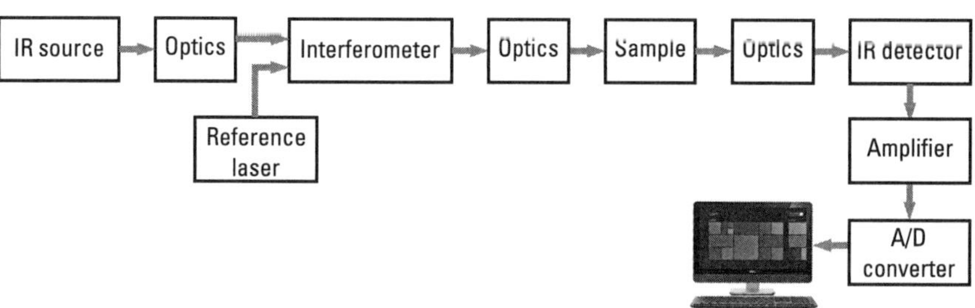

Figure 7.1 Block schematic arrangement of FTIR spectrometric setup.

The thermal response is observed and imaged by an infrared focal plane array detector as the laser wavelength is tuned across the resonant absorption bands of the species under examination. The thermal response is then used to determine the type and concentration of the explosive substance. Figure 7.2 shows the simplified setup of resonant infrared photothermal imaging spectroscopy setup. The concept has been experimentally demonstrated for detection and identification of trace explosives such as RDX, TNT, and ammonium nitrate on transparent, absorbing, and reflecting substrates.

7.3.2 Raman Spectroscopy

Raman spectroscopy offers another method for standoff detection of explosive agents. It has been extensively used for many years as a standard analytical tool for identification of chemical agents in the laboratory environment. The basis of detection in this case is the shift in the wavelength caused by inelastic Raman scattering by the target molecule. The inelastic scattering of impinging photons where some energy is lost to (or gained from) the target molecule returns scattered light with a higher (or lower) wavelength depending on whether energy was lost to (or gained from) the target molecule. The difference is dictated by the energy of vibrational modes of the target molecule and therefore constitutes

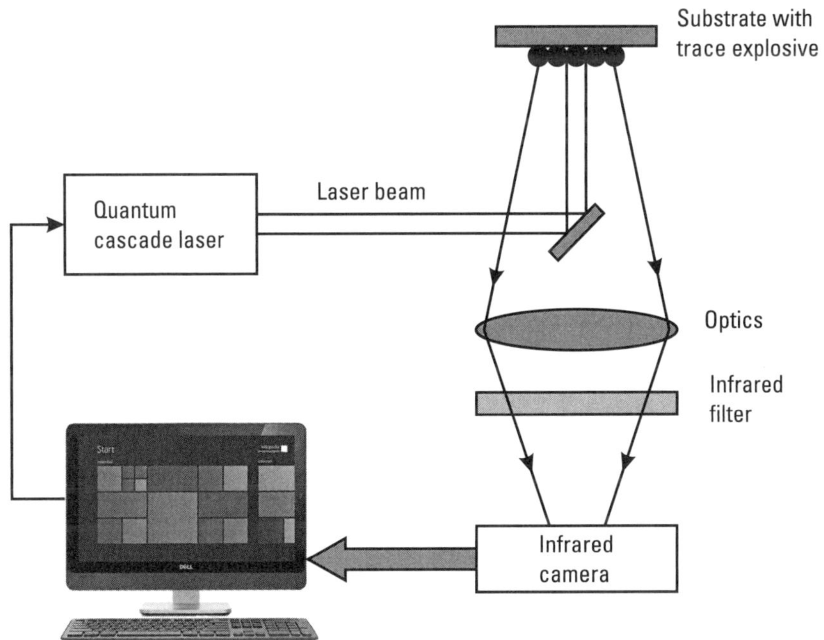

Figure 7.2 Resonant infrared photothermal imaging spectroscopy setup.

the fingerprint or the basis of identification. Complex mixtures are identified using algorithms for pattern recognition.

A major drawback of the Raman technique is its extremely poor sensitivity caused by the fact that Raman scattering occurs for one in about 10^7 photons impinging on the sample. Weak return signal intensity of Raman spectroscopy limits its use for trace detection as it also makes it sensitive to ambient light and fluorescence from the sample itself or other chemicals in the vicinity. The fluorescence masks the Raman signal. These problems are overcome by the use of resonant Raman spectroscopy. With a tunable laser, the wavelength can be chosen to match or nearly match a resonant absorption in the target molecule, leading to intensity enhancement of the order of 10^6. The problem of fluorescence masking a Raman signal can be overcome by use of either infrared or ultraviolet radiation. Infrared radiation does not have sufficient energy to cause fluorescence and ultraviolet radiation will cause fluorescence in the visible vicinity, which is well separated from the Raman signal. Figure 7.3 shows a typical Raman spectroscopy setup.

7.3.3 Laser-Induced Breakdown Spectroscopy

Laser-induced breakdown spectroscopy focuses a high-energy laser beam on the trace sample to break down a small part of the sample into a microplasma of

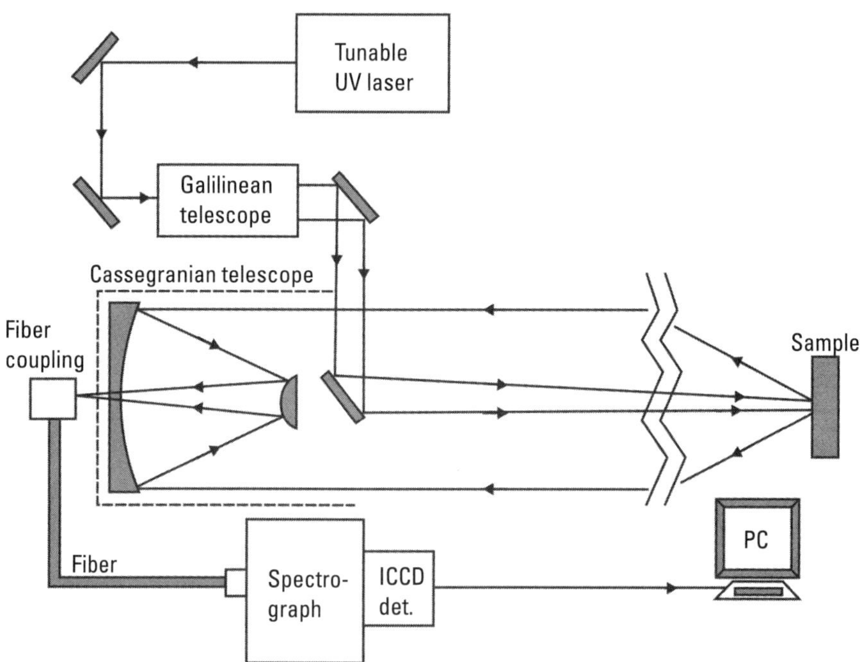

Figure 7.3 Raman spectroscopy setup.

excited ions and atoms. Plasma formation is accompanied by emission of a continuum radiation, which does not contain any useful information. Within a very small time, of the order of microseconds, the plasma expands at supersonic velocities and cools. It is at this point that the characteristic atomic emission lines from ionic, atomic, and small molecular species are observed. These emission lines after being detected by a spectrograph are used for fingerprinting and subsequent identification of explosive substances. Use of a time-gated ICCD camera allows rejection of continuum radiation. Figure 7.4 shows a typical laser-induced breakdown spectroscopy setup. One of the challenges in the use of LIBS is to assess its efficacy to detect and identify explosive species in a real environment that is replete with many interfering substances. One way to get the desired selectivity is to use double-pulse LIBS. In double-pulse LIBS, the first pulse is used to create a laser-generated vacuum and the second pulse transmitted a few microseconds later generates the return signal. Double-pulse LIBS is also observed to improve sensitivity in addition to enhancing selectivity. Selectivity can be further improved by adding temporal resolution to the LIBS emission analysis.

A wavelength of 1,064 nm has been widely employed for LIBS systems. Due to serious eye hazards posed by a 1,064-nm wavelength, scientists have also tried one that is 266 nm. A 266-nm wavelength has a 600 times higher MPE limit as compared to a 1,064 nm wavelength. It also allows the designer to build Raman capability into the system.

Lawrence Berkeley National Laboratory of Department of Energy in the United States has done pioneering work in developing laser ablation technologies such as LIBS for detection of explosives and other hazardous chemicals.

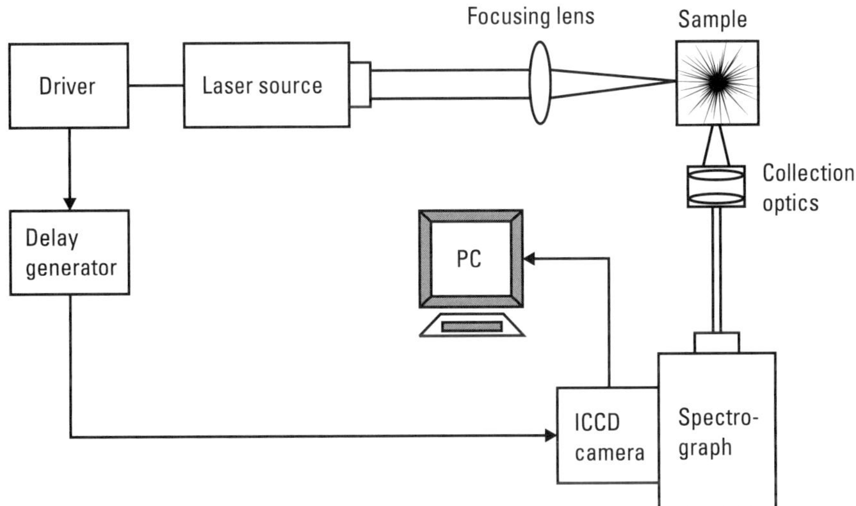

Figure 7.4 Laser-induced breakdown spectroscopy setup.

Applied Spectra, founded by one of the scientists of the Berkeley lab, has developed a portable LIBS-based explosive detection system. A military prototype of the system was reportedly field-tested on more than 100 samples in 2008. The detector was able to discriminate with 85% accuracy whether the tested samples contained residues of several types of explosives from standoff distances between 30 to 50m or whether the composition was of materials such as rock, wood, metal, or plastics.

7.3.4 Laser-Induced Fluorescence Spectroscopy

LIF spectroscopy is a type of spectroscopic technique in which the sample under examination is illuminated with a laser beam, thereby causing molecules in the sample to be electronically excited from a ground state to a higher energy state. As the molecules relax back to a lower energy level in the ground state, there is spontaneous emission of photons called fluorescence. Fluorescence photon has less energy than the one that was absorbed. The fluorescence emission wavelength is therefore longer than the excitation wavelength. While laser-induced fluorescence can generate very intense signatures, this technique is limited to materials that have strong electronic transitions, generally in the ultraviolet range. The resulting technique is known UV-LIF. The fluorescence photons are measured to provide the signal that tells about the type of molecules constituting the sample. Laser-induced fluorescence is an important tool for combustion diagnostics and studying decomposition of explosives. UV-LIF spectroscopic technique has been utilized for the detection of biological materials.

Photodissociation-laser–induced fluorescence (PD-LIF) employing photodissociation followed by laser-induced fluorescence has been found to be a potent technique for standoff detection of a variety of explosive substances, including ammonium nitrate, urea nitrate, RDX, DNT, TNT, and PETN.

Detection of explosive substances using PD-LIF spectroscopy is a three-step process. In the first step, a tunable pulsed ultraviolet laser with output wavelength band covering the absorption bands of targeted explosive materials illuminates the solid polyatomic explosive substance, thereby vaporizing and dissociating the explosive material to form gas phase diatomic NO fragments. In the next step, these vibrationally excited NO fragments get optically pumped to a higher energy state. NO fragments in the higher energy state fluoresce while relaxing to a lower energy state in the third and final step. A photodetector, which is typically a solar-blind PMT with narrowband filters, is used to detect fluorescence photons. Filters are used to suppress the scattered laser light. The wavelength of the fluorescence photons in this case is lower than that of the photons that pumped them. The wavelength of the fluorescence photons and the wavelength of excitation photons required to induce fluorescence are both very precise, which allows use of a narrow wavelength band source and narrowband detector. Excitation at 236 nm for instance produces 226-nm fluorescence photons. This further allows detection of NO fragments with a high degree of sensitivity and specificity. The advantage of gaseous diatomic NO molecule is that it has a much more distinct spectrum as compared to a solid-phase polyatomic molecule. This can be used for fingerprinting the explosive substance. As well, a relatively strong fluorescence signal allows use of eye-safe levels of excitation laser intensities. Another advantage of PD-LIF spectroscopy technique is that it is not susceptible to false alarms from traditional fluorescence processes while the fluorescence photons typically are at longer wavelengths.

7.3.5 Laser Photo Acoustic Spectroscopy

LPAS is a form of infrared spectroscopy where the detection is done with a photoacoustic sensor such as a microphone or a piezoelectric sensor. Laser photoacoustic spectroscopy, quartz-enhanced LPAS (QE-LPAS) in particular, has emerged as a promising tool for detection and identification of hazardous chemicals, biological warfare agents, and explosive substances from standoff distances of a few meters to a few tens of meters. Figure 7.5 shows a simplified schematic arrangement of a QE-LPAS setup. A tunable mid-infrared laser such as a quantum cascade laser covering the absorption spectra of the explosive substances of interest is used to irradiate the trace explosive sample in the form of trace particles adsorbed to a surface or trace vapors. The irradiating laser is pulsed at a frequency equal to resonant frequency of the quartz crystal detector.

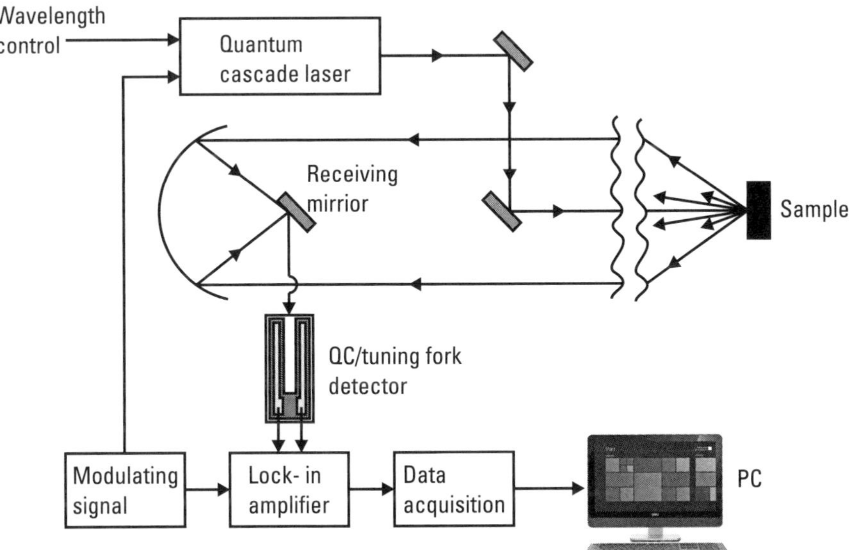

Figure 7.5 Quartz-enhanced laser photoacoustic spectroscopy.

When the laser wavelength matches the absorption wavelength of the explosive substance under examination, a part of the absorbed laser energy is converted into translation energy of the molecules by radiationless transitions, which is primarily responsible for heat production within the sample. The periodic heating of the sample caused by the modulation of laser energy results in a pressure modulation. It is these pressure fluctuations that are acoustically detected by a quartz crystal detector. The pressure wave modulated at a frequency equal to the resonant frequency of a quartz crystal detector drives it into oscillations, generating a periodic electrical signal.

The coupling medium plays an important role in addition to the thermal and optical properties of the sample. The detected signal is fed to the signal input of lock-in amplifier whose other input is fed with the reference signal used to modulate the laser radiation. The detected signal is fed to a PC platform through a suitable data acquisition interface. With the help of suitable software and with a wavelength control signal and detected signal as the inputs, we get the absorption spectrum of the explosive substance in terms of intensity versus wavelength plot. The spectrum shows absorption dips corresponding to wavelength/s unique to the explosive sample under examination. The large Q-factor of the quartz crystal detector allows detection of lower trace concentrations from larger standoff distances for a given laser power and coupling medium characteristics. This is further helped by availability of reasonable power quantum cascade laser sources. Standoff distance achievable in a given LPAS setup also depends on the nature of the surface to which trace particles of explosive substance are adsorbed. An aluminum surface, for example, gives a larger standoff range than a wood surface with other parameters remaining constant. Laser photoacoustic spectroscopy using pulsed lasers and piezoelectric detection of generated acoustic waves allows measurement of an absorbance of less than 10^{-7}. Reportedly, an explosive concentration less than 100 ppb has been detected from a standoff distance of more than 20m using an LPAS setup.

7.4 Representative Explosive Detection Systems

Different bulk, trace, and standoff explosive detection techniques were discussed in the preceding sections. Some of these techniques were capable of not only detecting the presence but also identifying the type of explosive substance. The following sections briefly outline the salient features of some representative explosive detection systems in each of the different categories.

7.4.1 Bulk Detection Systems

Of all the bulk detection techniques discussed in the preceding sections, X-ray technology is the most commonly used one for baggage and package screening.

A large number of manufacturers offer these machines, including those for fixed installation and portable machines for mobile use. Computed tomography, nuclear quadruple resonance, and fluoroscopy techniques have also been used for baggage, mail, and package screening. Millimeter-wave and terahertz imaging have been used for personnel screening.

The *Hi-SCAN 6040i* from Smiths Detection Inc. is an entry-level X-ray scanner designed for the screening of personal belongings and cabin baggage at airport checkpoints and critical infrastructure. The *eXaminer 3DX* from L3 Communications Security and Detection Systems (Figure 7.6) uses 3-D continuous-flow computed tomography technology. With its reliable threat detection, higher mean time between failure (MTBF), faster mean time to repair (MTTR), high inline throughput, and low false alarm rate, it is well-suited to challenging baggage inspection and handling requirements for air terminals. The *MAILGUARD* (Figure 7.7) from Control Screening LLC is a compact X-ray security screening system small enough to fit on a desktop. It uses fluoroscopy and is designed for detection of contraband material concealed in small items such as wallets, electronic gadgets, mail, mailbags, and parcels. The *QScan QR-160* from Quantum Magnetics Inc. detects explosives using nuclear quadrupole resonance technology. It is designed to screen mail, parcels, and personal items and can be used in stand-alone mode and together with X-ray systems.

7.4.2 Trace Detection Systems

Different trace explosive detection technologies were briefly discussed earlier. Of these, IMS, GC, MS, SAW, chemiluminescence (CL), and thermal redox are common. Quite often, more than one technology is used to build sensors with improved performance. Examples include GC-IMS, GC-IMS/CL, GC/

Figure 7.6 eXaminer 3DX X-ray machine.

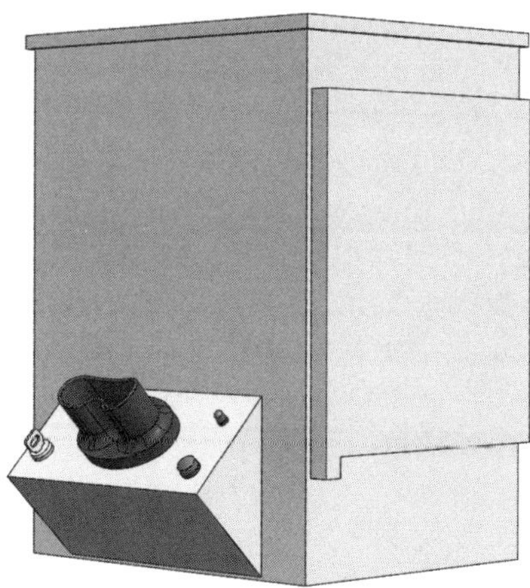

Figure 7.7   MAILGUARD X-ray scanner.

CL, GC/MS, and GC/SAW sensors. In addition, there are optical spectroscopic techniques that are used for the detection of trace explosives.

The *EVD-3000+* (Figure 7.8) from Scintrex Trace Corporation is a hand-held explosive detector that uses thermo-redox technology to detect vapors and particle traces of plastic and high-vapor pressure explosives, C-4, TNT, dynamite, PETN, Semtex, RDX, black powder, and nitroglycerine. The *E-3500* is also a portable device capable of detecting all threat explosives, including International Civil Aviation Organization (ICAO) taggants, military plastics, and TATP. Its operation is based on chemiluminescence technology. It also does not use any external carrier gas or radioactive source. Both systems can detect traces of both particulates and vapors, allowing for noninvasive searches of luggage, mail, vehicles, documents, and containers.

The *ZNose Model 4200* from Electronic Sensor Technology is a portable odor analyzer that uses GC/SAW technology. With its potential to detect and test all types of vapors, it can identify traces of biological, chemical, and organic

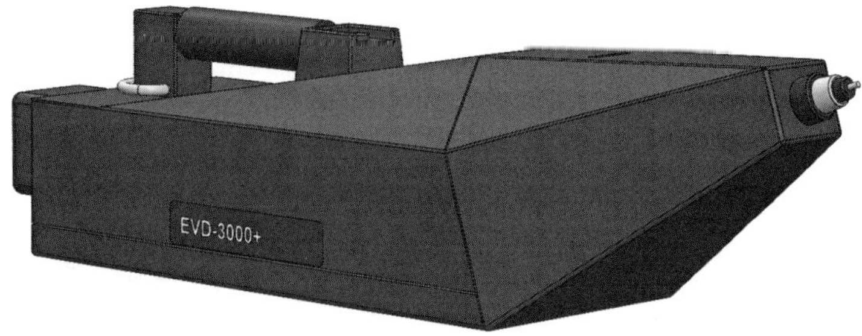

Figure 7.8   EVD-3000+ trace explosive detector.

compounds with accuracy and speed. It finds application in preventing homeland security threats by detecting chemical vapors and odors produced by explosives and biological/chemical weapons.

The *EGIS Defender* is a portable and compact explosive detector that combines high-speed gas chromatography (HSGC) with microdifferential ion mobility spectrometry (microDMx) to provide detection of plastic, commercial, and military explosives; TATP; HMTD; and enhanced AN-FO as well as ICAO marker compounds and narcotics with a high sensitivity of picogram level. Another important feature of the *EGIS Defender* is its ability to respond to new and emerging threats through future software upgrades. The *IONSCAN SENTINEL-II* (Figure 7.9) from Smiths Detection Inc. is a noncontact, walk-through personnel portal designed specifically for screening people for trace amounts of explosives and drugs. It is well suited to high throughput applications such as airport security checkpoints, customs, and building access control. It also uses ion mobility spectrometry technology.

7.4.3 Stand-Off Detection Systems

Although there are several spectroscopic technologies, discussed earlier in Section 7.3, with the potential for trace detection of explosives from stand-off distances, not all of them have reached a level of technological maturity to be transformed into field-usable systems. Most commercial spectroscopic explosive sensors use Raman and resonant Raman spectroscopy. Some field prototypes of LIBS-based standoff explosive detectors, like the one developed by Applied Spectra discussed earlier, have been reportedly field-tested. Other technologies, such as infrared photothermal imaging, photodissociation laser-

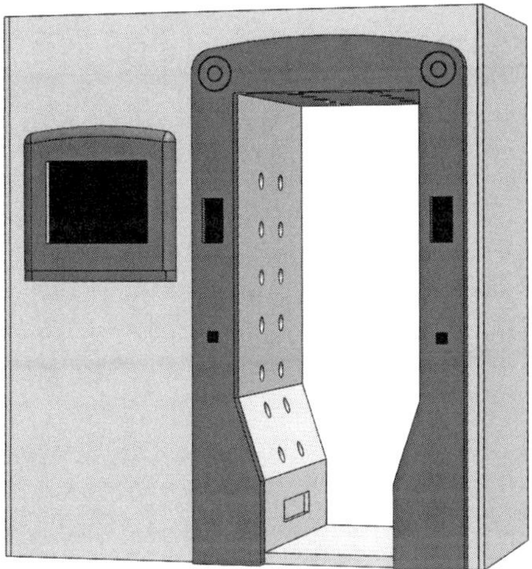

Figure 7.9 IONSCAN SENTINEL-II walk-through portal.

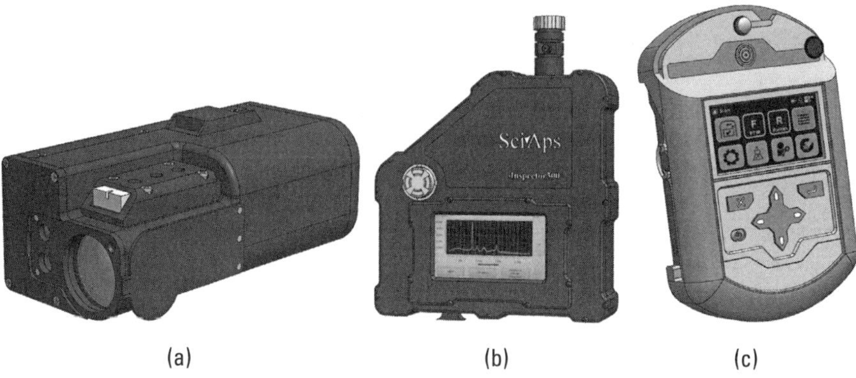

(a)   (b)   (c)

Figure 7.10   Raman explosive sensors.

induced fluorescence, and laser photoacoustic spectroscopy, are still at a level of experimental prototypes.

The *ObserveR*™ (Figure 7.10(a)) from SciAps, Inc. is portable Raman system that is designed for standoff detection and identification of explosives, toxic industrial chemicals (TICs). and toxic industrial materials (TIMs) from distances ranging from 0.1 to 3m. It has an onboard library of Raman signatures of many explosive samples, TICs, and TIMs, and provision for a large number of new entries that are user-programmable. The *Inspector-300* and the more sophisticated *Inspector-500* are the other portable Raman detectors operating, respectively, at 785 and 1,030 nm for the same purpose from the same company. The *Inspector-300* (Figure 7.10(b)) and *Inspector-500* are particularly suited to detection of samples that produce high levels of fluorescence. The *ReporteR*, a hand-held Raman system from SciAps, Inc., is a miniaturized version of the Inspector-300 system. The *Advantage-series* benchtop Raman systems from SciAps, Inc. offering four variants operating at 532, 633, 785, and 1,064 nm are the other systems.

The *Gemini Analyzer* (Figure 7.10(c)) from Thermo Scientific is a hand-held sensor that integrates complementary and confirmatory chemical identification techniques of FTIR and Raman spectroscopy in a single instrument to detect explosives and chemical warfare agents. The Gemini analyzer includes an extensive library that enables operators to identify unknown explosives, chemical warfare agents, and industrial chemicals and precursors, thereby enabling detection of a broader range of samples than either technique alone.

Selected Bibliography

Baudelet, M. (ed.), *Laser Spectroscopy for Sensing: Fundamentals, Techniques and Applications*, Cambridge, UK: Woodhead Publishing Limited, 2014.

Cremers, D. A., and L. J. Radziemski, *Handbook of Laser Induced Breakdown Spectroscopy*, Chichester, UK: Wiley-Blackwell, 2006.

Demtroder, W., *Laser Spectroscopy Volume-1: Basic Principles*, Fourth Edition, Berlin: Springer, 2008.

Demtroder, W., *Laser Spectroscopy Volume-2: Experimental Techniques*, Fourth Edition, Berlin: Springer, 2008.

Marshall, M., and J. C. Oxley (eds.), *Aspects of Explosive Detection*, Amsterdam: Elsevier Science, 2008.

Sayeedkia D. (ed.), *Handbook of Terahertz Technology for Imaging, Sensing and Communications*, Oxford, UK: Woodhead Publishing Limited, 2013.

Sun, Y., *Field Detection Technologies for Explosives*, Hertfordshire, UK: ILM Publications, 2010.

Woolard, D. L., J. O. Jens, R. J. Hwu, and M. S. Shur, *Terahertz Science and Technology for Military and Security Applications*, Hackensack, NJ: World Scientific, 2007.

Yinon, J., *Counterterrorist Detection Techniques of Explosives*, Oxford, UK: Elsevier Science, 2007.

8

Detection of CBRN Threats

CBRN incidents, those caused by weaponized forms of CBRN materials and those occurring due to either human error or natural or technological reasons, have similar short- and long-term consequences. This necessitates that first responder personnel are equipped with specialized equipment along with decontamination and countermeasure systems for detection, identification, and mitigation of such threats. In this chapter, beginning with a brief introduction to different types of chemical, biological, and radiation threats, we confine our discussion to the different technologies used for detection of chemical and biological warfare agents and radiation detectors. Optronic technologies are discussed in greater detail than the nonoptronic technologies.

8.1 CBRN Threats

There are certain chemical, biological, radiological, and nuclear materials that pose serious hazards to human life and property. These are collectively known as CBRN materials. CBRN threats are those originating from both weaponized and nonweaponized forms of these materials. Because of their potential for causing mass casualties and enormous damage to property in a short span of time, both man-made and accidental occurrence of CBRN threats pose the gravest danger to national security. The following sections discuss different types of CBRN threats along with the extent of damage these threats are capable of inflicting to human life and property. A brief history of the use of CBRN agents in the past is also presented toward the end of this section.

8.1.1 Types of CBRN Threats

Due to advances in technology, it has become relatively much easier to produce hazardous substances than it was even a few years ago. This has significantly increased the risk of hazardous chemical, biological, radiological, and nuclear incidents in military or even civil space with potentially devastating consequences for society at large and for the environment. As the name suggests, the domain of CBRN threats includes chemical and biological agents, and radiological and nuclear materials. Chemical substances, including chemical warfare agents, hazardous industrial chemicals, and even household chemicals affect people through poisoning or injury. Biological agents such as dangerous bacteria or viruses and biological toxins when released cause diseases such as fever, malaise, fatigue, shortness of breath vomiting, abdominal pain, and visual disturbance on a mass scale. Exposure to harmful radioactive materials also causes illness. Thermal and blast effects arising from nuclear detonation lead to mass casualties and destruction to property on an enormous scale.

8.1.2 Chemical Warfare Agents

Chemical warfare agents (CWA) are toxic chemical substances used with the intent of killing, causing injury, or incapacitating the adversary's armed forces during war, and both the armed forces and civil population during acts of terrorism. These are broadly classified as nerve agents, blistering agents, cyanides, and toxic industrial chemicals that are extremely hazardous materials capable of causing a catastrophic medical disaster leading to mass destruction.

Nerve agents adversely affect the nervous system by affecting transmission of nerve impulses. They belong to the group of organo-phosphorus compounds. Nerve agents considered important from the viewpoint of chemical warfare include *tabun* (O-ethyl dimethyl amido phosphoryl cyanide) with the American denomination GA, *sarin* (isopropyl methyl phosphono fluoridate) with the American denomination GB, soman (pinacolyl methyl phosphono fluoridate) with the American denomination GD, cyclohexyl methyl phosphono fluoridate, with the American denomination GF, and O-ethyl S-diisopropyl aminomethyl methylphosphono thiolate with the American denomination VX. All nerve agents are easily dispersible. They are highly toxic and have rapid effects both when absorbed through the skin and via respiration. Consumption of liquids or foods contaminated with nerve agents may cause poisoning.

Blistering agents, also called *vesicants,* are toxic compounds that cause severe skin and eye injuries like the ones caused by burns. On inhalation, they affect the upper respiratory tract and the lungs, producing pulmonary edema. There are two forms of blistering agents: mustards and arsenicals with *sulfur mustard* being the most important blistering agent. Nitrogen mustards and lewisites are the other agents in this class. The image shown in Figure 8.1 illustrates the effects of use of mustard gas.

Cyanides or *cyanogenic agents*, also referred to as *blood agents*, are cyanide group of chemicals. They prevent the normal utilization of oxygen by body tissues, thereby adversely affecting bodily functions. These agents are also known as systemic agents as they inhibit certain specific enzymes. Sodium and potassium cyanides, hydrogen cyanide, and cyanogen chloride are the chemical warfare agents in this class. Sodium or potassium cyanides are white-to-pale yellow salts that can be easily used to poison food or drinks. Cyanide salts can be disseminated as a contact poison when mixed with chemicals that enhance skin penetration but may be detected since most people will notice if they touch wet or greasy surfaces contaminated with the mixture. Hydrogen cyanide and cyanogen chloride are liquids that turn into a gas near room temperature. Both hydrogen cyanide and cyanogen chloride need to be released in high concentration levels to be effective. Exposure to cyanides may produce nausea, vomiting, anxiety, palpitations, hyperventilation, and vertigo that may lead to a state of unconsciousness, coma, and death.

Several toxic industrial chemicals (TICs) are potential chemical warfare agents. Though they are not as toxic and hazardous as nerve and blister agents and cyanides, lower toxicity is compensated for by using larger quantities. Chlorine and phosgene gases produce effects similar to that of mustard agent. Organophosphate pesticides such as parathion are in the same chemical class as nerve agents.

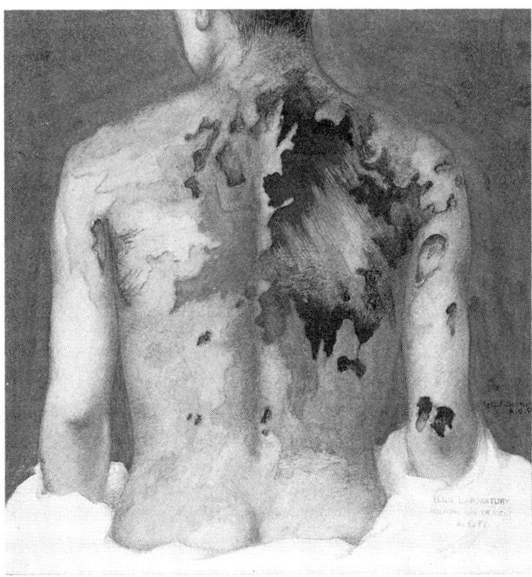

Figure 8.1   Mustard gas effect. (Source: Wikimedia Commons.)

8.1.3 Biological Warfare Agents

Biological warfare agents (BWAs) are microorganisms such as bacteria, viruses, fungi, and toxins produced by them. These agents when dispersed in a biological warfare or bioterrorist attack pose grave health hazards being capable of causing large-scale mortality, morbidity, and incapacitating a large population. The effects of these agents are not instantaneous and the symptoms appear in the affected population in a time span ranging from few hours to weeks. The bioagents are self-replicating and only a small quantity needs to be released to cause a disease outbreak. The collage of pictures shown in Figure 8.2 illustrates only a fraction of the devastating effects of a biological warfare agent attack.

Biological warfare agents, as outlined above, are classified as *bacteria, viruses,* and *toxins.* Common bacterial bioagents include *Bacillus anthracis* (causing *anthrax*), *Yersinia pestis* (causing *plague*), *Brucella melitensis* (causing *brucellosis*), *Brucella abortus* (causing *premature abortion of fetus* in cattle), *Burkholderia Mallei* (causing *glanders*), and *Burkholderia pseudo mallei* (causing *melioidosis*). Bacterial agents are generally disseminated in aerosol form. Anthrax-causing *Bacillus anthracis* is the most common form of bacterial agent. The anthrax causing bacterium often penetrates the body via wounds in the skin and may also infect humans as aerosol or ingestion. The resistance of the anthrax spores to harsh environmental conditions like heat and humidity, disinfectants, and ultraviolet radiation makes anthrax the most important biological warfare agent. Anthrax is usually fatal unless antibiotic treatment is begun within hours of inhaling anthrax spores. In a biological warfare or bioterrorist attack, the *plague* bacilli are inhaled as aerosol directly, resulting in the pneumonic plague, which is highly contagious and spreads from person to person through airborne droplets. Human brucellosis is a multisystem disease that may be present with a broad spectrum of clinical manifestations. Its diagnosis requires microbiological confirmations by means of isolation from blood culture or demonstration of the presence of specific antibodies by serological tests. Brucellosis is a zoonotic

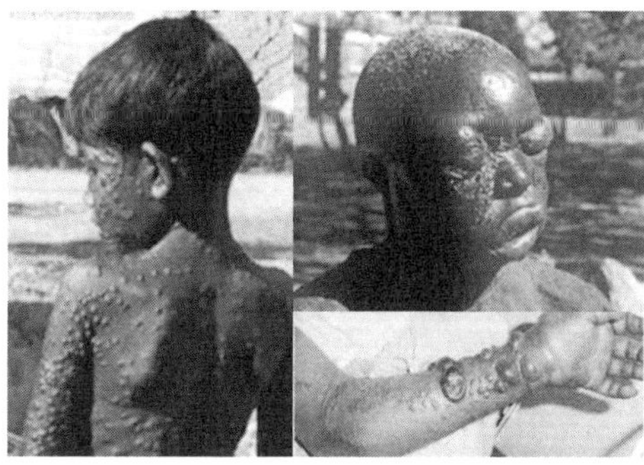

Figure 8.2   Effects of a biological warfare agent attack.

infection transmitted from animals to humans by ingestion through infected food products, direct contact with an infected animal, or inhalation of aerosols. The disease is known by various names, including Mediterranean fever, Malta fever, gastric remittent fever, and undulant fever. *Glanders* is a disease of equines (i.e., horses, mules, and donkeys). The disease can be present as either a cutaneous or systemic disease. Cutaneous infection causes nodules with accompanying localized lymphadenitis. The systemic illness usually manifests itself either as brancho or lobar pneumonia. Infection in humans can be fatal in the absence of antimicrobial therapy. Melioidosis, also called Whitmore's disease, is an infectious disease affecting humans or animals. It is predominately a disease of tropical climates. *Burkholderia mallei* bacteria, which causes melioidosis, are found in contaminated water and soil. It spreads to humans and animals through direct contact with the contaminated source.

Common viruses include *Variola virus* (causing *smallpox*), *Ebola virus* (causing *Ebola hemorrhagic fever*), and *Marburg virus* (causing *Marburg hemorrhagic fever*). Route of infection in the case of viruses is also aerosol. *The variola virus* is the causative agent of smallpox. Though smallpox was eradicated worldwide in 1977, it has regained interest because of its potential as a bioterrorism agent. Smallpox is considered one of the most serious bioterrorist threats. British soldiers reportedly used it as a biological weapon by distributing smallpox-infected blankets to American Indians in the French and Indian Wars from 1754 to 1767. As well, in the 1980s, the Soviet Union developed variola as an aerosol biological weapon intended for use in intercontinental ballistic missiles. *Ebola virus disease* (EVD) or *Ebola hemorrhagic fever* is a serious condition that is quite often fatal. Ebola infection is highly contagious and is transmitted through direct contact with the blood, body fluids, and tissues of infected people or animals. The disease is often characterized by the abrupt onset of fever, intense weakness, muscle pain, headache and sore throat. The virus causes severe bleeding and organ failure that can lead to death. *Marburg virus disease* affects both humans and non-human primates and is kind of viral hemorrhagic fever like the Ebola virus disease. The disease is highly contagious and is transmitted from person to person by exposure to blood and other bodily secretions. The early symptoms of the Marburg virus disease include fever, chills, headaches, and muscle aches, which might subsequently lead to hemorrhagic fever and death.

Common toxins belonging to class of bioagents include *Clostridium botulinum* (causing *botulism*), *Staphylococcus aureus* (causing *Staphylococcal enterotoxin type B*), *Ricin* (causing *Ricin toxin*), and *Trichothecene* (causing *trichothecene T2 toxin*). Ricin is a plant toxin and trichothecene belongs to fungus family. Toxins are spread through contaminated food and water. Botulism is a rare but serious disease that is caused by bacterium called *Clostridium botulinum* occurring naturally in soil. There are several kinds of botulism, including *foodborne botulism* that comes from eating toxin-contaminated foods and *wound botulism* caused by a wound infected with the bacteria making the toxin. Infant botulism is caused by consuming the bacteria spores from soil or honey. Symptoms include double or blurred vision, drooping eyelids, slurred speech, difficulty swallowing, dry mouth, and muscle weakness. *Staphylococcus aureus* is a group of bacteria that can cause numerous diseases. It is also known as *Staph* infection, which is caused either due to direct infection or production of toxins by the bacteria. Boils, impetigo (a kind of bacterial skin infection), food poisoning, cellulitis (a bacterial infection of the skin and tissues beneath it), and toxic shock syndrome are all examples of diseases that can be caused by Staphylococcus. Ricin is a toxin protein occurring naturally in castor beans. Ricin-caused illness is not contagious and cannot be transmitted from person to person through casual contact. However, coming in contact with people who have ricin on their body or clothes could make you vulnerable. The U.S. military had reportedly experimented with using ricin as a possible warfare agent. As well, some terrorist organizations have reportedly used ricin as a warfare agent in the 1980s in Iraq. Trichothecene is a type of toxin produced by fungus. It is one of the most notorious mycotoxins because of its extremely high toxicity and also because it is very difficult to destroy. Exposure to trichothecene mycotoxins causes symptoms such as dry eyes, tiredness, fatigue, vomiting, diarrhea, abdominal pain, mental impairment, rash, and bleeding.

8.1.4 Radiological and Nuclear Devices

This category of CBRN threats includes RDDs and INDs. Radiological dispersal devices are designed to disseminate radioactive material and the radiation produced by the material then becomes the cause of contamination, destruction, and injury. Potential materials used in RDDs include americium-241, californium-242, cesium-137, strontium-90, iridium-132, plutonium-238, polonium-210, radium-226, and cobalt-60.

There are two types of radiological devices: *passive RDDs* and *active RDDs*. While a passive RDD uses unshielded radioactive material that is dispersed manually at the targeted location, an active RDD combines conventional explosives with radioactive material and employs explosive force of detonation to disperse the radioactive material. Active RDD is also known as a dirty bomb. Note that active RDDs are conventional bombs and not yield-producing nuclear devices. There are also RDDs known as *atmospheric RDDs*, where radioactive material is converted into a form easily transportable by air currents. There are many unaccounted sources of radioactive materials around the world, providing an easy access to rogue nations and/or terrorist groups to obtain these materials for RDDs. Radiological dispersal devices such as dirty bombs are attractive to terrorists because the materials necessary to build the weapons are relatively easy to acquire and the technology is simple. Not only there is an easy access to unaccounted sources of radioactive materials, radioactive materials with potential for use in dirty bombs are used in hundreds of medical, industrial, and academic applications. Explosive RDDs might initially kill a few people in the immediate area of the blast; they are primarily used to create panic and terror in the targeted population, thereby producing psychological rather than physical harm.

An IND is designed to produce a yield-producing nuclear explosion. Unlike radiological dispersal devices that can be made by using almost any radioactive material, improvised nuclear devices require fissile material such as highly enriched uranium or plutonium. A nuclear device in the hands of a rogue or terrorist group would be fabricated by using either illegally obtained fissile nuclear weapons material or built from the components of a stolen nuclear weapon. If the IND failed to produce nuclear yield, the result would be similar to that of an RDD in which fissile material was dispersed locally. Production of nuclear yield results in catastrophic loss of life, destruction of infrastructure, and contamination of a very large area. The effects include blast injuries including primary, secondary and tertiary blast injuries, thermal/burn injuries including flash burns, flame burns and eye injuries, and radiation injuries caused during the incident and also through long-term effects.

8.1.5 Chemical versus Biological Warfare

The differences between chemical and biological warfare lie primarily in the type of agents that make up the chemical or biological weapon, modes of dissemination, or dispersal of those agents and the effects produced as a consequence of use of the weapon. The key points of distinction between the two types of weapons are briefly outlined below.

1. Chemical warfare uses man-made chemical agents whereas the agents used in biological warfare have a natural origin. All chemical warfare agents including nerve agents such as tabun, sarin, and soman, blister agents such as sulfur mustard, blood agents such as hydrogen cyanide, and toxic industrial chemicals including chlorine and phosgene gases are man-made. On the other hand, biological warfare agents such as viruses, bacteria, and fungus are actual living organisms. The toxins used in biological warfare are those originating from these living organisms.

2. Chemical warfare agents and chemical weapons can be built in a large-scale industrial production setup and are relatively cheaper to produce. In contrast, biological warfare agents and weapons are difficult to develop, are far more expensive. and are made in small quantities.

3. All chemical warfare agents with the exception of sarin gas have a specific odor and taste. Biological warfare agents are odorless and tasteless.

4. Chemical warfare agents are disseminated as aerosols or liquids. Biological warfare agents are disseminated as aerosols in air or as contamination in water and food.

5. In terms of effects, chemical warfare agents can penetrate skin. The effects of chemical agents are often immediate. Generally, chemical agents tend to present an immediately noticeable effect. Cause of death or injury is massive burning or poisoning. Most biological agents do not penetrate skin. Physical effects are delayed and the cause of death is disease such as anthrax or smallpox. Most biological agents will take days before symptoms appear

6. Use of biological warfare agents causes much higher fatality. One biological weapon can kill 100 to 10,000 times as many people as compared to one chemical weapon of the same weight.

8.1.6 History of Chemical and Biological Weapon Use

The use of chemical and biological warfare agents is not a phenomenon of recent history, although it poses a much graver danger today due to the increased threat of terrorism. There have been incidents of use of chemical and biological agents in different centuries since 1000 B.C. Incidents of chemical agent attacks started much earlier than those of biological agent attacks. It is not possible to highlight each and every occurrence of a chemical and biological attack since then; a summary of major incidents is presented here in chronological order.

1. Arsenic smoke was reportedly used by the Chinese in 1000 B.C. Solon of Athens poisoned the drinking water of Kirrha during a siege of a city in 600 B.C. and Hannibal of Carthage reportedly hurled clay pots full of Viperidae, a family of venomous snakes, onto the decks of enemy ships in a sea battle in 184 B.C.

2. There were incidents of hurling the bodies of victims of smallpox or the plague over city walls during the twelfth century. In the fifteenth century, Leonard da Vinci reportedly proposed an arsenic-based anti-ship weapon and the Spanish reportedly offered wine adulterated with the blood of leprosy patients to the French near Naples. Again in 1650, the Polish artillery general Kazimierz Siemienowicz fired spheres filled with the saliva of rabid dogs at his enemies.

3. In 1763, in an early instance of biological warfare, British officers reportedly came up with a plan to distribute smallpox-infected blankets to American Indians besieging the Fort Pitt, Pennsylvania. This started an epidemic.

4. Several incidents of the use of chemical warfare agents during World War I have been reported. Biological warfare was generally not as successful. The focus was mainly on infecting enemy livestock with anthrax or glanders. In 1914, German artillery fired around 3,000 105-mm shells filled with a lung irritant in dianisidine chlorosulfate against British troops. The German forces could not see the effect of using chemical agent as the shells exploded, thereby destroying the chemical agent. Subsequent to that, in late 1914, German scientist Fritz Haber came up with the idea of creating a cloud of poison gas by using thousands of cylinders filled with chlorine gas, a choking agent. These were deployed in April 1915 during the battle for Ypres, France. The Germans had relied on prevailing wind conditions for gas dispersal, which was responsible for injury and death of a large number of German soldiers carrying out the attack. Around 6,000 French casualties were reported, caused primarily by asphyxia and tissue damage in lungs. A large number were also blinded. In 1915, Allied troops also launched their own chemical attacks using chlorine gas. This began a race for use of chemical agents with greater toxicity, which is evidenced by use of diphosgene gas by Germany and cyanide gas by France. In July 1917, Germany introduced mustard gas, which burned the skin as well as the lungs. An estimated number of 90,000 soldiers were killed and another 1.3 million soldiers were injured due to chemical attacks during World War I.

5. Another important incidence of chemical warfare was in 1919 during the Russian civil war (November 1917–October 1922) when the British military used mustard gas against Red Army soldiers in Russia. This was followed by the Red Army using chemical agents against peasant rebels and civilians in 1921 in the Tambov area of Russia.

6. During the Third Rif War in Spanish Morocco (1921–1927), the Spanish Army of Africa used mustard gas repeatedly against Rif tribes in an attempt to put down the Riffian Berber rebellion during 1924–1925.

7. Incidents of the use of multiple biological agents by the Japanese military in Ping Fan Manchuria on war prisoners during 1932–1945 in an attempt to test biological weapons on humans have also been reported. An estimated 1,000 people died and another 2,000 were injured.

8. The Soviet military carried out multiple mustard gas attacks on Chinese soldiers in Sinkiang during 1934–1937. During the second Italo-Ethiopian war (1935–1936), the Italian military carried out multiple attacks with sulfur mustard gas against Ethiopian soldiers in Ethiopia, causing as many as 15,000 casualties. Italian forces initially dropped sulfur mustard air bombs and then replaced them by aerial spray tanks.

9. In another major use of chemical weapons, the Japanese military extensively used mustard gas and other chemical agents against Chinese soldiers and civilians during a war with China between 1938 and 1945, causing 10,000 casualties and injuring another 72,000. The Japanese military also carried out attacks with cholera and other bioagents on 11 Chinese cities between 1940 and 1942, killing 2400 people and injuring another 10,000, including some casualties among Japanese soldiers.

10. Between 1941 and 1945, Nazi Armed Forces in Germany carried out extensive chemical weapon attacks with Zyklon-B and carbon monoxide on concentration camp prisoners in Germany, Austria, and Poland, killing millions of prisoners.

11. During the Yemen War of 1963 through 1967, Egypt used CN tear gas, mustard blistering gas, and phosgene asphyxiant in support of South Yemen against royalist troops in North Yemen. Some reports claim that Egypt also used an organophosphate nerve agent against Yemeni Royalist forces. An estimated number of 1,400 people were killed in the attacks.

12. Laotian and Vietnamese militaries used chemical agents against civilians and soldiers in Laos-Vientiane, Xiangkhoang, and Louangphrabang provinces between 1975 and 1983, killing around 6,500 people. The Vietnamese military carried out chemical attacks in western Kampuchea between 1978 and 1983, killing more than 1,000 soldiers and civilians.

13. The Soviet and Afghan militaries carried out multiple chemical agent attacks in Afghanistan, killing more than 3,000 civilians and rebel soldiers between 1979 and 1981.

14. The Iraqi military carried out extensive attacks with chemical warfare agents during 1983–1988, killing an estimated 21,000 Iranian soldiers and civilians in Iran and Iraq.

15. In a major attack of chemical agents using cyanides, mustard blistering agents, and nerve agent in 1988, Iraqi military killed estimated 5,000 and injured 8,000 Kurdish Iraqi civilians in Halabja, Iraq.

16. In another shocking incident of chemical agents attack in March 1995, Asahara's cult, Aum Shinrikyo, released sarin nerve agent during the Monday morning rush hour in one of the world's most crowded subway systems in Tokyo, Japan. Members of the cult used the tips of their umbrellas to puncture plastic bags filled with liquid sarin on five crowded cars before hurrying off at subway stops and leaving their fellow riders trapped with the toxic gas. The incident killed 12 people while thousands sought medical attention.

17. In yet another mysterious incident in October 2001, a large number of anthrax-laden letters were received in the offices of the Florida-based tabloid *The Sun* and ABC, CBS, and NBC in New York. Several people were reported to have developed cutaneous anthrax. Anthrax was also found in the New York office of Gov. George Pataki. In the same month, letters containing anthrax arrived at the Senate mail-room. The incidents killed five people and another 10,000 U.S. residents required extensive medication for possible anthrax exposures. Reportedly, anthrax used in the attacks was of weapon grade. The perpetrator(s) of these attacks were never identified.

18. Between 2003 and 2013, Islamist terrorists reportedly carried out a large number of chemical and biological weapons attacks predominantly in Iraq targeting U.S. soldiers, policemen, school children, and civilians. The agents used in the attacks included sulfur mustard, chlorine gas, sarin nerve gas, chemical toxins, pesticides, rat poison, and other agents.

19. Between 2013 and 2014, the Syrian military reportedly used chlorine gas and sarin nerve gas in Damascus, Kafr Zita in Hama and Talmenes in Idlib against civilians, killing more than 1,500 including a large number of children.

20. Between and 2014 and 2016, there have been several chemical weapons attacks with sulfur mustard gas and chlorine gas at different places in Iraq.

21. There have been at least 34 confirmed cases of chemical attacks in Syria, mostly using chlorine gas, some sarin gas, and other unspecified chemical agents from 2013–2017. In one such attack with sarin gas in Ghouta, 1,400 people were killed. On April 7, 2018, in a suspected and not yet confirmed chemical weapons attack with a toxic gas in the rebel-held town of Douma near Damascus, an estimated 42 people were killed due to suffocation and hundreds were reported to be suffering from burning eyes, breathing problems, and indications of white foam coming out of their mouths and nostrils.

8.2 Detection of Chemical Warfare Agents

Although chemical agents, which include chemical warfare agents and toxic industrial chemicals, have been used during conventional warfare against armed forces, it is because of the increasing use of these agents in terrorist activities that the threat has extended to civilian population. Therefore, there needs to be in place a capability to rapidly detect, identify, and monitor chemical agents for efficient use of both military and civilian defense resources. Timely assessment of the extent and severity of chemical hazards helps in identifying clean and contaminated areas and providing the required inputs to first responders to implement precautionary measures and initiate countermeasures. The following sections discuss important parameters in the detection of chemical agents, technologies available for their detection with a focus on optical technologies, and representative commercial chemical agent detection systems.

8.2.1 Detection Parameters

Parameters that need to be considered while selecting a suitable detector for a given threat situation include detection sensitivity, selectivity, response time, false alarm rates, portability, setup, warmup and recovery times, operating and environmental conditions, user interface, calibration requirements, and cost. Some of these parameters have been discussed in Chapter 7 with reference to detection of explosive agents. For the sake of continuity and in the context of detection of chemical warfare agents, each of these parameters is briefly described.

Sensitivity, also called *limit of detection* (LOD), is the lowest detectable concentration of the chemical agent. It is also a measure of the ability of the detector to discriminate between small differences in the concentration of the agent under analysis. A more sensitive detector therefore produces a large change in output signal intensity for a small change in concentration of the chemical agent. Sensitivity depends on a number of factors, which includes the type of chemical agent and operating and environmental conditions. Sensitivity is generally measured in mg/m^3 and $\mu g/m^3$ or ppm/ppb/ppt; $1\ g/m^3$ equals 1 ppm. A typical value of sensitivity ranges from a few tens of $\mu g/m^3$ to a few tens of mg/m^3.

Selectivity is the ability of a detector to discriminate between the targeted chemicals and other compounds that may be present in the sample. While a highly selective detector responds only to targeted chemicals, a less selective detector responds to a larger number of chemicals including both chemical agents and nontoxic substances. Selective detectors can detect only a limited number of compounds and they respond to chemicals that possess properties to those of chemical agents, thus producing a false positive response. Nonselective and selective detectors may be used in field conditions; the former for surveying the suspected area as a broad spectrum early warning system and the latter for identifying or discriminating potential chemical agent threats from other nonhazardous compounds present.

Response time is the time taken by a detector to respond to the targeted chemical agents. It is the accumulation of time periods required for collection and analysis of the sample and the time required to provide feedback. The response time typically varies from a few seconds to few tens of seconds primarily depending on the detection technology. Ideally, the detector should respond in real time or nearly real time, which is important to prevent exposure of targeted population to chemical agents. Detection technologies available today are more suited to monitoring or presumptive CA identification in the event of an incidence of chemical agent attack.

There are two types of false alarms: false positives, where the detector indicates the presence of a chemical agent when there is none, and false negatives, where the detector fails to detect a chemical agent when one is present. False positive alarms usually occur when the targeted chemical compound is in the presence of an interferent having a chemical composition similar to the hazardous chemical agent. Repeated false positives may lead to future real alarms being ignored. False negatives pose greater risks as the failure to produce an alarm when one is present may lead to dangerous situations.

Portability is important from the viewpoint of field operation. Whether detection equipment is portable or not should include the portability aspect of support equipment required for its operation.

Setup, warmup, and *recovery times,* respectively, are the time periods required to power up a detector, get ready for analysis after it has been turned on, and for the display to return to the baseline "no response" value after being removed from the agent.

The performance of current chemical agent detectors does get affected by *environmental conditions* such as temperature, humidity, wind, dust, and contamination concentration in the air to lesser or greater degree depending on the technology of detection. This information helps in identifying a suitable detector for the intended operating environment.

User interface is another important aspect of the suitability of a given detector for the intended use. Factor such as to how the information generated will be used, the required level of training of users, and the environment in which a detector is to be used need to be paid due attention. A user-friendly interface that displays results of detection unambiguously is highly desirable, more so for first responders.

Calibration requirements define the periodicity with which the equipment needs to be calibrated to guarantee operation as per claimed performance specifications. Calibration is usually carried out using a known no-toxic chemical as a simulant of the targeted compound.

Cost is another factor, which should include maintenance costs and costs of consumables.

8.2.2 Review of Detection Technologies

Technologies that have been extensively explored for detection of CWAs and TICs were outlined in Chapter 1. Except for photo ionization detection, flame photometry, and flame ionization detection, other techniques, including photo-acoustic spectroscopy and active infrared detection based on FTIR spectroscopy were discussed in further detail in Chapter 7 with reference to explosive detection. The discussion is equally valid for detection of chemical agents. Passive infrared detection and DIAL techniques are discussed in the following sections.

8.2.3 Passive Infrared Detection

Passive or active infrared detection uses spectral signatures of the chemical agent in the gas or vapor phase to detect, identify, and determine concentration from a standoff distance. Depending on the technology and algorithms employed, some infrared sensors uniquely identify a chemical and measure its concentration. Others only identify the family the chemical belongs to and provide a measure of relative concentration. While active sensors use their own source of infrared light, passive sensors rely on thermal contrast or temperature differences between the chemical cloud and the background. In the case of passive infrared detection, since all objects with temperatures above absolute zero emit energy in the infrared region, different chemicals emit or absorb light at specific narrow wavelength bands, creating a unique spectrum or fingerprint that can be used to identify the chemical. Chemicals hotter than the background emit infrared energy and those colder than the background absorb infrared energy at the same wavelengths, thereby creating chemical-specific spectra, which are then compared with the spectra stored in the library of the equipment to identify the chemical. In essence, detection and identification of chemicals relies on the uniqueness of the chemical signature or spectrum measured and whether that signature can be separated from the background interference. Figure 8.3(a) shows a simplified schematic arrangement illustrating the operational concept of passive infrared detection. Different categories of passive infrared detection sensors include single-band imagers requiring more than one sensor for different chemicals or chemical families, multispectral imagers capable of detecting a wider range of chemicals, and hyperspectral imagers that with their high-resolution spectra can detect and identify a large range of different chemicals. The main difference between multispectral imaging and hyperspectral imaging lies in the number and spectral width of bands. Multispectral imagery generally

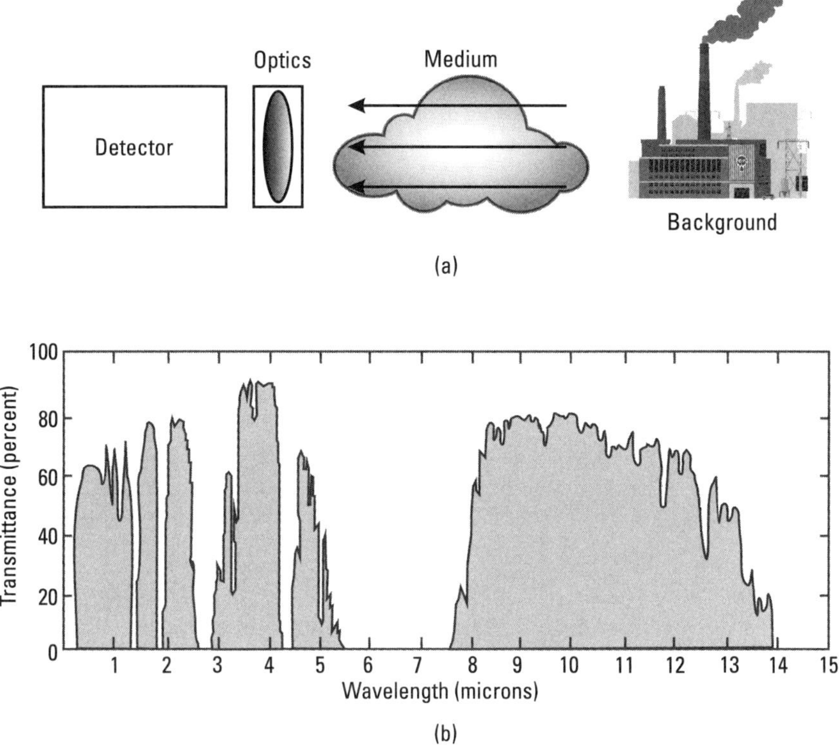

Figure 8.3 Passive infrared detection: (a) schematic arrangement, and (b) atmospheric transmission windows.

refers to 3 to 10 bands. Each band is obtained using a different sensor. Hyperspectral imaging on the other hand has hundreds or thousands of much narrower bands. In general, it comes from an imaging spectrometer. For standoff detection using passive or active infrared detection methodology, it is essential that the characteristic absorption wavelengths corresponding to targeted chemical agents lie within the transmission windows of the atmosphere. These are shown in Figure 8.3(b) and include three small transmission windows in the medium-wave infrared (MWIR) from about 2 to 5 μm with some breaks, and a large window in the long-wave infrared (LWIR) from about 8 to 14 μm. Systems using the 2- to 5-μm band are called MWIR sensors and those using the 8- to 14-μm band are called LWIR sensors.

8.2.4 Differential Absorption Lidar

DIAL is the most frequently used technique used for detection of chemical warfare agents along with detection of toxic gases and pollutants. It uses two wavelengths; one corresponding to the wavelength of peak absorption of the targeted molecule and the other corresponding to the weak absorption of the targeted molecule. The ratio of the two received backscattered signals measures the concentration of the targeted chemical warfare agent. Figure 8.4 shows the block schematic arrangement of a DIAL system. Two laser pulses, called ON and OFF wavelengths, are transmitted toward the targeted area of interest in the atmosphere that is suspected to be contaminated with a potentially harmful chemical warfare agent. As outlined earlier, the ON wavelength corresponds to the peak absorption of the suspected harmful species and the OFF wavelength is slightly detuned from the ON wavelength to encounter significantly lower absorption from the same species. Note that the choice of ON and OFF wavelengths is unique for a given chemical species and is determined by using a tunable laser that scans the area of interest by transmitting pairs of these ON and OFF wavelengths. Obviously, the ON wavelength, called λ_{ON}, encounters maximum absorption with the result that the corresponding backscattered signal is relatively weak compared to the λ_{OFF} that encounters weak absorption and produces a relatively stronger backscattered signal. This forms the basis of knowing atmospheric constituents at that time. The ratio of the two backscattered signals gives the concentration of the molecule of interest.

Many chemical warfare agents have significant absorption bands in the 3- to 4-μm and 9- to 11-μm bands. CO_2 lasers emitting in the 9- to 11-μm band have been commonly used for the detection of the majority of chemical agents. Some DIAL systems do make use of both these bands and therefore

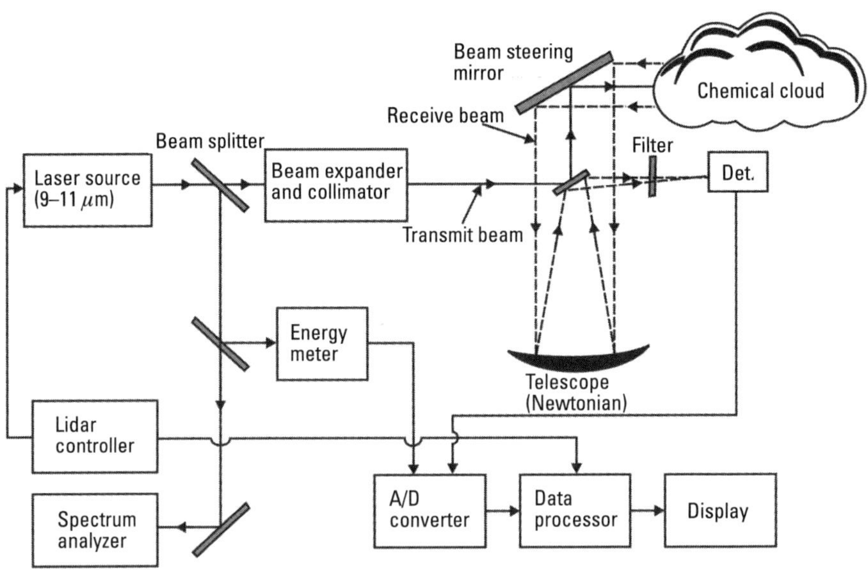

Figure 8.4 Differential absorption lidar.

employ both tunable mid-infrared as well as CO_2 laser sources. The schematic arrangement of the DIAL system shown in Figure 8.4, however, makes use of only a CO_2 laser. The transmit laser beam after suitable collimation is directed toward the atmosphere in the desired direction with the help of a scanning gimbal mirror. The backscattered signal is received by the receiving telescope and is further focused on to the detector subsystem. Interference filter is used to block the undesired radiation and allow only the radiation of wavelength of interest to pass through it. The detected signal is then digitized and fed to the data processor to extract the desired information on the type and concentration of the chemical species.

8.2.5 Representative Systems for Detection of Chemical Agents

There are a large number of detectors available for detection of chemical warfare agents and toxic industrial chemicals using the different technologies outlined in Section 8.2.2. Some representative systems include a *chemical agent monitor* (CAM) and *RAID-M* (ion mobility spectrometry). The *AP2C* and its updated version, *AP4C portable chemical contamination control detectors,* from PROEN-GIN (flame photometry), the *FirstDefender* and *FirstDefender-XL* from Ahura Corporation (Raman spectroscopy), *HAZMATCAD, ChemSentry-150C,* and *CW Sentry Plus* (SAW technology), *MiniRAE-2000, MiniRAE-3000, ppbRAE family, MultiRAE Plus* and *ToxiRAE Plus,* all from RAE Systems, and the *Toxic Vapour Analyzer Model: TVA-1000B* from Thermo Scientific (photo ionization detection).

Most information on lidar complexes specifically designed for detection and identification of chemical warfare agents is classified and not available in public domain. Some atmospheric and meteorological lidar systems in addition to monitoring atmospheric pollutants and meteorological parameters are also capable of stand off detection of chemical warfare agents such as organophosphates. One such system is the *mobile lidar complex* from Laser Systems, Russia.

Some of the representative passive infrared detectors of chemical warfare agents and toxic industrial chemicals include single-band imagers types *GF-300/320* and *GF-304* from FLIR, the multispectral imager *Second Sight-MS* from Bertin, hyperspectral scanners *RAPID Plus* from Bruker Detection Systems, *SIGIS-2* from Bruker Optics and *iMCAD* from Mesh Inc., and hyperspectral imagers *HI-90* from Bruker Optics and *Firefly* from Mesh Inc. The salient features of representative lidar systems and passive infrared detectors intended for chemical agent detection are briefly discussed next.

1. *Mobile lidar complex* combines an *aerosol lidar,* a *wind lidar,* a *long-wave (LW) DIAL lidar,* and a *short-wave (SW) DIAL lidar* with maximum detection ranges of 20, 5, and 3 km, respectively. The system

is capable of detecting hydrocarbons, freon gases, and organophosphates, and so forth. The scanning lidar offers an azimuth field of view of +/− 180° and an elevation field of view of −15° to +90°.

2. The *GF-304* and *GF-300/320* are mainly intended for detection of gas leaks. The GF-304 is used for detection of refrigerant gases in the wavelength band of 8 to 8.6 μm. The detector used is a cooled quantum-well infrared photodetector.

3. The FLIR GF300/GF320 infrared camera is capable of detecting methane and volatile organic compound (VOC) fugitive emissions from the production, transportation, and use of oil and natural gas. It employs a cooled indium antimonide detector.

4. The *Second Sight MS* is a passive, long-range chemical and toxic gas cloud detector based on infrared spectral imaging technology. It is designed for early warning and real time visualization of all suspicious gas clouds that absorb in band III of infrared. It can identify the family of the gas and measure its path length concentration.

5. The *RAPID Plus* is a second-generation passive infrared standoff detector that can automatically detect, identify, and monitor all known CWAs and important TICs from remote distances up to 5,000m.

6. *SIGIS-2* (Figure 8.5) is a scanning infrared gas imaging system based on the combination of an infrared spectrometer with a single detector element and a scanner system. It allows identification, quantification, and visualization of potentially hazardous gas clouds from long distances.

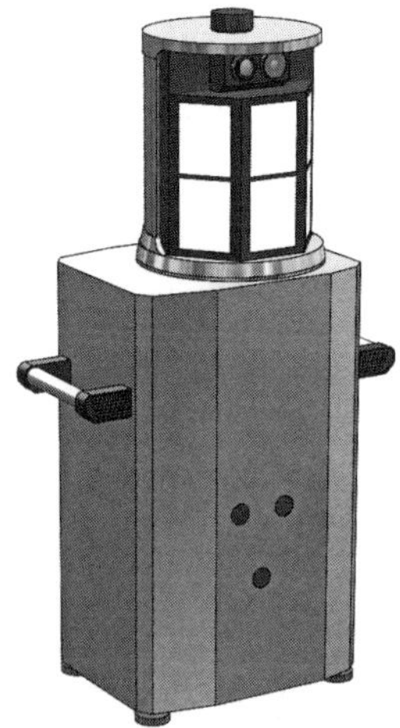

Figure 8.5 SIGIS-2 scanning infrared gas imaging system.

7. The improved Mobile Chemical Agent Detector (*iMCAD*) is an advanced standoff sensor that uses combination of passive FTIR instrument with thermal imaging and visible cameras. It can detect, identify, and map chemical weapon vapors, chemical aerosols, toxic industrial chemicals, and biological particles at distances of up to 6,000m.

8. The *HI-90* (Figure 8.6) is a hyperspectral imaging system based on the combination of a Michelson interferometer and a focal plane array detector. The system allows standoff identification, quantification, and visualization of potentially hazardous gases from long distances of up to several kilometers.

9. The *Firefly* is a FTIR hyperspectral imaging sensor that is intended for standoff visualization and identification of chemical vapor plumes.

8.3 Detection of Biological Warfare Agents

Biological warfare agents, which include pathogenic microorganisms and toxins generally of microbial, plant, or animal origin, are a grave threat to a nation not only because of their potential to cause massive disease outbreaks striking a large population over a wide geographical spread and thereby inflicting heavy casualties, but also because of the social, economic, and psychological impact they have on the entire nation. The ease with which they can be disseminated, their contagiousness, their ability to replicate in a host organism, and the fact that they remain undetected for days, weeks, and sometimes months before the onset of disease and its subsequent confirmation through clinical tests makes them far more dangerous. Biological agent detection systems that can detect in almost real time are therefore required so that the first responders do not have

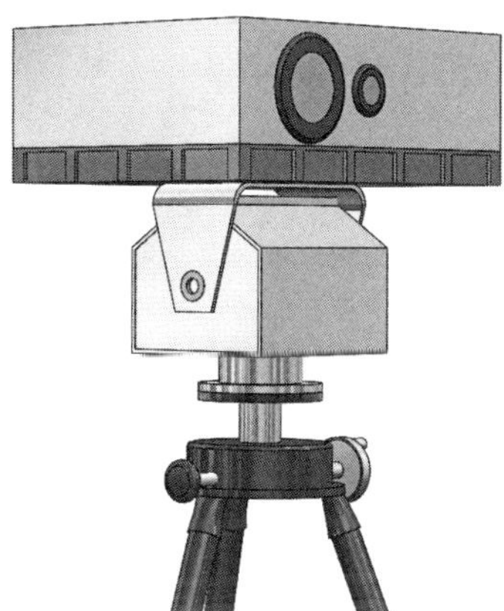

Figure 8.6 HI-90 hyperspectral imaging system.

to wait for the appearance of symptoms to initiate measures to stop its spread. Early detection and identification of biological warfare agents during a deliberate biological incident is essential to initiate corrective emergency responses for efficient management of such incidents.

8.3.1 Detection Parameters

Detection parameters in the case of biological warfare agents are broadly no different from those discussed earlier in the case of chemical warfare agents. In this case too, sensitivity or limit of detection, selectivity, and response time are the most critical detection parameters. Sensitivity in the case of biological warfare agent detectors needs to be even higher than what might be acceptable in the case of chemical detectors as relatively much smaller concentrations of biological agents could be fatal. As an example, a concentration of 100 particles per liter of anthrax-causing bacterium called *Bacillus anthracis* or only 10 particles per liter of tularemia-causing bacterium called *Francisella tuarensis* can infect a person. Discrimination of pathogens from other nonhazardous biological and nonbiological components in the environment present in normally much higher concentration than the targeted biological agent is also very important to achieve a low false-positive rate. Rapid detection is very crucial for an effective intervention by first responders. Ideally, detection should take place in real time, but current technology does not allow for that.

8.3.2 Review of Detection Technologies

Biological warfare agent detection technologies as outlined in Chapter 1 are broadly classified as point and standoff detection technologies. There are a number of detection and identification techniques in each of the two categories. Point detection techniques are further classified as specific point detection techniques capable of detecting and identifying the biohazard and nonspecific point detection techniques that can only detect and not identify the biohazard. Standoff detection systems are capable of detecting the presence of biological warfare agents at a standoff distance from the point of release. Common nonspecific point detection techniques include particle sizers, fluorescence-based systems, viable particle size samplers, and virtual compactors. Important specific point detection techniques include molecular biology, flow cytometry, and mass spectrometry and immunoassay technologies. Standoff detection systems are based on lidar technology. UV-LIF lidar is the potential candidate for detection of biological warfare agents from standoff distances. An overview of various optronic and nonoptronic technologies was presented in Chapter 1. A UV-LIF lidar-based system is described in the following sections.

8.3.3   UV-LIF Lidar

The standoff detection of biological warfare agents is based on the concept of the LIF effect. Laser-induced fluorescence is the emission from atoms or molecules after they have been excited to go to higher energy levels by excitation by another laser. The emission takes place at a wavelength that is higher than the wavelength of the laser light exciting it. Biological warfare agent molecules mainly constitute aromatic amino acids and coenzymes. Aromatic amino acids such as tryptophan, tyrosine, and phenylanine absorb laser radiation at 280- to 290-nm bands and fluoresce in 300- to 400-nm bands. It is therefore possible to detect biological warfare agents by using excitation by UV laser of suitable wavelength. As well, discrimination of biological agents can be achieved only from the LIF signal because the fluorescence cross section for particles in the range of 1 to 10 μm are sufficiently large to make single-particle interrogation feasible.

Figure 8.7 shows the block schematic arrangement of typical monostatic UV-LIF lidar system. The principle components of this lidar system include a frequency tripled or quadrupled ND-YAG laser emitting at 355 nm (frequency-tripled output) and 266 nm (frequency-quadrupled output), respectively, a telescope used for both transmission and reception, a photomultiplier tube used for recording the backscattered signal and the biofluorescence signal, a spectrograph with a gated ICCD array for recording the dispersed fluorescence spectra, and the lidar controller. The fourth harmonic at 266 nm is transmitted toward the biological cloud. The telescope receives the backscattered and biofluorescence signals, which are then fed to the PMT channel. The backscattered signal is received by the gated PMT channel and is present whenever there is a cloud along the beam path. This channel is used to measure the distance to the cloud. A solar-blind PMT channel is used to receive the fluorescence signal and is activated only when there is a suspicious cloud. A spectrograph with a gated ICCD is used to identify the nature of the biomolecule responsible for the fluorescence. The bandwidth of the receiver channel is usually kept small to disallow unwanted background radiation from entering the channel.

Early UV-LIF lidar systems used excitation at near 350 nm. This wavelength had a fundamental problem in that many innocuous fluorophores, amino acids, and other substances are also excited at that wavelength, which made discrimination of threats challenging. Use of a wavelength in the 280-nm region along with a wavelength in the 350-nm region enabled more effective weeding out of harmless targets and better identification of harmful bioaerosols.

8.3.4   Representative Systems for Detection of Biological Agents

While the technologies for detection and identification of chemical warfare agents are more mature, the same is not true for detection and identification

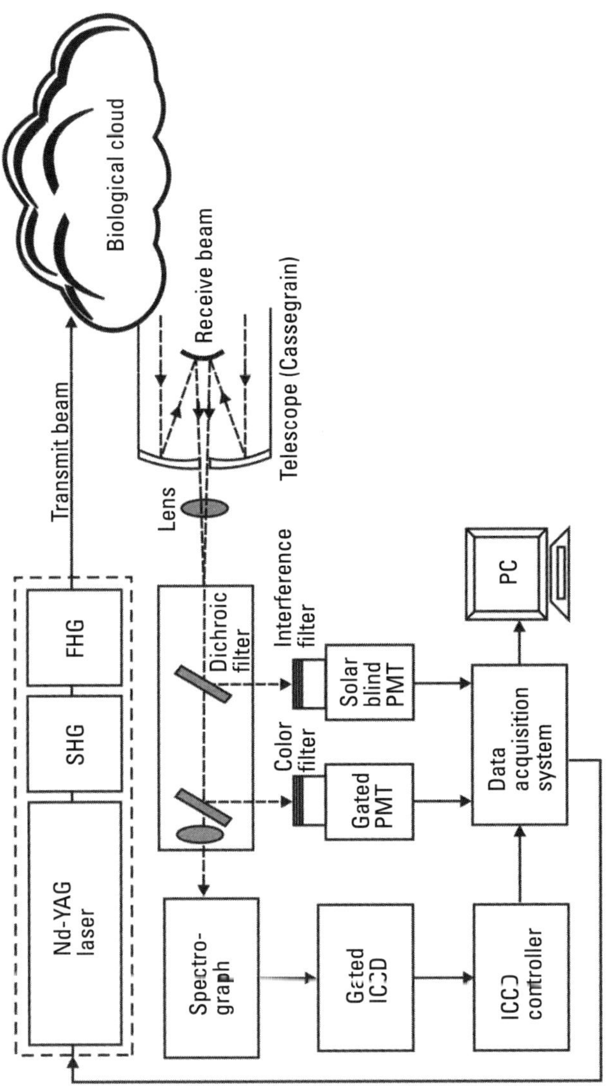

Figure 8.7 UV-LIF lidar.

of biological warfare agents. As a consequence of this, while there are a large number of field-usable chemical agent detectors exploiting various technologies, the number of biological agent detectors is much smaller. As well, not all bioagent detection technologies are mature enough to be converted into field-worthy systems. Some better known bioagent detectors include the *Bio-seeq Plus biological agent identifier* from Smiths Detection, Inc. utilizing linear after the exponential polymerase chain reaction (LATE PCR) technology, *IBAC-2 biological detection system* from FLIR using UV-LIF technology, *MAB biological agent monitor* from PROENGIN using flame spectrophotometry technology, *AP4C-FB* combined chemical and biological detector from PROENGIN, and *Smart Bio Sensor (SBS)* from Smiths Detection, Inc. based on fluorescence spectroscopy. Each is briefly described below.

1. The *Bio-Seeq PLUS* is a field-portable, high-precision bioagent identifier utilizing LATE PCR technology to detect trace levels of both viral and bacterial biological warfare agents including anthrax (pX01 and pX02), plague, smallpox, tularemia, Q-fever, and brucellosis through DNA replication. LATE PCR technology is associated with reduced false positives and improved overall reliability of results.

2. The *IBAC-2* is a continuous, real-time monitoring system that responds in less than 60 seconds in the event of the presence of an airborne biothreat. It is based on UV-LIF technology to detect all four classes of biothreats including spores, viruses, bacteria, and toxins in concentration levels of less than 100 agent-containing particles per liter of air (ACPLA).

3. The *MAB* (*Biological Alarm Monitor*) uses flame spectrophotometry technology to analyze particles in the atmosphere and produces an alarm on detection of suspect bioagents including bacteria, fungi, viruses, and biotoxins with a concentration level above a specified threshold. The MAB biological monitor has a sensitivity of 20 ACPLA and it responds in nearly real time with a response time typically of the order of 10 to 15 seconds. Figure 8.8 shows a MAB biological monitor.

4. *AP4C-FB* is a chemical and biological agent detector that combines the chemical detection features of the chemical agent detector AP4C-F and MAB biological agent monitor. *AP4C-VB*, another chem-bio agent alarm system from PROENGIN, is designed for integration on mobile platforms.

5. The *Smart Bio Sensor* (*SBS*) uses fluorescence spectroscopy to detect and classify biological agents or airborne toxins, which includes bacterial and fungal spores, viruses, and biotoxins. The detector is immune

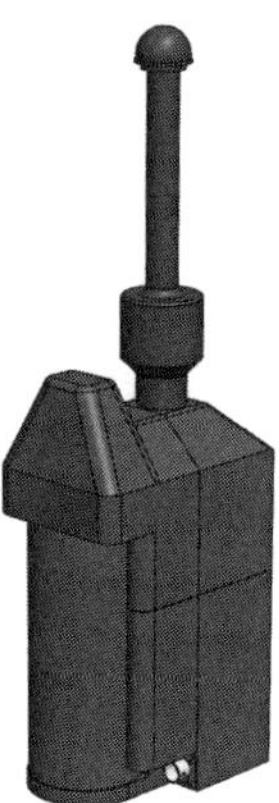

Figure 8.8   MAB biological agent monitor.

to ambient biologicals and offers near-real-time detection with a response time of approximately 1 minute. The SBS is designed for field use in varied military environments, including integration on vehicular platforms where it is used to alert troops to the presence of biological agents or biotoxins and classify the potential threat. The SBS can also be used to protect critical infrastructure where it is used to alert emergency responders of a biological incident.

Most information on development of systems for standoff detection of hazardous bioaerosols is classified. Some of the experimental systems are reported in the public domain. Two known systems include the *UV-LIF Lidar system* of the Norwegian Defence Research Establishment and the *Long Range Biological Stand-off Detection system* (LR-BSDS).

1. Figure 8.9 shows a simplified schematic of the Norwegian biological lidar. It is a breadboard system using only commercially-off-the-shelf (COTS) components. Major components include a pulsed Nd-YAG laser emitting 150-mJ, 10-Hz pulses at 355 nm, a 1,200-mm focal length Newtonian telescope, a PMT for receiving elastic scatter, and a gated ICCD-coupled spectrograph for receiving inelastic scatter.

2. LR-BSDS is used to detect, track, and map large-area aerosol clouds at ranges up to 30 km. The system has three major components including a 100-Hz operationally eye-safe laser transmitter, a 24-inch receiving telescope, and a transferred electron-intensified photodiode detector with an information processor integrated into a frame. This system provides information about the cloud configuration such as size, shape, and relative intensity and cloud location including range,

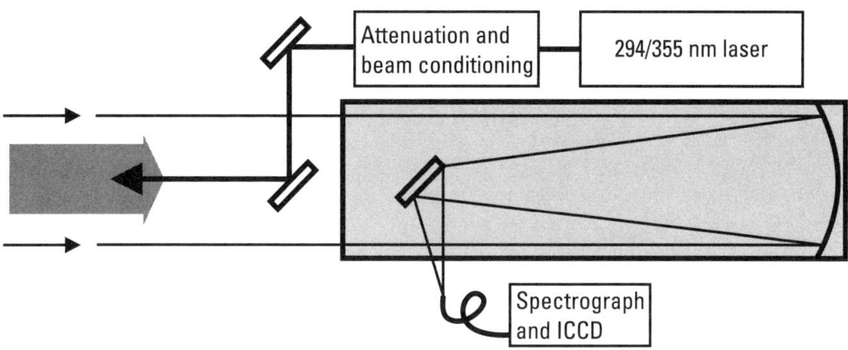

Figure 8.9 Norwegian biological lidar.

width, height, height above ground, and drift rate. The LR-BSDS is intended to be a corps-level asset that provides early warning and information on the location and configuration of aerosol cloud formation. This information is used to enhance contamination avoidance efforts and cue other biological detection measures.

8.4   Radiation Detection Devices

Radiation detectors are used to detect and identify the presence of specific types of radiation including alpha, beta, gamma, and neutron radiations in the environment, on the surface of people in the event of external contamination, inside people in the event of internal contamination, and also radiation received by people during exposure to an external radiation source while working in a radiation area. Different categories of radiation detectors, each suited to a specific application, are used for detection of specific radiation, for measurement of specific levels or ranges of radiation energy, for measurement of exposure to X-rays or gamma radiation such as Roentgens per unit time, and for measuring accumulated dose and current dose rate, respectively, in gray and gray per unit time. Radiation detection devices are broadly categorized as *radiation survey meters* and *dosimeters*. These are briefly discussed next.

8.4.1   Radiation Survey Meters

Radiation survey meters are portable instruments used to detect, identify, and quantify external or ambient ionizing radiation fields and also detect and monitor personnel, equipment, and facilities for radiation and radioactive contamination. These meters are used by emergency responders including law enforcement agencies, firefighters, health care professionals, and other professionals trained to handle hazardous materials in response to radiological incidents or during interception and prevention of movement of hazardous materials. There are small radiation detectors called *radiation pagers* worn by personnel and used to provide audio, visual, or vibration alarm warning about the presence of an elevated radiation level. Radiation pagers are used as personal alarming devices for early detection and are designed to respond only to highly penetrating ionizing radiation such as gamma and neutron but not to alpha and beta radiations. The alarm may be used to alert the emergency responders equipped with relatively larger and more sensitive radiation survey meters to locate the radioactive source or material or quantify the radiation level to assess the scale of the radiological problem. Some radiation survey meters have a readout unit with interchangeable probes for detecting different types of ionizing radiation. There are two types of radiation survey meters: one that use *gas-filled detectors* and one that use *scintillation detectors.* Common gas-filled detectors used in radiation survey meters include Geiger-Mueller (GM) tubes, ion chambers, and proportional counters. Radiation survey meters based on gas-filled detectors and scintillation detectors are briefly described in next.

8.4.1.1   Geiger Counters

Geiger counters are configured around a Geiger-Mueller tube, which is a metal cylinder filled with inert gas such as neon or argon at low pressure and sealed at one end by a ceramic or mica window. A thin metal wire, usually of tungsten, runs down the tube with one of its ends connected to positive terminal of a power supply. The negative terminal of the power supply is connected to the curved surface of the metal tube. The wire and the curved surface act as anode and cathode, respectively. Figure 8.10 shows the schematic arrangement of a Geiger counter.

The radioactive particles entering the tube ionize the gas to produce positive ions and electrons. Negatively charged electrons are instantly attracted to the anode, while the positive ions are repelled by the large positive charge and flow toward the cathode. As the electrons move down the gas, they collide into more atoms, thereby initiating a chain reaction of ionization called Geiger discharge that produces more ions and electrons. A large number of electrons arriving at the anode generate a pulse of electricity that is measured on a counter. Counts rate or frequency measured in counts per second (cps) gives indication of the strength of the radiation field. Different units, such as milli-Roentgens per hour (mR/hr) or micro-Sieverts per hour (μS/hr), are currently used to display the severity of radiation in a given area. Most Geiger counters have an associated amplifier and loudspeaker to produce a sound click every time there is an electrical pulse. Before the counter can detect any more radiation, it needs to be restored to its original state through a process called quenching. This is achieved by releasing a halogen gas that attracts free ions. A large value resistance can also be used to oppose the flow of current. Geiger counters can

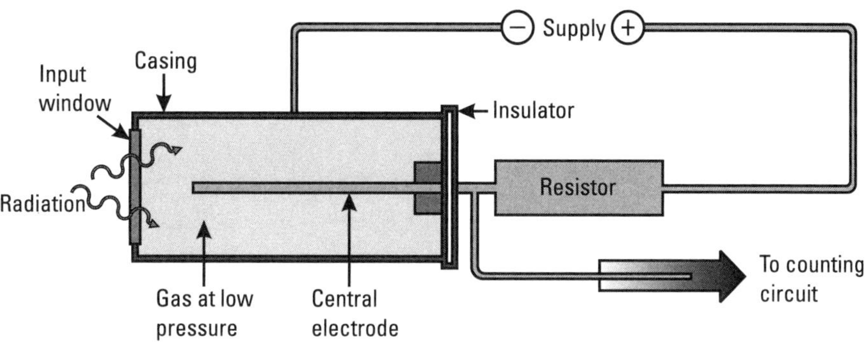

Figure 8.10 Geiger counter.

be used to measure the intensity of alpha, beta, gamma radiation, and X-rays. Geiger counters have a limitation that they produce an output pulse of the same magnitude regardless of the energy of the incident radiation. As a consequence, they cannot distinguish between different radiations.

8.4.1.2 Ionization Chambers

Ionization chambers, like Geiger counters, are also gas-filled radiation detectors. They are used for measuring intensity of beta, gamma, and X-ray radiation. It consists of a gas-filled chamber with two electrodes called anode and cathode. The electrodes may be in the form of parallel plates as in the case of a parallel plate ionization chamber or a cylindrical arrangement with a coaxially located internal anode wire. The ionization chamber only uses the discrete charges created by each interaction between the incident radiation and the gas and does not involve the multiplication mechanisms present in other radiation instruments, such as the Geiger-Mueller counter or the proportional counter. These positive and negative charges while reaching respective electrodes cause an observable pulse of current to flow through the external circuit. Ionization chambers offer a uniform response to radiation over a wide range of energies and are preferred for measurement of high levels of gamma radiation. Figure 8.11 illustrates the operational principle of an ionization chamber.

8.4.1.3 Proportional Counters

Proportional counters, like Geiger counters, utilize the phenomenon of gas multiplication to increase the pulse amplitude. Virtually all proportional counters use a wire anode of a small diameter placed inside a larger cylindrical chamber that acts as cathode and also serves to enclose the gas. However, in contrast to a Geiger counter where the amplitude of the output pulse is same irrespective of the energy of incident radiation, in a proportional counter, each original free electron that is formed along the track of the particle creates its own individual

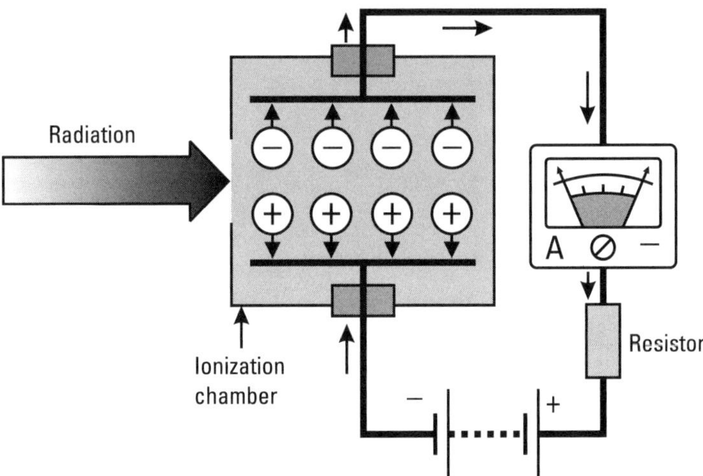

Figure 8.11   Ionization chamber.

avalanche and different avalanches created by different original electrons are of the same size with the result that the final total charge remains proportional to the number of original ion pairs formed along the particle track. The proportionality between the size of the output pulse and the amount of incident energy allows discrimination between different types of particles. Proportional counters are commonly used for measurement of alpha and beta activity. They are also used for neutron detection, and to some extent for X-ray spectroscopy.

8.4.1.4   Scintillation Counters

Scintillation counters measure ionization radiation by using a scintillator that fluoresces when coming in contact with ionizing radiation, a PMT detector to convert fluorescence photons to proportional electrical current pulses, and electronics to process the signal to measure and count radiation. Figure 8.12 illustrates the constructional features of a scintillation counter. The scintillation phosphors are either inorganic crystals such as sodium iodide, zinc sulphide, or lithium iodide, or organic crystals such as stilbene or anthracene, or plastic phosphors. There are also liquid scintillation counters. These use a mixture of solvent and fluors. The energy in the ionizing radiation is transferred to the solvent, which in turn transfers it to the fluors to excite them to the unstable higher-energy state. The energy released as the fluors fall back to the stable state is detected to measure the radiation.

8.4.2   Dosimeters

Dosimeters are radiation detection devices used to measure the level of exposure to the ionizing radiation of personnel, including radiographers, those working

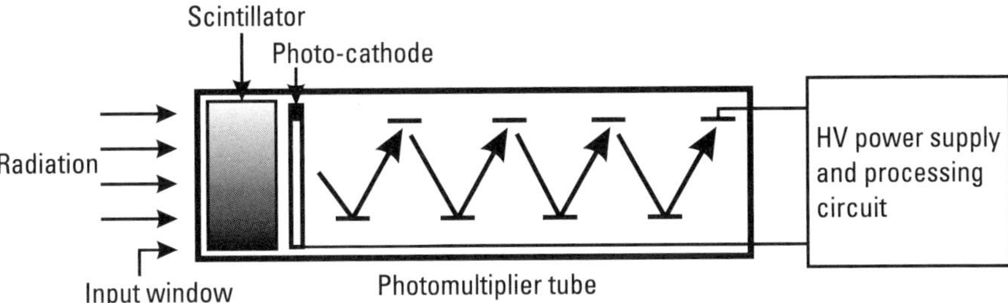

Figure 8.12   Scintillation counter.

in nuclear facilities, doctors using radio therapy, and also those working in laboratories using radionuclides. Different types of dosimeters including personal dosimeters and operational dosimeters were outlined in Chapter 1. There are personal dosimeters that are worn to obtain a whole body radiation dosage and also specialist types that can be worn on the fingers or clipped to headgear to measure the localized body irradiation for specific activities. Note that radiation doses received following internal exposure from the incorporation of radioactivity cannot be measured by dosimeters.

In a *film badge* type of personal dosimeter, the density of the developed photographic film indicates the received radiation dose. However, the same dose may produce different density changes due to changes in film properties. This drawback can be overcome by comparing with the density change of an identical film exposed to a known radiation dose to cancel out variations resulting from differences in film properties or development procedures. Another drawback of a film badge dosimeter is the limited dynamic range between underexposure and overexposure. The dynamic range may be extended by fitting the holder containing film badge with a set of small metallic filters to cover selected regions of the film.

Thermoluminescent dosimeters are based on the use of crystalline materials such as lithium fluoride in which ionizing radiation creates electron-hole pairs. Electron-hole pairs formed by the incident radiation are quickly captured and immobilized. The population of trapped charges accumulates in the material during the exposure time of radiation. The trapped charge is more or less permanently stored if the exposure is carried out at normal temperatures. The temperature of crystal is raised to initiate rapid release of trapped charges, leading to emission of photons by recombination of liberated electrons with trapped holes, or alternatively by recombination of liberated hole with a trapped electron. The total intensity of the emitted light is proportional to the original population of trapped charges, which in turn is proportional to the radiation dose accumulated over the exposure period. The light output is measured by using a PMT detector. Since the readout process effectively empties all the traps, the material can be recycled for repeated use.

Ionization chamber dosimeters are ionization chambers used in conjunction with a high gain, negative feedback operational amplifier with a standard resistor or capacitor connected in the feedback path as shown in Figure 8.13. While a resistor in the feedback is used to measure the chamber current; a capacitor in the feedback path is used to measure charge collected over a fixed time interval. The principle of operation of ionization chamber was discussed in Section 8.4.1.2.

8.4.3 Representative Radiation Detectors

A large number of companies worldwide offer different categories of radiation detectors. Most devices of a given type have similar specifications. The salient features of some representative devices are briefly mentioned below.

1. The *GCA-07-DL* from Images Scientific Instruments Inc. is a hand-held digital Geiger counter capable of detecting radiation energy levels of alpha particles above 3 MeV, beta radiation above 50 KeV, and X-ray and gamma radiation above 7 KeV. It offers counting and radiation measurement resolution figures of one count per minute (CPM) and 0.001 mR/hour, respectively. Count and radiation ranges are one CPM–1,000 counts per second (CPS) and 0.001–1000 mR (imperial measurements) or 0.01–10,000 μSv/hour (metric), respectively. The DTG-01 from the same company is a desktop Geiger counter with similar features and specifications.

2. The *ion chamber survey meter type 451P* from Fluke Biomedical is a handheld pressurized ion chamber survey meter offering μR (micro-

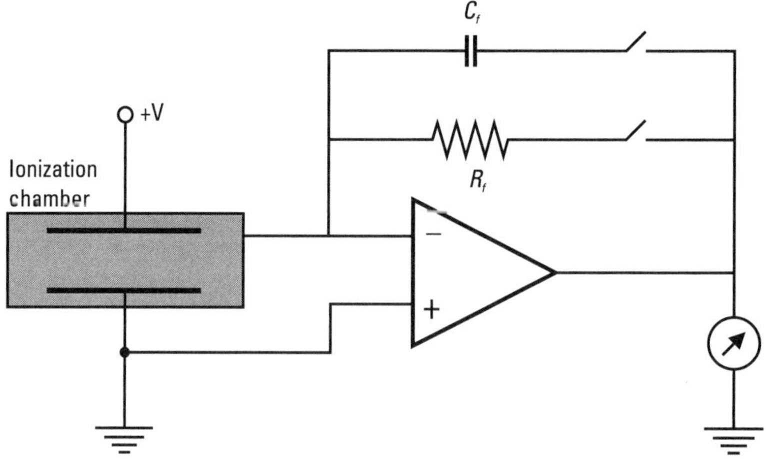

Figure 8.13 Ionization chamber dosimeter.

Roentgen) resolution. It detects beta radiation above 1 MeV and X-ray and gamma radiations above 25 keV. The operating range is 0 to 5 R/hour. Response time is in the range of 1.8 to 5 seconds depending on radiation level.

3. The *RI-02* from MIRION Technologies is designed for highly stable and accurate measurement of dose rates and integrated dose of beta, gamma, and X-ray radiation. It offers a measurement range of 0.1 mR/hr to 100 R/hr (1 μSv/hr–1 Sv/hr) and up to 1,000 R (10 Sv) of integrated dose. The radiation energy level detection range is 20 keV to 2 MeV.

4. The *RAD-60/62 electronic dosimeters* from MIRION Technologies are precise radiation measuring instruments for reliable detection and registration of gamma and X-rays for ensuring the personal safety of the user. The RD-60 is suitable for a broad range of everyday radiation monitoring purposes in stand-alone conditions. The RAD-62 is an integrated solid-state dosimeter with a full range of functions for use in the automatic dose management systems. Both dosimeters are characterized by integrated dose range of 0.1 mR to 999 R (1 μSv–9.99 Sv). The dose rate and energy response ranges are 0.5 mR/h to 300 R/h (5 μSv/h–3 Sv/h) and 60 keV to 6 MeV, respectively.

5. The *DMC-3000 Personal Electronic Radiation Dosimeter* features superior gamma and X-ray energy response. Key performance specifications include an X-ray and gamma radiation energy range of 15 keV to 7 MeV, an integrated dose range of 1 μSv to 10 Sv (0.1 mR–1,000

Figure 8.14   DMC-3000 electronic dosimeter with add-on neutron module.

R), and a dose rate range of 0.1 μSv/h to 20 Sv/h (0.01 mR/h–2,000 R/h). It features programmable alarms with visual LED, audible, and vibrating alarm indicators and simple two-button navigation. Various add-on modules including a beta module, neutron module, and telemetry module can be attached to the DMC-3000 dosimeter to expand the detection and communications capabilities of the dosimeter. Figure 8.14 shows the DMC-3000 dosimeter with an add-on neutron module.

Selected Bibliography

Accetta, J. S., *The Infrared and Electro-optic Systems Handbook,* Volume 7, Bellingham, WA: SPIE International Society for Optical Engineering, 1993.

Baudelet, M. (ed.), *Laser Spectroscopy for Sensing: Fundamentals, Techniques and Applications,* Cambridge, UK: Woodhead Publishing Limited, 2014.

Demtroder, W., *Laser Spectroscopy Volume 1: Basic Principles,* Fourth Edition, Berlin: Springer, 2008.

Demtroder, W., *Laser Spectroscopy Volume 2: Experimental Techniques,* Fourth Edition, Berlin: Springer, 2008.

Ellison, D. H., *Handbook of Chemical and Biological Warfare Agents,* Second Edition, Boca Raton, FL: CRC Press, 2007.

Gupta, R. C. (ed.), *Handbook of Toxicology of Chemical Warfare Agents,* Second Edition, London: Academic Press, 2015.

Khan, M., *Electrochemical Detection of Chemical Warfare Agents,* Latvia: LAP Lambert Academic Publishing, 2013.

Knoll, G. F., *Radiation Detection and Measurement,* Chichester, UK: John Wiley & Sons, 2010.

Sferopoulos, R., *A Review of Chemical Warfare Agent (CWA) Detector Technologies and Commercial-Off-The-Shelf Items,* No. DSTO-GD-0570, Victoria, Australia: Defence Science and Technology Organisation Victoria (Australia), Human Protection and Performance Division, 2009.

Sun, Y., and K. Y. Ong, *Detection Technologies for Chemical Warfare Agents and Toxic Vapours,* Boca Raton, FL: CRC Press, 2004.

Waynant, R., and M. Ediger (eds.), *Electro-Optics Handbook,* New York: McGraw-Hill, 2000.

Weitkamp, C. (ed.), *Lidar: Range Resolved Optical Remote Sensing of the Atmosphere,* New York: Springer, 2014.

Glossary

Acoustic weapons (less-lethal)   An acoustic weapon is a type of less-lethal weapon that employs audible sound, infrasound, or ultrasound to achieve the intended effect. Three broad categories of less-lethal acoustic weapons include *acoustic-optical devices* such as stun grenades, *acoustic generators* that project sound in audible, ultrasonic, or infrasonic frequency ranges at a decibel level generally exceeding 85 dB, and *vortex generators* that project a vortex of air at high speed.

Active countermeasures   Active countermeasures further comprise of soft-kill countermeasures and hard-kill countermeasures. Soft-kill countermeasures change the electromagnetic, acoustic, or other forms of signatures of the platform to be protected, thereby adversely affecting the tracking or sensing capability of the incoming threat. In the case of hard-kill active countermeasures, high-energy lasers are used to damage the front-end optics of the sensor system.

ANSI Z136.1 laser safety standard   The ANSI Z136.1 laser safety standard document provides guidance for the safe use of lasers and laser systems by defining control measures for each of seven laser hazard classifications.

Avalanche photodiodes   Avalanche photodiodes (APDs) are high-speed; high-sensitivity photodiodes utilizing an internal gain mechanism that functions by applying a relatively higher reverse bias voltage than that is applied in the case of PIN photodiodes.

Biological warfare agents   Biological warfare agents are microorganisms such as bacteria, viruses, fungi, and toxins produced by them.

Blistering agents   Blistering agents, also called *vesicants,* are chemical warfare agents. These are toxic compounds that cause severe skin and eye injuries like the ones caused by burns. On inhalation, they affect the upper respiratory tract and the lungs producing pulmonary edema. Sulfur mustard and lewisites are examples of blistering agents.

Bolometer   A type of thermal sensor in which a sensing element is a resistor with a high-temperature coefficient. A bolometer is different from a photo-conductor; in a photoconductor a direct photon-electron interaction causes a change in the conductivity of the material, whereas in a bolometer the increased temperature and the temperature coefficient of the element causes the resistance change.

Bulk method of detection of explosives   Bulk explosive detection involves the detection of a macroscopic mass of explosives material through measurement of material characteristics such as mass, density, and effective atomic number, also called Z-number. X-ray scatter, neutron, and γ-based techniques, magnetic techniques, millimeter-wave imaging, and terahertz spectroscopy are important bulk detection methods.

Bullet camera   Bullet cameras are shaped like a rifle bullet shell.

Carbon dioxide laser   A carbon dioxide (CO_2) laser is the most widely used and diversely exploited type of gas laser. The laser medium in this case is a gas mixture of CO_2, helium (He), and nitrogen (N_2). It produces laser radiation in two principle wavelength bands centered on 10.6 and 9.4 μm.

CBRN threats   CBRN threats are those originating from both weaponized and nonweaponized forms of chemical, biological, radiological, and nuclear materials.

Charge-coupled device (CCD)   A CCD is a light-sensitive device comprised of an array of thousands to millions of light-sensitive elements in the form of MOS capacitors called pixels etched onto a silicon surface. The image is stored in the form of charges in different MOS capacitors. The charge is read out by suitable electronics and then converted into a digital bit pattern, which represents the image.

Chemical warfare agents   Chemical warfare agents are toxic chemical substances used with the intent of killing, causing injury, or incapacitating adversary armed forces during war and both armed forces and civilian population during acts of terrorism. These are broadly classified as nerve agents, blistering agents, cyanides, and toxic Industrial chemicals.

Chemiluminescence (explosive detection)   Chemiluminescence is the emission of light from a chemiluminescent reaction. Chemiluminescence is used for detection of explosives by measuring the amount of emitted infrared light in a chemiluminescence reaction. The infrared light is a measure of amount of NO present in the material, which in turn is indicative of the amount of the original nitrogen-containing explosive material.

CMOS sensor   A CMOS sensor is a light-sensitive device that records the intensities of light as variable charges similar to a CCD sensor. Unlike CCD sensors, in a CMOS sensor, each pixel has its own charge-to-voltage convertor, amplifier, and pixel select switch. This is called active pixel sensor architecture in contrast to passive pixel sensor architecture used in a CCD sensor. As well, the sensor often also includes on-chip amplifiers, noise-correction, and analog-to-digital conversion circuits and other circuits critical to pixel sensor operation. The chip in this case outputs digital bits.

Coherence   Light is said to be coherent when different photons (or the waves associated with those photons) have the same phase and this phase relationship is preserved with respect to time.

Collisional cascading   See *Kessler syndrome*

Colorimetric chemical agent sensor   A colorimetric chemical agent sensor uses a sorbent substrate such as paper or paper ticket to which a reagent has been applied. When the targeted chemical comes into contact with the substrate, it reacts with the reagent to produce a distinctive color change. This allows identification of the chemical agent.

Computed tomography   A computed tomography (CT) scan uses X-ray radiation to produce a three-dimensional image of the object under screening or examination.

Deflagration   Deflagration is an exothermic process in which the transmission of decomposition reaction is based on thermal conductivity. Deflagrations are thermally initiated reactions propagating at subsonic speeds.

Detectivity (photosensor)   Detectivity of a photosensor is the reciprocal of its NEP. A sensor with a higher value of detectivity is more sensitive than a sensor with a lower detectivity value. Detectivity, like NEP, depends on noise bandwidth and sensor area.

Detonation   Detonation is a reaction involving chemically unstable molecules that when energized, instantaneously split into many small pieces, subsequently recombining into different chemical products releasing very large amounts of thermal energy. The speed of reaction in this case is supersonic with 1,000 m/s considered as the minimum speed that distinguishes deflagration from detonation.

Dielectrometry   Dielectrometry is an imaging technique that uses a low-energy microwave field to irradiate the object under screening. It is commonly used for personnel screening.

Differential absorption lidar   Differential absorption lidar (DIAL) uses two wavelengths; one corresponding to the wavelength of peak absorption of the targeted molecule and one corresponding to the weak absorption of the targeted molecule. The ratio of the two received backscattered signals measures the concentration of the targeted chemical warfare agent.

Digital night vision   Digital night vision differs from conventional night vision in the sense that while in a conventional night vision device, available light is collected through the objective lens and focused onto an intensifier. Most digital night vision devices process and convert the optical image into an electric signal through a highly sensitive CCD image sensor.

Direct detection (thermal imaging)   In the case of direct detection, the detector element translates the photons directly into electrons. The charge accumulated, the current flow, or the change in conductivity is proportional to the radiance of objects in the scene.

Directed-energy weapons   A directed-energy weapon (DEW) system primarily uses directed energy in the form of a concentrated beam of electromagnetic energy in the targeted direction to cause intended damage to the enemy's equipment, facilities, and personnel. Intended damage could be lethal or nonlethal. Based on the wavelength of the directed electromagnetic energy, these are classified as high-power microwave (HPM) weapons, millimeter-wave (MMW) weapons, and laser weapons.

Discreet camera   Discreet cameras use discreet or covert housings. Such cameras are generally used either by architects seeking to achieve certain aesthetics within a space or in situations where surveillance is done without the knowledge of targeted individuals, such as in sting operations.

Divergence (laser beam)   Divergence is an indicator of the spread in the laser beam spot as it travels away from the source. It is a function of the wavelength (λ) and size of output optics.

Dome camera   Dome cameras have a domelike shape.

Dosimeter   Dosimeters measure the amount of energy deposited by ionizing radiation over a given period and therefore are used to estimate the effective dose received by the human body through exposure to external ionizing radiation while working in radiation area.

Dynamic range (explosive detection)   Dynamic range is the explosive concentration from LOD to maximum concentration that can be reliably detected.

Electrical pumping   Electrical pumping uses electrical discharge and is common in gas lasers. The excited electrons in the gas discharge plasma transfer their energy to the lasing species either directly or indirectly through the atoms or molecules of another element.

Electrical weapon (less-lethal)   An electrical weapon usually works by sending a high-voltage, low-current electrical discharge through the body of the targeted person. The discharge voltage typically varies from 1 MV to a few tens of megavolts. The discharge current is limited to less than 5 mA. The electrical shock interferes with the communication between the brain and the muscles, causing involuntary muscle contractions and impairment of motor function.

Electrified water cannon   A type of electrical weapon that requires a direct contact with the body to produce the intended effect.

Erbium:glass laser   An erbium-doped glass laser produces output at 1,540 nm, which is attractive as an eye-safe alternative to neodymium-doped YAG- (or glass) based military laser rangefinders and laser target designators.

Erbium:YAG laser   An erbium-doped YAG laser produces output at 2,940 nm. This wavelength holds promise for medical applications in the field of plastic surgery due to its extremely large absorption by water in tissue.

External cavity diode laser   An external cavity diode laser is a semiconductor diode laser whose resonator is completed with one or more optical components outside the diode laser chip.

False alarm rate (explosive detection)   There are two types of false alarms: false negative and false positive. A *false positive alarm* is said to occur when the system gives an alarm indicating the presence of an explosive even in the absence of one. A *false negative alarm* is said to occur when the system fails to detect an explosive when one is present. The *false alarm rate* is the number of times the system gives a positive or negative false alarm divided by total number of tests performed.

Fiber laser   Fiber laser is a type of solid-state laser where the gain medium is a glass fiber doped with rare earth element ions such as neodymium (Nd^{3+}), erbium (Er^{3+}), ytterbium (Yb^{3+}), thulium (Tm^{3+}), holium (Ho^{3+}), or praseodymium (Pr^{3+}).

Flame ionization detection (detection of chemical warfare agents)   Flame ionization detection (FID) technology is in principle similar to photoionization detection (PID) technology in the sense that an analyte in both cases is ionized, with the difference that FID sensors use a hydrogen flame as the ionization source rather than UV radiation.

Flame photometry (detection of chemical warfare agents)   Flame photometry is an atomic spectroscopy technique that uses the characteristic emission spectrum of the atoms for fingerprinting of chemical warfare agents as they return to lower energy states.

Flicker noise   Flicker noise, or 1/f noise, occurs in all conductors where the conducting medium is not a metal and exists in all semiconductor devices that require a bias current for their operation. Its amplitude is inversely proportional to the frequency. Flicker noise is usually predominant at frequencies below 100 Hz.

Flow cytometry (detection of chemical warfare agents)   Flow cytometry is a technology widely used to analyze the physical and chemical characteristics of particles in a fluid as it passes through at least one laser beam. A flow cytometer counts and measures the size of particles dispersed after liquid phase concentration using a laser diffraction system.

Fluorescence technologies (detection of biological warfare agents)   Fluorescence technologies involve excitation of molecular components of a biological agent with light, usually in the ultraviolet (UV) region of the spectrum. The fluorescence-based devices exploit the properties of endogenous fluorophores (i.e., fluorophores growing or originating from within the organism), to detect biological agents through bioluminescence.

Fluoroscopy   Fluoroscopy is an X-ray imaging technique that uses low-energy X-rays to form a still or a dynamic image of the object under screening. The equipment operates on detection of X-ray radiation transmitted through the object.

FM-CW laser range finder   The FM-CW laser range finding technique is like the one followed in the case of its radar counterpart (FM-CW radar). In this, the frequency of a narrow line width laser is modulated with a ramp or sinusoidal signal; collimated, and then transmitted toward to the target. The received signal corresponding to the reflected specular or diffused laser beam is mixed with the reference signal representing the transmitted laser beam to produce a beat frequency, which is then used to compute range.

Geiger counter   A type of radiation survey instrument that is commonly used for detection of beta particles and gamma rays.

Generation 0 night vision devices   Generation 0 night vision devices are based on image conversion rather than image intensification. A night vision device is primarily comprised of a photocathode that convertd incident photons into electrons. The electrons are accelerated toward an anode by applying a positive potential to the anode. The device uses active infrared illumination.

Genearion 1 night vision devices   Generation 1 night vision devices are an adaptation of Generation 0 technology. A major deviation in Generation 1 night vision devices from Generation 0 devices is in the absence of the infrared source used in the case of the latter devices to provide scene illumination. Generation 1 devices depend on ambient light provided by the moon and stars.

Generation 2 night vision devices   Generation 2 night vision devices use a microchannel plate (MCP) for electron multiplication, leading to a significant increase in device sensitivity.

Generation 3 night vision devices   Generation 3 night vision devices have two distinctive changes from the technology used in Generation 2 devices. These include the use of a gallium arsenide photocathode and an ion barrier coating on the MCP.

Generation 3+ devices   Generation 3+ devices offer improved performance specifications over Generation 3 devices. Two important features associated with Generation 3+ night vision devices are automatic gated power supply system and a thinned ion barrier layer.

Generation 4 night vision devices   Generation 4 night vision devices were conceived to use filmless and gated technology. The proposal was to remove the ion barrier film from the MCP introduced in Generation 3 devices. Introduction of automatic gated power supply for the photocathode enabled the device to adapt instantaneously to light level fluctuations from a low light level to a high light level or from a high light level to a low light level. Removal of the ion barrier was also intended to reduce the halo effect seen around bright spots or light sources.

Generation-recombination noise   Generation-recombination noise is caused by the fluctuation in current generation and the recombination rates in a photosensor. This type of noise is predominant in photoconductive sensors operating at infrared wavelengths.

Hard-kill systems (laser countermeasures)   Hard-kill systems are laser countermeasure systems that are capable of inflicting physical damage to the front-end optics of any electro-optic system.

Helium-neon laser   The active medium of a He-Ne laser is a gas mixture of helium and neon, which is predominantly helium with only 10% to 20% neon. It produces laser radiation in the visible (543.5 nm, 632.8 nm) and infrared wavelengths (1153 nm, 3391 nm).

Heterojunction lasers (semiconductor lasers)   In the case of heterojunction lasers, the active layer and either one or both of the adjacent layers are of a different material. If only one of the adjacent layers is of a different material, it is called a simple heterojunction and if both are different, it is called a double heterojunction.

Holographic sight   A holographic sight is a nonmagnifying sight. In the case of a holographic weapon sight, the shooter looks through a glass window with the laser transmission hologram of a reticle image built into it. The shooter sees a reticle image superimposed on the field of view at a distance when the recorded hologram is illuminated by a collimated light beam from a laser diode.

Homojunction lasers (semiconductor lasers) In the case of homojunction semiconductor lasers, all layers are of the same semiconductor material. One such example is a GaAs/GaAs laser.

Image intensification Image intensification or enhancement is used in image intensifier tube based night vision devices. It works on the principle of collecting small quanta of light reflected off the target scene to be viewed in visible and near-infrared bands of electromagnetic spectrum in low light conditions. The collected photons are amplified through the processes of photon-electron conversion, electron multiplication, and electron-photon conversion.

Image intensifier tube An image intensifier tube works on the principle of image intensification or enhancement.

Immunoassay An immunoassay is a test that relies on biochemistry to measure the presence and/or concentration of an analyte. Immunoassays technology allows detection and identification of biological agents using the principle of specific antigen/antibody interaction.

Infrared imaging scanner Infrared imaging scanner depends for its operation on the infrared radiation emitted by the human body and the objectionable objects concealed therein, if any, due to the temperature difference of different parts of the body and the concealed objects.

Infrared spectroscopy (detection of chemical warfare agents) In the case of infrared spectroscopy, infrared radiation, generally in the mid-IR region of electromagnetic spectrum of 2.5 to 15 μm, is passed through the sample. The radiation is partly absorbed and partly transmitted. The wavelength spectrum of detected radiation represents the molecular absorption or transmission and hence the fingerprint of the sample.

Intensified CCD Intensified CCD (ICCD) exploits the optical amplification provided by an image intensifier to overcome limitations of the basic CCD sensor. An intensified CCD primarily is comprised of an image intensifier tube whose light output is coupled to a CCD sensor. Two important features of an intensified CCD are high optical gain and gated operation.

Ionization chamber A type of radiation survey instrument that uses a gas-filled detector.

Ion-mobility spectrometry (detection of chemical warfare agents)   IMS is an analytical technique that allows ionized analyte molecules to be distinguished on the basis of their mass, charge, and mobility in the gas phase.

IP camera   An IP camera is a networked digital video camera capable of transmitting data over a fast Ethernet link.

Iron sight   A type of weapon sight that is comprised of a system of shaped alignment markers having a front sight, usually a block or post, and a rear sight with a notch. While aiming, the front sight and the notch in the rear sight need to be aligned and also be on the target.

Irradiance   Irradiance, also referred to as power density, is defined as the power per unit area of the laser radiation falling on the target. It is expressed as watt/m^2.

Johnson noise   Johnson noise, also known as Nyquist noise or thermal noise, is caused by the thermal motion of charged particles in a resistive element. The RMS value of the noise voltage depends on the resistance value, temperature, and the system bandwidth.

Kessler syndrome (space debris)   The Kessler syndrome refers to a scenario where collisions between objects produce a cascade effect due to each collision generating space debris that further increases the probability of more collisions.

Kinetic energy weapons (less-lethal)   Less-lethal kinetic energy weapons achieve the desired effect by transferring kinetic energy from the weapon to the targeted person or material object. Common kinetic energy weapons include kinetic impact projectiles such as rubber and plastic bullets, beanbag rounds, pellet rounds, foam rounds, and sponge rounds, and water cannons. Kinetic impact projectiles are intentionally designed to inflict pain and incapacitate an individual without penetrating the body.

Laser-blinding shell   A laser-blinding shell generates extremely intense flashes of light on activation. The shell contains explosive material together with inert gases like argon, neon, and xenon. When the explosive material is ignited, high-pressure, high-temperature gases are generated leading to formation of plasma at high temperature, accompanied by emission of an extremely intense flash. The flash generates omnidirectional and directional radiation in the wavelength band from ultraviolet to infrared.

Laser countermeasures   Laser countermeasures constitute a subset of electro-optic countermeasures employed against systems operating in the optical spectrum of electromagnetic radiation from ultraviolet to infrared. These are categorized as passive countermeasures and active countermeasures. Active countermeasures are further classified as soft-kill and hard-kill countermeasures.

Laser dazzler   A laser dazzler is a nonlethal weapon that emits a high-intensity laser beam in the visible band, usually in the blue-green region, to temporarily impair the vision of an adversary without causing any permanent or lasting injury or adverse effect to the subject's eyes.

Laser decoy   A laser decoy is used to mimic the laser signatures in terms of wavelength and pulse repetition frequency code of the actual target to be protected.

Laser fence   A laser fence, also called *laser wall*, uses multiple pairs of a laser source and sensor to create an invisible laser fence along the perimeter of the infrastructure to be protected. Intrusion of this line of sight between source and sensor is used to activate an alarm or relay the information immediately to a nearby post for action.

Laser grenade   See *Laser-blinding shell*.

Laser-induced breakdown spectroscopy (explosive detection)   A laser-induced breakdown spectroscopy (LIBS) is a trace detection technique used for standoff detection of explosive materials. It focuses a high-energy laser beam on the trace sample to break down a small part of the sample into plasma of excited ions and atoms. The plasma emits light that is characteristic of emissions from ionic, atomic, and small molecular species. These light emissions are detected by a spectrometer to identify the elemental composition.

Laser-induced fluorescence spectroscopy (explosive detection)   Laser-induced fluorescence (LIF) spectroscopy is a type of spectroscopic technique in which the sample under examination is illuminated with a laser beam. The resultant fluorescence photons are measured to detect explosives.

Laser microphone   A laser microphone is a surveillance device that can be used to eavesdrop on suspect and rogue elements from a distance of a few hundred meters.

Laser photoacoustic spectroscopy   Laser photoacoustic spectroscopy (LPAS) is a form of infrared spectroscopy where the detection is done with a photoacoustic sensor such as a microphone or a piezoelectric sensor. Quartz-enhanced LPAS is an improved form of LPAS used for standoff detection of explosives.

Laser scanner   A laser scanner, also called a lidar sensor, employs the time-of-flight principle to create an invisible IR barrier and trigger an alarm or a specified event once the IR plane is broken.

Laser sight   A laser sight, also known as a *laser aiming aid,* is an aiming device that uses a small, low-cost, low-power semiconductor diode laser modules for target aiming and pointing, particularly during nighttime operations.

Laser triangulation (laser range finder)   Laser triangulation laser range finders use simple laws of trigonometry to compute distance to the target.

Laser wall   See *Laser fence.*

Lead salt semiconductor laser   A lead salt semiconductor laser is a P-N junction diode laser that consists of a single crystal of lead telluride (PbTe), lead selenide (PbSe), lead sulfide (PbS), or their alloys with themselves or with strontium selenide (SnSe), strontium telluride (SnTe), cadmium sulphide (CdS), and other materials.

Lidar sensor   See *Laser scanner.*

Light-dependent resistors   See *Photoconductor.*

Limit of detection (LOD)   LOD is the smallest amount of chemical, biological, or explosive agent that the detector system will respond to with an alarm.

Mass spectrometry (explosive detection)   Mass spectrometry is an analytical technique. It detects explosives based on conversion of the explosive sample into gaseous ions, with or without fragmentation, which are then characterized by their mass to charge ratios and relative abundances.

Maximum permissible exposure (MPE)   MPE is defined as the highest power density measured in W/cm^2 (in the case of a continuous-wave light source) or energy density measured in J/cm^2 (in the case of a pulsed light source) that poses a negligible risk of causing damage to an exposed surface of the body. It is measured at the cornea or skin for a given wavelength and exposure time. The MPE is specified as power or energy per unit surface based on the power or energy that can pass through a fully open pupil (0.39 cm^2) for visible and near-infrared wavelengths.

Microchannel plate (MCP)   The MCP is a thin glass disc about 0.5-mm thick consisting of an array of millions of tilted glass channels, each of about 5- to 6-μm diameter bundled in parallel. The number of stages to be used in the MCP depends on the required value of the gain. The strip current that flows through the MCP decides the dynamic range or linearity of the image intensifier tube.

Millimeter-wave imaging scanner   The operational principle of a millimetre-wave imaging scanner is the same as that of a backscattered X-ray scanner except that the former uses electromagnetic radiation called millimeter waves lying in the 30- to 300-GHz band of the spectral region between radio waves and infrared waves and not X-rays.

Mode locking   The phenomenon of mode locking can be considered to be a process that locks together in phase a cluster of photons. The mode-locking element transmits this cluster every time it passes through it while bouncing back and forth between the cavity mirrors. The repetition rate of the mode-locked pulses is equal to the round-trip transit time of the resonator cavity.

M^2 value M^2 value is a measure of beam quality. (M^2) and therefore is defined as the ratio of the divergence of the real beam to that of a theoretical diffraction-limited beam of the same waist size with a Gaussian beam profile (TEM_{00} mode). It is also referred to as the beam propagation ratio as per the ISO-11146 standard.

Monochromaticity   Monochromaticity refers to single frequency or wavelength property of the radiation. Laser radiation is monochromatic, and this property has its origin in the stimulated emission process by which laser emits light.

Nd:glass laser   A Nd:glass laser is a solid-state laser in which the active medium is neodymium ions in a glass host. The host materials are silicate, phosphate, and fused silica glasses with silicate and phosphate being more common.

Nd:YAG laser   A Nd:YAG is the most important and most widely used of all neodymium-doped lasers because of its high gain and good thermal and mechanical properties. The active medium is neodymium-doped yttrium aluminum garnet where trivalent neodymium replaces trivalent yttrium.

Nd:YLF laser   A type of neodymium-doped solid-state laser. The active medium is neodymium-doped yttrium lithium fluoride.

Nd:YVO$_4$ laser   A type of neodymium-doped solid-state laser. The active medium is neodymium-doped yttrium vanadate. This laser material is important because of several properties that make it particularly attractive for laser diode pumping. These include a large stimulated emission cross section and a strong broadband absorption around 808 nm.

Nerve agents   Nerve agents are chemical warfare agents that adversely affect the nervous system by affecting transmission of nerve impulses. They belong to the group of organophosphorus compounds. Tabun, sarin, and soman are examples of nerve agents.

Nominal ocular hazard distance (NOHD)   The distance from the light source at which the intensity or the energy per surface unit becomes lower than the maximum permissible exposure (MPE) on the cornea as dictated by American National Standards Institute (ANSI) standards for eye safety.

Nonlinear wave mixing (explosive detection)   Nonlinear wave mixing is a standoff technique for detection of explosive agents. In this, two laser beams are made to overlap in the region of presence of the explosive agent. Molecules present in the overlapping region interact with laser beams and the chemical information is transmitted to a detector as a laser-like beam.

Nuclear magnetic resonance (explosive detection)   The nuclear magnetic resonance (NMR) technique interrogates the nuclei of the material for explosive detection. NMR distinguishes between different chemical species based on magnetic resonance signals.

Nuclear quadrupole resonance (explosive detection)   Nuclear quadrupole resonance (NQR) detects explosives based on nitrogen quadrupole detection when the quadrupole nuclei present in explosive material are exposed to a pulsed radio frequency field.

Optical pumping   Optical pumping employs optical radiation and is used for lasers with a transparent active medium. Solid-state and liquid dye lasers are typical examples. The most commonly used pump sources are flash lamps in the case of pulsed and arc lamps in the case of continuous-wave solid-state lasers.

Panoramic sight   Panoramic sight is a type of artillery sight with a large field of view. It permits the gunner to sight in all directions without moving his or her head while laying an artillery piece for direction.

Particle sizer (detection of biological warfare agents)   The operational principle of a particle sizer is based on determining the relative number of particles in a predetermined size range.

Passive countermeasures   Passive countermeasures for platform protection include use of armour, camouflage, fortification, and other protection technologies such as a self-sealing fuel tank.

Passive infrared detection (detection of chemical warfare agents)   In the case of passive infrared detection, since all objects with temperatures above absolute zero emit energy in the infrared region, different chemicals emit or absorb light at specific narrow wavelength bands, creating a unique spectrum or fingerprint that can be used to identify the chemical agent.

Periscopic sight   A periscopic sight enables an operator to observe intended surroundings in the absence of direct line of sight. It also allows the operator to remain under cover or behind armor or in a submerged location while viewing the scene around him or her.

Phase shift laser range finder   In the phase shift technique of range finding, a laser beam with sinusoidal power modulation is transmitted toward the target and the diffused or specular reflection from the target is received. The phase of the received laser beam is measured and compared with that of the transmitted laser beam to compute time of flight and hence the range.

Photodiode   Photodiodes are junction-type semiconductor light sensors that generate current or voltage when the P-N junction in the semiconductor is illuminated by a light of sufficient energy. Photodiodes are mostly constructed using silicon, germanium, indium gallium arsenide (InGaAs), lead sulfide (PbS), and mercury cadmium telluride (HgCdTe).

Photo ionization detection (detection of chemical warfare agents)   Photo ionization detection (PID) technology relies on the ionization of molecules for detection of chemical agents.

Photomultiplier tube   Photomultiplier tubes (PMTs) are extremely sensitive photosensors operating in the ultraviolet, visible, and near-infrared spectra. PMTs have an internal gain of the order of 10^8 and can even detect a single photon of light. They are constructed from a glass vacuum tube that houses a photocathode, several dynodes, and an anode. When the incident photons strike the photocathode, electrons are produced as a result of the photoelectric effect.

These electrons accelerate toward the anode and in the process electron multiplication taken place due to the secondary emission process from the dynodes.

PIN photodiodes   In PIN photodiodes, an extra-high resistance intrinsic layer is added between the P and N layers. This has the effect of reducing the transit or diffusion time of the photo-induced electron-hole pairs that in turn results in improved response time. PIN photodiodes feature low capacitance, thereby offering high bandwidth, making them suitable for high-speed photometry as well as optical communication applications.

Polymerase chain reaction technology (detection of biological warfare agents)   The polymerase chain reaction (PCR) is one of the most commonly used molecular biology techniques in clinical laboratories for identification of microorganisms, thereby allowing detection of biological agents such as bacteria and bacteria spores or viruses.

Proportional counter   A proportional counter is a type of radiation survey meter using a gas-filled detector. Like the Geiger counter, it utilizes a multiplication mechanism to increase the pulse amplitude.

Protocol IV for blinding-laser weapons   Protocol IV for blinding-laser weapons gives guidelines for prohibiting the use and transfer of blinding-laser weapons. It was adopted on October 13, 1995 during the Vienna Convention.

PTZ camera   PTZ stands for pan, tilt, and zoom. PTZ cameras with their pan/tilt and zoom features allow them to monitor large areas with a single security camera.

Pulsed fast neutron analysis   Pulsed fast neutron analysis, also known as pulsed fast neutron activation, uses ultrashort fast neutron pulses of the order of nanoseconds that interact with the nuclei of interest to generate a characteristic gamma ray emission, which is measured for detection of explosives. Use of ultrashort fast neutron pulses also allows determining the location of detected explosive material.

Pyroelectric sensors   Pyroelectric sensors are characterized by spontaneous electric polarization, which is altered by temperature changes as light illuminates these sensors. Pyroelectric sensors are low-cost, high-sensitivity devices that are stable against temperature variations and electromagnetic interference. Pyroelectric sensors only respond to modulating light radiation and there will be no output for a CW incident radiation.

Q-switching   Q-switching is the mechanism of producing short laser pulses with a pulse width of the order of a few nanoseconds.

Quantum cascade laser (QCL)   QCL are compact high-power wavelength-agile semiconductor lasers that emit in the mid-infrared to far-infrared wavelength band. While conventional semiconductor lasers are all interband devices, quantum cascade lasers are unipolar devices and the laser emission in this case occurs across the intersubband, also called intraband transitions of electrons in the conduction band.

Quantum efficiency   Quantum efficiency is defined as ratio of the number of photoelectrons released to the number of photons of incident light absorbed. It is the percentage of input radiation power converted into a photocurrent.

Radiation survey meters   Radiation survey meters are portable radiation detection and measurement devices used to detect and measure external or ambient ionizing radiation fields.

Raman spectroscopy (explosive detection)   Raman spectroscopy is a trace detection technique used for standoff detection of explosive agents. The basis of detection in this case is the shift in the wavelength caused by inelastic Raman scattering by the target molecule. The inelastic scattering of impinging photons where some energy is lost to (or gained from) the target molecule returns scattered light with a higher (or lower) wavelength depending on whether energy was lost to (or gained from) the target molecule. The difference is dictated by the energy of vibrational modes of the target molecule and therefore constitutes the fingerprint or the basis of identification.

Red dot sight   See *Reflector sight.*

Reflector sight   In a reflector sight, also called reflex sight, the shooter looks through a partially reflecting glass element to see at infinity a projection of a reflective or luminous aiming point or some other image placed at the focus of a lens or a curved mirror with the result that anything at the focus will look like as if it is sitting in front of the viewer at infinity. When red light-emitting diodes are used to create an illuminated reticle, it is called red dot sight.

Reflex sight   See *Reflector sight.*

Resolution (explosive detection)   Resolution is the smallest concentration variation that can be detected when the concentration is continuously changed.

Response time (photosensors)   Response time is expressed as rise/fall time parameter in photoelectric sensors and as a time constant parameter in thermal sensors. Rise and fall times are the time durations required by the output to change from 10% to 90% and 90% to 10% of the final response, respectively. It determines the highest signal frequency to which a sensor can respond. Time constant is defined as the time required by the output to reach to 63% of the final response from zero initial value.

Response time (chemical or biological detection)   Response time is the time taken by a detector to respond to the targeted chemical or biological agents. It is the accumulation of time periods required for collection and analysis of the sample and the time required to provide feedback.

Responsivity (photosensors)   Responsivity is defined as the ratio of electrical output to radiant light input determined in the linear region of the response. It is measured in amperes per watt (A/W) or V/W if the photosensor produces a voltage output rather than a current output. Responsivity is a function of the wavelength of incident radiation and band gap energy. Spectral response is a related parameter.

Satellite laser ranging (SLR)   SLR is the technique of measuring range of an earth-orbiting satellite using a laser with the objective of determining the orbital parameters of satellites and their variation from predicted values and through this accurately determine the temporal variation of the earth's center of mass.

Schottky photodiodes   In Schottky-type photodiodes, a thin gold coating is sputtered on to the N-material to form a Schottky effect P-N junction. Schottky photodiodes have enhanced UV response.

Scintillation counter   A type of radiation survey instrument used for the measurement of alpha, beta, and neutron particles. A scintillation counter is comprised of a scintillator that generates photons in response to incident radiation, a photomultiplier tube that converts photons into electrical signals, and processing electronics that extracts the desired result.

Selectivity (chemical or biological detection)   Selectivity is the ability of a detector to discriminate between the targeted agent and other interferents that may be present in the sample.

Selectivity (explosive detection)   Selectivity is defined as the ability of the explosive detection system to detect a specific explosive molecule in the presence of interferents.

Semiconductor laser Semiconductor lasers use a semiconductor material as the active medium. The optical gain in this case is usually achieved by a process of stimulated emission at an interband transition triggered by prevailing conditions of high carrier density in the conduction band.

Sensitivity (chemical or biological detection) Sensitivity, also called limit of detection or LOD, is the lowest detectable concentration of the chemical or biological agent. It is also a measure of the ability of the detector to discriminate between small differences in the concentration of the agent under analysis.

Sensitivity (explosive detection) Sensitivity is defined as the minimum quantity of an explosive material that the system can detect under a given set of conditions.

Shot noise Shot noise in a photosensor is caused by the discrete nature of the photoelectrons. It is related to the statistical fluctuation of both dark current and the photocurrent. It depends on the average current through the photosensor and system bandwidth.

Slope efficiency Slope efficiency is determined by the slope of the characteristic I-V curve above the threshold current and is measured in mW/mA (or W/A). Slope efficiency is strongly dependent on temperature and decreases with an increase in temperature.

Sniper detector A sniper detector is an electro-optic device that is capable of detecting and identifying battlefield optoelectronic sighting systems. The device operates on the principle of the cat-eye effect.

Soft-kill systems (laser countermeasures) Soft-kill systems are laser countermeasures systems that are capable of causing only a temporary disability of electro-optic devices and optoelectronic sensors deployed by an adversary. The pulse energy level in these systems is in the range of a few hundreds of millijoules to a few joules.

Spatial coherence Spatial coherence is preservation of phase across the width of the beam. It tells about the correlation in phase of different photons transverse to the direction of travel.

Spontaneous emission Spontaneous emission is the phenomenon in which an atom or molecule undergoes a transition from an excited higher-energy level to a lower level all by itself without any outside intervention or stimulation and in the process emits a resonance photon. The rate of spontaneous emission process is proportional to the related Einstein coefficient.

Stimulated emission   In the case of stimulated emission, there first exists a photon called a stimulating photon that has energy equal to the resonance energy ($h\nu$). This photon perturbs another excited species (atom or molecule) and causes it to drop to the lower energy level, in the process emitting a photon of the same frequency, phase, and polarization as that of the stimulating photon. The rate of stimulated emission process is proportional to the population of the higher excited energy level and the related Einstein coefficient.

Stun baton/belt/gun   A type of electrical weapon that requires a direct contact with the body to produce the intended effect.

Surface acoustic wave sensor (detection of chemical warfare agents)   A surface acoustic-wave (SAW) device employs an acoustic wave guided along the surface of a piezoelectric crystal coated with a chemically sensitive polymer. The chemicals present in the chemical vapors undergo sorption by the polymer on the surface of the piezoelectric substrate, which alters the surface wave propagation on the substrate in terms of its amplitude and frequency. The signals are processed to identify chemicals sorbed from the sample and also determine the concentration of the chemical by assuming that sorption equilibrium is reached.

SAW sensor (explosive detection)   A SAW sensor detects explosive materials by measuring the change in the resonant frequency of the SAW sensor caused by absorption of chemical vapors into chemically selective coatings on the sensor surface.

TEA CO_2 laser   A transversely excited atmospheric pressure (TEA) CO_2 laser is a high-power pulsed CO_2 laser.

Telescopic sight   Telescopic sight is a type of weapon sight that generally is comprised of a combination of lenses and reflective surfaces within a metal sleeve. The ocular lens that the shooter looks through is smaller and has lower magnification. The lens at the telescope's far end, called an objective, increases magnification.

Temporal coherence   Temporal coherence is preservation of the phase relationship with time.

Terahertz band (imaging)   Terahertz radiation (0.3–3 THz) can penetrate through a wide variety of dielectric materials such as fabric, paper, plastic, leather, and wood, is nonionizing and has minimal effects on the human body, has very large absorption due to water, and is highly reflected from metals. This forms the basis of using terahertz imaging, including detection of concealed weapons and hidden explosives.

Thermal damera Thermal cameras operate on heat radiated by an object rather than the visible light reflected off the object.

Thermal imaging Thermal imaging is a night vision technology that works on the principle of detecting temperature difference between the objects in the foreground and those in the background. The information on the temperature difference available in the form of infrared energy is collected by the thermal imaging device and converted into an electronic image.

Thermal neutron analysis (explosive detection) Thermal neutron analysis, also known thermal neutron activation, depends for its operation on the interaction of a low-energy neutron beam with the nitrogen nuclei of the matter under inspection, leading to absorption of the thermal neutron and subsequent emission of a high-energy gamma ray with characteristic energy of 10.8 MeV. Detection of any emitted 10.8 MeV gamma rays indicates the presence of nitrogen, indicating a high probability of the object containing explosive material.

Thermo-redox (explosive detection) Thermo-redox technology of explosive detection is based on thermal decomposition of explosive material to release NO_2 molecules, which are subsequently reduced and detected.

Thermal sensors Thermal sensors absorb radiation, which produces a temperature change that in turn causes a change in the physical or the electrical property of the sensor. In other words, thermal sensors respond to change in their bulk temperature caused by the incident radiation. Thermocouple, thermopile, bolometer and pyroelectric sensors belong to the category of thermal sensors. Thermal sensors lack the sensitivity of photoelectric sensors and are generally slow in response but have a wide spectral response.

Thermal sight Thermal sight operates on the principle of infrared thermography that converts thermal information and the associated infrared wavelengths emitted from the object into an image.

Thermocouple sensors Thermocouple sensors are based on the Seebeck effect (i.e., the temperature change at the junction of two dissimilar metals generates an EMF proportional to the temperature change). Commonly used thermocouple materials are bismuth-antimony, iron-constantan, and copper-constantan.

Thermopile sensors   A thermopile sensor is a series connection of a number of thermocouples. The responsivity of a single thermocouple is very low and therefore to increase the responsivity, several junctions are connected in series to form a thermopile.

Throughput (explosive detection)   Throughput is the rate at which a sample can be analyzed.

Time-of-flight laser range finder   In a time-of-flight laser range finder, a narrow pulse width laser beam is transmitted toward the intended target. The target range is measured from the time taken by the laser pulse to travel to the target and back.

Trace detection (explosive detection)   Trace detection methods require only trace amounts of explosives either in gas phase or particle form for detection. Important trace detection methods include ion mobility spectrometry (IMS), electronic noses, cavity ring down spectroscopy, surface plasmon resonance (SPR), and surface enhanced Raman spectroscopy (SERS).

Ultraviolet laser induced fluorescence technology (detection of biological warfare agents)   Ultraviolet laser induced fluorescence (UV-LIF) is a promising lidar technology to allow fast standoff detection of biological warfare agents. UV-LIF lidar allows gross discrimination between biological agents and background noise, taking advantage of the intrinsic fluorescence of biological molecules.

Vacuum photodiode   A vacuum photodiode is comprised of an anode and a cathode placed in a vacuum envelope. When irradiated, the cathode releases electrons that are attracted by the positively charged anode, thus producing a photocurrent proportional to the light intensity.

Vertical cavity surface-emitting laser (VCSEL)   In the case a VCSEL, the optical cavity is along the direction of flow of injection current. The laser beam in this case emerges from the surface of the wafer rather from its edges. The two mirrors are either grown epitaxially as a part of the diode structure or grown separately and then bonded to the semiconductor chip having the active region.

Vertical external cavity surface-emitting laser (VECSEL)   A VECSEL is a variant of VCSEL where the resonator is completed with a mirror placed external to the diode structure, thus introducing a free-space region in the resonant cavity.

Viable particle size sampler (detection of biological warfare agents)   A viable particle size sampler or impactor operates by accelerating an air flow through a nozzle before deflecting it against an impact surface maintained at a fixed distance.

Virtual impactor (detection of biological warfare agents)   Virtual impactors belong to the broad category of viable particle size samplers. A virtual impactor separates particles by size into two airstreams. The impaction surface of a conventional impactor is replaced with a virtual space of stagnant or slow moving air. Large particles are captured in a collection probe rather than impacted onto a surface.

Virtual PTZ camera   A virtual PTZ camera, also called a 360-degree camera or sometimes an ePTZ (electronic PTZ) camera, is comprised of several high-resolution fixed cameras in a single, usually dome, housing. The images captured by individual cameras are stitched together to provide a full 360-degree view.

Webcam   A webcam, short for web camera, is a digital video camera directly or indirectly connected to a computer or computer network via the internet.

X-ray scanner   An X-ray scanner uses the intensity of transmission and scattering of X-rays through different materials to construct an image of hidden objects.

About the Author

Dr. Anil K. Maini is an engineering graduate in electronics and communication from Punjab Engineering College (now PEC University) and Ph.D. in electronics and communication from Amity University Uttar Pradesh (India). He served Defence Research and Development Organisation (DRDO) in various capacities from March 1978 to June 2014. At the time of his superannuation, he was holding the post of outstanding scientist and director of laser science and technology center, Delhi, a premier R&D DRDO under Ministry of Defence, Government of India. He has more than 36 years of R&D experience as a scientist, team leader, project director and director in diverse disciplines related to Electronics, Lasers and optoelectronics technologies and systems. His contributions have led to a number of innovations in the development of different types of electronics systems, optronic sensors and directed energy laser systems for applications in defense and homeland security. The key products and technologies developed under his guidance include a wide range of high voltage electronics systems, family of laser systems for homeland security applications including non-lethal laser dazzlers for antiterrorist and counterinsurgency applications, laser ordnance disposal system for safe neutralization of explosives, sniper locator sensors, laser-based explosive detectors, laser fence, laser systems for electro-optic countermeasures (EOCM) applications, and optoelectronic simulators and sensors for testing electro-optically guided precision strike munitions. He has authored and co-authored 18 books, more than 150 research papers and technical articles in national and international journals and magazines and is inventor and coinventor in nine patents. currently, he is working as a consultant (*Defence Technologies*).

Index

Smart Structures and Materials, Brian Culshaw

Substrate Surface Preparation Handbook, Max Robertson

Surveillance and Reconnaissance Imaging Systems: Modeling and Performance Prediction, Jon C. Leachtenauer and Ronald G. Driggers

Wavelength Division Multiple Access Optical Networks, Andrea Borella, Giovanni Cancellieri, and Franco Chiaraluce

For further information on these and other Artech House titles, including previously considered out-of-print books now available through our In-Print-Forever® (IPF®) program, contact:

Artech House
685 Canton Street
Norwood, MA 02062
Phone: 781-769-9750
Fax: 781-769-6334
e-mail: artech@artechhouse.com

Artech House
16 Sussex Street
London SW1V 4RW UK
Phone: +44 (0)20 7596-8750
Fax: +44 (0)20 7630-0166
e-mail: artech-uk@artechhouse.com

Find us on the World Wide Web at: www.artechhouse.com